HIGHLY EVOLVED CLOSE
BINARY STARS: FINDING CHARTS

Advances in Astronomy and Astrophysics

A series of books and monographs which aims to supply professionals and serious graduate students with the intellectual tools necessary for the appreciation of the present status of topics at the forefront of current research, and provides a framework upon which future developments may be based.

Series Editors:
V.G. Gurzadyan, Yerevan Physics Institute, Armenia
S. Inagaki, Kyoto University, Japan
G. Meylan, European Southern Observatory, Gärching bei München, Germany

Volume 1
Highly Evolved Close Binary Stars: Catalog
Highly Evolved Close Binary Stars: Finding Charts
A.M. Cherepashchuk, N.A. Katysheva, T.S. Khruzina, S.Yu. Shugarov

Additional Volumes in Preparation
Volume 2
Basic Physics of Accretion Disks
S. Kato, S. Inagaki, J. Fukue, S. Mineshige

This book is part of a series. The publisher will accept continuation orders which may be cancelled at any time and which provide for automatic billing and shipping of each title in the series upon publication. Please write for details.

HIGHLY EVOLVED CLOSE BINARY STARS: FINDING CHARTS

**A.M. Cherepashchuk, N.A. Katysheva,
T.S. Khruzina, S.Yu. Shugarov**

Sternberg Astronomical Institute, Moscow State University, Russia

GORDON AND BREACH PUBLISHERS

Australia • Canada • China • France • Germany • India • Japan • Luxembourg
Malaysia • The Netherlands • Russia • Singapore • Switzerland • Thailand
United Kingdom

Copyright © 1996 by OPA (Overseas Publishers Association) Amsterdam B.V. Published in The Netherlands under license by Gordon and Breach Science Publishers SA.

All rights reserved.

No part of this book may be reproduced or utilized in any form or by any means, electronic or mechanical, including photocopying and recording, or by any information storage or retrieval system, without permission in writing from the publisher. Printed in Singapore.

Emmaplein 5
1075 AW Amsterdam
The Netherlands

British Library Cataloguing in Publication Data

Highly evolved close binary stars: finding charts. –
  (Advances in astronomy & astrophysics; 1)
  1. Double stars 2. Astrographic catalog and chart 3. Astronomy
  – Charts, diagrams, etc.
  I. Cherepashchuk, A.M.
  523.8'41'0223

  ISBN 90-5699-014-4 (Hardcover)
  ISBN 90-5699-015-2 (Softcover)

CONTENTS

Introduction	vii
Description of the Atlas	viii
The General List of Stars	1
Massive Close Binary X-Ray Sources	8
Low-Mass X-Ray Close Binaries	29
Precataclysmic Close Binary Systems	56
AM Canus Venaticorum Type Stars	62
Symbiotic Binary Stars	65
Cataclysmic Variables. Novae, Dwarf Novae and Related Objects	79
Intermediate Polars	204
AM Herculis Type Stars (polars)	223
References	239

INTRODUCTION

Due to great progress in space and ground based astronomy, a new class of astronomical objects – X-ray binary systems – has been discovered. The nature and evolution of these objects have been understood after pioneering works devoted to the investigation of accretion processes onto relativistic objects (neutron stars, black holes) and an evolutionary scenario has been calculated for close binary systems of different masses (see the first part of our *Catalog* where all references are presented).

A huge amount of observational data concerning X-ray binaries and other types of highly evolved close binary stars (HECBSs), i.e., CBSs after the first mass exchange containing very peculiar objects (Wolf-Rayet stars, white dwarfs, neutron stars, black holes) has been obtained up to now. These observational data are collected in our *Catalog of Close Binary Stars On Late Evolutionary Stages* published by Moscow University Press in 1989 and in the first part of the present *Catalog*. The Catalog can be useful for theoreticians as well as for observers. Due to their spectacular appearance, highly evolved CBSs are very interesting and promising objects, especially for observational investigations. In this connection, the availability of finding charts and knowledge of comparison stars are highly desirable for the observational investigations of HECBSs. This information was partially published in the first version of our *Catalog* (1989) and in the *Atlas of Cataclysmic Variables. U Geminorum Stars* published by T.S. Khruzina and S.Yu. Shugarov in 1991 with Moscow University Press.

Here we are publishing the finding charts and comparison stars for numerous HECBSs which are included in the first part of the present *Catalog of HECBSs*. These data will be useful not only for professional astronomers but also for members of amateur societies such as the American Association of Variable Star Observers (AAVSO), British Astronomical Society (BAA), New Zealand Astronomical Society (RASNZ) and numerous others.

In our work we have used data published in "Atlas des Étoiles Variables du Type U Geminorum" (Brun A., Petit M., 1959, *Peremennye Zvezdy*, **12**, 78), "Catalog of Variable Star Charts" published by the American Association of Variable Star Observers and "Charts of Southern Variables" which were published by F.M.Bateson and his collaborators, "Atlas of Photographic Charts of Cataclysmic Stars" published by G.Williams (1983, *Ap.J.Suppl.Ser.*, **53**, 523). Also we have used data published by N.Vogt and F.M.Bateson (1982, *Astr.Ap.Suppl.Ser.*, **48**, 383) and A.Bruch, F.-J.Fischer and U.Wilmsen (1987, *Astr.Ap.Suppl.Ser.*, **70**, 481).

The second part of our *Catalog* contains finding charts (in many cases with comparison stars) for 335 HECBSs, the basic data for which are published in the first part of the present *Catalog*. To simplify the identification of some faint binaries, we give two finding charts in different scales: as a rule, about $10' \times 10'$ together with that of about 1 square degree.

We hope that our book will help astronomers and amateurs in detailed observational investigations of HECBSs. We are grateful to Nikolai N. Samus for his assistance during the preparation of our manuscript.

DESCRIPTION OF THE ATLAS

All binary systems given in the *Atlas* are divided into groups using the classification of binary stars according to descriptions given in *Part I* of our *Catalog*. The general list of the stars having an identification chart in the *Atlas* is given in the table on page 1. In some cases, two charts are presented: the first chart (top) shows the surrounding star field with the size of about 1 square degree, the second one (bottom) is more detailed, showing a region with $\sim 10' \times 10'$ size.

In the subsequent chapter, information about binary stars included in the *Atlas* is given. For each star, the following observational data are presented.

1. One or two finding charts. For each case, dimensions in acrminutes are given. North is top, east is left. In many cases comparison stars, used by different observers, are also indicated. As a rule, stars having only visual or photographic stellar magnitudes are marked by a number. The comparison stars having photoelectric magnitudes are marked by Latin letters. Comparison stars' magnitudes in a corresponding system are given on the same page.

2. The type of the system.

3. Equatorial coordinates of the HECBS for the equinox 2000.0.

4. Spectrum of the system, or the spectrum of the secondary (less evolved) component, or that of both components.

5. Stellar magnitudes of the binary corresponding to the brightest maximum and the faintest minimum. Symbols V, B and so on denote magnitudes given in corresponding bands of the wide-band photometry. Letters Pg and Vis denote photographic and visual light variability limits.

6. Orbital period of the variable star (in days) with corresponding epoch of maximum or minimum light (if they are known). Figures in brackets denote a value of error in the last decimal. In Remarks, it is indicated from what kind of observations - photometric or spectroscopic - this period value was obtained.

7. Stellar magnitudes of comparison stars.

8. Remarks containing additional information about the system.

9. In square brackets, a list of references to articles giving finding charts and additional information about the variable is presented. The complete list of references is given at the end of the book. The designations of cited sources are as follows.

The journals in Russian

Abast.Bull. — Bulletin of the Abastuman Astrophysical Observatory
AF — Astrofizika, Erevan
Astron.Zh. — Astronomicheskii Zhurnal
Astron.Circ. — Astronomicheskii Tsirkular (Russian Academy of Sciences)
DAN — Doklady, Russian Academy of Sciences
IK — Izvestiya Krymskoi Astrofizicheskoi Observatorii
NIAS — Nauchnye Informatsii Astronomicheskogo Soveta, Soviet Academy of Sciences
Pis'ma Astron.Zh. — Pis'ma v Astronomicheskii Zhurnal
UFN — Uspekhi Fizicheskikh Nauk
PZ — Peremennye Zvezdy (Variable Stars)
PZ(P) — Peremennye Zvezdy (Prilozheniya) (Variable Stars, Supplement)
SCAO — Soobsheniya Spezialnoi Astrofizicheskoi Observatorii

Other journals

AAp.Tr. — Astronomy and Astrophysics Transactions
Acta Astr. — Acta Astronomica
Acta Astr.Sinica — Acta Astronomica Sinica
Acta Ap.Sinica — Acta Astrophysica Sinica
Adv.Astr.Ap. — Advances in Astronomy and Astrophysics
Adv.Space Res. — Advances in Space Research
AJ — The Astronomical Journal
AN — Astronomische Nachrichten
Ann.Rev.Astr.Ap. — Annual Review of Astronomy and Astrophysics
Ann.Uppsala Astr.Obs. — Uppsala Astronomiska Observatoriums Annaler
Ap.J. — The Astrophysical Journal
Ap.J.Suppl.Ser. — The Astrophysical Journal. Supplement Series
Ap.Lett. — Astrophysical Letters
Ap.Space Sci. — Astrophysics and Space Science
Ap.Space Sci.Libr. — Astrophysics and Space Science Library
Astr.Ap. — Astronomy and Astrophysics
Astr.Ap.Rev. — Astronomy and Astrophysics Review
Astr.Ap.Suppl.Ser. — Astronomy and Astrophysics. Supplement Series
Astr.Ges.Mitt. — Mitteilungen der Astronomischen Gesellschaft
Astr.Rep. — Astronomical Reports
BAAS — Bulletin of the American Astronomical Society
BAC — Bulletin of the Astronomical Institutes of Czechoslovakia
BAF — Bulletin de l'Association Française d'Observateurs d'Étoiles Variables
Bergd Abh. — Astronomische Abhandlungen der Hamburger Sternwarte, Bergedorf
BSAM — Bulletin de la Station Astrophotographique de Mainterne
Comm.Ap. — Comments on Astrophysics
DAO Preprint — Preprint of the Dominion Astrophysical Observatory, Victoria, B.C.
DOB — Documentation des Observateurs, Bulletin, Paris
ESO Mess. — European Southern Observatory Messenger

Harv.Ann. — Annals of the Astronomical Observatory of Harvard College
Harv.Ann.Card — Harvard College Observatory Announcement Card
IAUC — Circular of the International Astronomical Union
IBVS — Information Bulletin on Variable Stars. Commission 27 and 42 of the IAU, Budapest, Konkoly Observatory
JAAVSO — The Journal of the American Association of Variable Star Observers
J.Brit.Astr.Ass. — The Journal of the British Astronomical Association
JO — Journal des Observateurs
Kod.Obs.Bul. — Kodaikanal Observatory Bulletins. Series A, Bangalor
KVB — Kleine Veröffentlichungen der Remeis-Sternwarte Bamberg
Lick Obs.Publ. — Lick Observatory of the University of California Publications
Lowell Bull. — Bulletins of the Lowell Observatory
McCormick.Publ. — Publications of the Leander McCormick Observatory of the University of Virginia
Mem.S.A.It. — Memorie della Società Astronomica Italiana
Messenger — ESO Messenger
MNASSA — Monthly Notes of the Astronomical Society of South Africa
MNRAS — Monthly Notices of the Royal Astronomical Society
MVS — Mitteilungen über veränderliche Sterne, Berlin-Babelsberg und Sonneberg
Nature Phys.Sci. — Nature Physical Science
NZAS Bull. — Bulletin of Variable Star Section, Royal Astronomical Society of New Zealand
NZAS Publ. — Publications of Variable Star Section, Royal Astronomical Society of New Zealand
PASJ — Publications of the Astronomical Society of Japan
PASP — Publications of the Astronomical Society of the Pacific
Padova Publ. — Publicazioni dell'Osservatorio Astronomico di Padova
Preprint ESO — European Southern Observatory Scientific Preprint
Proc.ASA — Proceedings of the Astronomical Society of Australia
Publ.Bei.Astr.Obs. — Publications of the Beijing Astronomical Observatory, China
Publ.Dom.Ap.Obs. — Publications of the Dominion Astrophysical Observatory, Victoria. B.C.
Phys.Rep. — Physics Reports
Rev.Mex.Astr.Ap. — Revista Mexicana de Astronomia y Astrofisica
Ric.Astr. — Specola Astronomica Vaticana Richerche Astronomiche
J.Roy.Astr.Canada — The Journal of the Royal Astronomical Society of Canada
Sky Tel. — Sky and Telescope
Sonn.Mitt. — Mitteilungen der Sternwarte zu Sonneberg
South.Stars — Southern Stars, New Zealand
Sov.Astr. — Soviet Astronomy (from 1993 Astronomy Reports)
Sov.Astr.Lett. — Soviet Astronomy Letters (from 1993 Astronomy Letters)
Space Sci.Rev. — Space Science Review
TTB — Boletin de los Observatorios Tonantzintla y Tacubaya
VSS — Veröffentlichungen der Sternwarte in Sonneberg
Z.f.Ap. — Zeitschrift für Astrophysik

Catalog of HECBSs

THE GENERAL LIST OF STARS

Massive close binary X-ray sources

Name of the star				Page
	2S	0050–727	SMC X-3	8
	2S	0052–739	SMC X-2	9
γ Cas	4U	0054+60		12
V 662 Cas	2S	0114+650	LS1+65°010	10
	4U	0115–737	Sk 160/SMC X-1	11
V 635 Cas	4U	0115+634		23
V 615 Cas	4U	0236+610	LS1+61°303	13
BQ Cam	X	0331+53		14
	4U	0532–664	LMC X-4	17
V 725 Tau	A	0535+26	HDE 245770	20
	4U	0538–641	LMC X-3	16
	A	0538–66		13
	4U	0540–697	LMC X-1	15
	4U	0728–25		15
GP Vel	4U	0900–403	HD 77581/Vela X-1	12
	1E	1024–5732	Wack 2134	21
	1E	1048–5937		18
	A	1118–615	WRA 793/He 3-640	22
V 779 Cen	4U	1119–60	Cen X-3/KRZ star	20
V 830 Cen	1E	1145–6141		22
V 801 Cen	4U	1145–619	Hen 715/HD 102567	23
BP Cru	4U	1223–62	GX 301-2/WRA 977	28
V 850 Cen	4U	1258–61	GX 304-1	19
BR Cir	4U	1516–56	Cir X-1	24
QV Nor	4U	1538–52	Nor X-2	26
	2S	1553–542		25
	OAO	1657–415		21
V 884 Sco	4U	1700–377	HD 153919	27
	X	1907+097		27
V 1343 Aql	A	1909+04	SS 433	14
V 1357 Cyg	4U	1956+35	HDE 226868/Cyg X-1	19
	EXO	2030+375		26
V 1521 Cyg	4U	2030+407	Cyg X-3	28
	4U	2206+54		25

Low-mass X-ray binaries with a neutron star or a suspected black hole

Name of the star				Page
V 518 Per	GRO	J0422+32	XN Per 1992	29
	4U	0521–720	LMC X-2	39
	4U	0543–682	CAL 83	33
	4U	0547–711	CAL 87	30
V 1055 Ori	4U	0614+091		30
V 616 Mon	A	0620–00		31
UY Vol	EXO	0748–676		32
	2S	0918–549		32
V 395 Car	2S	0921–630		34
GU Mus	GRS	1124–684	XN Mus 1991	33
GR Mus	4U	1254–690		37
	X	1323–619		35
BW Cir	GS	1354–64	XN Cir 1987	34
V 822 Cen	A	1455–31	Cen X-4	35
KY TrA	4U	1524–617	TrA X-1	48
HL Lup	4U	1543–475		40
LU TrA	4U	1556–605		38
QX Nor	4U	1608–522		54
V 818 Sco	4U	1617-155	Sco X-1	53
KZ TrA	4U	1626–673		36
	4U	1630–47		37
V 801 Ara	MXB	1636–536		41
HZ Her	3U	1653+35	Her X-1	38
V 2134 Oph	MXB	1658–298		39
V 821 Ara	4U	1659–487	GX 339–4	40
	2S	1702–363	GX 349+2/Sco X-2	41
V 2107 Oph	H	1705–250	Nova Oph	42
	4U	1705–44		42
V 2216 Oph	4U	1728–169	GX 9+9	47
V 2116 Oph	4U	1728–247	GX 1+4	52
V 926 Sco	MXB	1735–444		55
V 4134 Sgr	4U	1755–338		44
	GRS	1758–258	GX 5–1	43
	4U	1811–17	GX 13+1	44
NP Ser	4U	1813–14	GX 17+2/Ser X-2	45
V 691 CrA	4U	1822–371		46
MM Ser	MXB	1837+05	Ser X-1	46
	XB	1905+000		47
V 1333 Aql	4U	1908+005	Aql X-1	48
V 1405 Aql	4U	1916–053		49
V 1408 Aql	4U	1957+115		49

Name of the star			Page
QZ Vul	GS 2000+25	Nova Vul 1988	50
V 404 Cyg	GS 2023+338	Nova Cyg 1989	51
V 1727 Cyg	4U 2129+470		52
V 1341 Cyg	4U 2142+380	Cyg X-2	53
V 665 Cas ?	1E 2259+586	G 109.1–1.0	54

Precataclysmic close binary systems

Name of the star		Page
RR Cae	LFT 349/BMP 31582/LHS 1660//WD 0419−487	56
V 664 Cas	PN HFG 1/Cas 1/HFG 1/136+5°	56
IN Com	PN LT5/BD+26°2405/SAO 82570/HD 112313	57
V 477 Lyr	PN Abell 46	57
UU Sge	PN Abell 63/PK 053−3°1	58
V471 Tau	BD+16°516/H 0349+17/G 7-23/WD 0347+171	59
BE UMa	PG 1155+492	60
GK Vir	PG 1413+015/WD 1413+015	61
	WD 0232+035/Feige 24	59
	EUVE 2013+40/RE 2014+400	61

Symbiotic binary stars

Name of the star		Page
Z And	HD 221650/IRAS 23312+4832	65
EG And	HD 4174/BD+39°167/SAO 36618/IRAS 00415+4024	65
V 1413 Aql	AS 338/SS 428	66
R Aqr	HD 222800/IRAS 23412−1533	66
UV Aur	HD 34842	67
T CrB	IRAS 15374+2603	67
BF Cyg	MWC 315	68
CH Cyg	HD 182917/BD+49°299/SAO 031632/IRAS 19232+5008	68
CI Cyg	IRAS 19483+3533	69
V 1016 Cyg	MWC 415/IRAS 19553+3941	69
V 1329 Cyg	HBV 475	70
AG Dra	BD+67°922/IRAS 16013+6656	70
YY Her	AS 297	71
V 443 Her	MWC 603	71
RW Hya	HD 117970	72
BX Mon	AS 150	72

Name of the star		Page
SY Mus	HD 100336	73
AR Pav	MWC 600	73
AG Peg	HD 207757/IRAS 21486+1223	74
AX Per	MWC 411	74
RX Pup	HD 69190/IRAS 08124−4133	75
CL Sco	AS 213	75
FG Ser	AS 296/SS 148/IRAS 18125−0019	76
HM Sge	IRAS 19396+1637	76
V 2601 Sgr	AS 313	77
V 2756 Sgr	AS 293	77
RR Tel	Hen 1811/IRAS 20003−5552	78
PU Vul	IRAS 20189+2124/Kuwano object	78

AM Canus Venaticorum type stars

	Name of the star	Page
AM CVn	HZ 29/EG 91	63
V 803 Cen	AE 1	62
GP Com	G 61−29/GR 389/LTT 18284/1ES(X)1303+1817	64
CP Eri		63

U Geminorum type stars

Name	Page	Name	Page	Name	Page
RX And	80	AC Cnc	181	EY Cyg	118
AR And	81	AR Cnc	101	V 792 Cyg	120
DX And	82	AT Cnc	102	V 1776 Cyg	184
LT And	86	CC Cnc	103	CM Del	123
HL Aqr	177	SV CMi	104	AB Dra	124
UU Aql	88	BV Cen	107	DM Dra	125
V 1315 Aql	179	MU Cen	108	ES Dra	186
WX Ari	179	V 442 Cen	110	AH Eri	126
SS Aur	91	V 485 Cen	111	U Gem	128
FS Aur	92	WW Cet	112	CF Gru	187
V 363 Aur	181	TT Crt	116	AH Her	131
Z Cam	96	RZ Gru	186	V 544 Her	132
SY Cnc	99	EM Cyg	117	LY Hya	188

Name	Page	Name	Page	Name	Page
X Leo	136	IP Peg	149	BC UMa	165
RU LMi	137	KT Per	151	CH UMa	167
CW Mon	142	AY Psc	153	CI UMa	168
HP Nor	143	BV Pup	155	EI UMa	200
CN Ori	144	V Sge	196	ER UMa	201
CZ Ori	145	UZ Ser	160	TW Vir	172
V 1193 Ori	192	SW Sex	198	HS Vir	202
BD Pav	146	KK Tel	162	VW Vul	174
V 345 Pav	191	RW Tri	199		
RU Peg	148	UX UMa	199		

SU UMa type stars

Name	Page	Name	Page	Name	Page
FO And	83	V 1251 Cyg	121	UV Per	150
KV And	84	V 1504 Cyg	122	TY Psc	152
LL And	85	AQ Eri	127	TY PsA	154
VY Aqr	178	AW Gem	129	RZ Sge	156
FO Aql	87	IR Gem	130	WZ Sge	158
TT Boo	94	VW Hyi	133	V 4140 Sgr	159
YZ Cnc	100	WX Hyi	134	EK TrA	163
OY Car	105	T Leo	135	SU UMa	164
HT Cas	106	RZ Leo	189	CY UMa	169
V 436 Cen	109	BR Lup	138	DV UMa	170
WX Cet	113	BK Lyn	190	CU Vel	171
Z Cha	114	AY Lyr	139	HV Vir	201
TV Crv	178	TU Men	140		
V 503 Cyg	119	EF Peg	147		

Nova-like stars

Name	Page	Name	Page	Name	Page
HV And	176	AT Ara	89	UZ Boo	95
PQ And	175	TT Ari	90	AF Cam	97
PX And	175	T Aur	180	BZ Cam	98
UU Aqr	176	KR Aur	93	V 425 Cas	182
V 794 Aql	177	QZ Aur	180	GO Com	115

Name	Page	Name	Page	Name	Page
V 394 CrA	182	RS Oph	193	RT Ser	197
V 751 Cyg	183	V 380 Oph	193	CT Ser	198
V 1668 Cyg	184	V 442 Oph	194	LX Ser	161
V 1974 Cyg	185	V 841 Oph	194	BZ UMa	166
HR Del	185	HX Peg	183	RW UMi	200
V 825 Her	187	V Per	192	SW Vul	173
DI Lac	188	FY Per	195	PW Vul	202
DO Leo	189	RR Pic	195	0903+41	203
BH Lyn	190	T Pyx	196	2136+11	203
MV Lyr	191	WY Sge	157		
BT Mon	141	U Sco	197		

Intermediate polars

	Name of the star	Page
V 603 Aql	N Aql 1918	204
AE Aqr		204
FO Aqr	H 2215−086	205
HL CMa	1E 0643−1648	206
BG CMi	3A 0729+103	207
TV Col	2A 0526−328	205
TX Col	H 0542−407	206
AL Com		208
SS Cyg		209
DO Dra	3A 1148+719/2A 1150+72/E 1140.8+7158	210
PQ Gem	RE J0751+14/RX J0751+14	221
DQ Her	N Her 1934	207
V 533 Her	N Her 1963	211
V 795 Her	PG 1711+336	211
EX Hya		212
BI Lyn	PG 0900+401	222
V 426 Oph		212
V 2301 Oph	1H 1752+081	213
GK Per	N Per 1901	213
TW Pic	H 0534−581	216
AO Psc	H 2252−035	214
CP Pup	N Pup 1942	214
V 347 Pup	4U 0608−49/LB 1800	215
V 348 Pup	1H 0709−360	215
VZ Pyx	H 0857−242	217

Name of the star		Page
V 1223 Sgr	3A 1851−312	217
V 1062 Tau	H 0459+246	218
SW UMa		219
KO Vel	1E 1013−477	216
	X 0022−7221 (47Tuc)	222
	1H 0551−819	220
	S 193	221

Polars or AM Herculis type stars

Name of the star		Page
BY Cam	H 0538+608/4U 0541+60/1H 0533+607	223
V 834 Cen	1E 1405−451/HEAO 1409−45	223
V 1500 Cyg	N Cyg 1975	224
EP Dra	1H 1907+690	231
EF Eri	2A 0311−227	224
UZ For	EXO 0333−255	225
CE Gru	Grus V1/Hawkins-V1	225
AM Her	3U 1809+50/2A 1814+4950	226
WW Hor	EXO 0234−5232	227
BL Hyi	H 0139−68	227
DP Leo	1E 1114+182	228
ST LMi	CW 1103+254	228
GQ Mus	N Mus 1983	229
V 2051 Oph		230
VV Pup	RE 0812−1254	231
MR Ser	PG 1550+191	232
AN UMa	PG 1101+453	233
DW UMa	PG 1030+590/FSB 1031+59	232
EK UMa	1E 1048.5+5421	233
SS UMi	PG 1551+719	235
QQ Vul	1E 2003+225	234
	EXO 0329−260/EXO 032957−2606.9	235
	RX J 0531−46	236
	1E 0830.9−2238	236
	RX J 0929.1−2404	237
	RE 1149+28	237
EV UMa	RE 1307+537	238
	V 2009−65.5	238

MASSIVE CLOSE BINARY X-RAY SOURCES

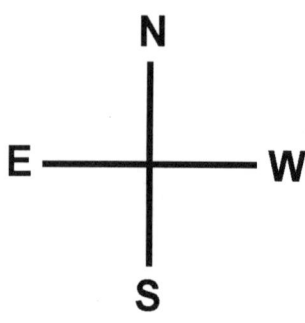

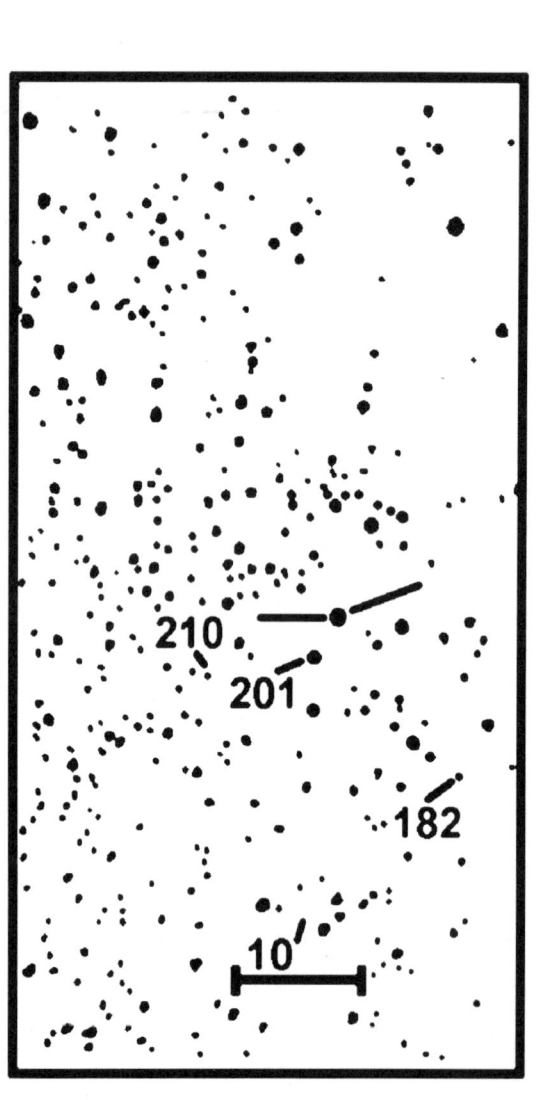

α (2000.0)=$00^h 52^m 10^s.5$
δ (2000.0)=$-72°26'27''.6$
Type: X-ray transient
Spectrum: O9 III–V
$Pg \sim 14.54$

Star	Pg
182	14.45
201	14.44
210	14.56

Be/X? The optical identification of SMC X-3 is based on accurate determination of the position of the system and on spectral characteristics of this star which are similar to optical spectra of Be/X transients in the Galaxy.

SMC X-3/2S 0050–727 [192]

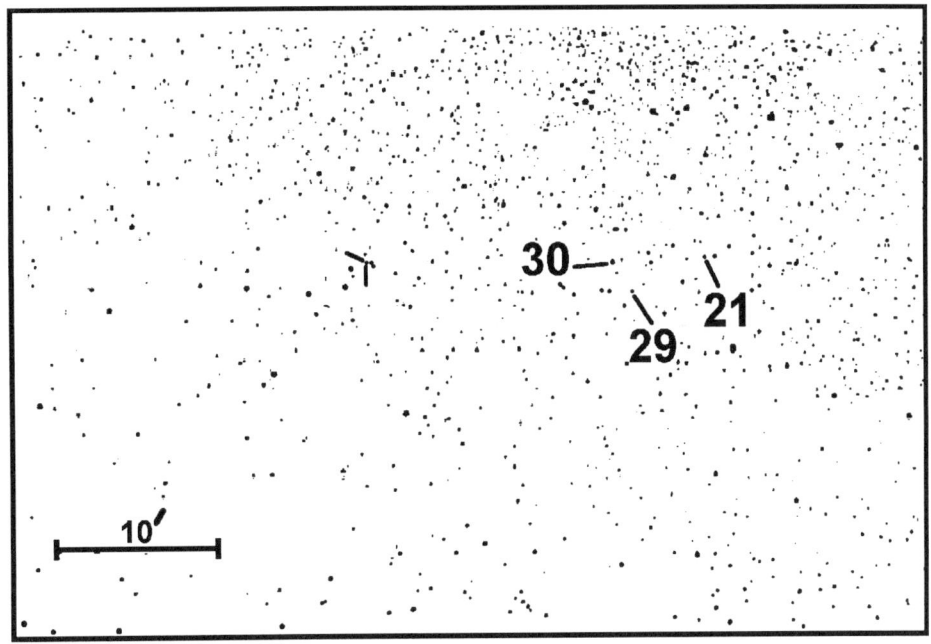

SMC X-2/2S 0052–739 [290]

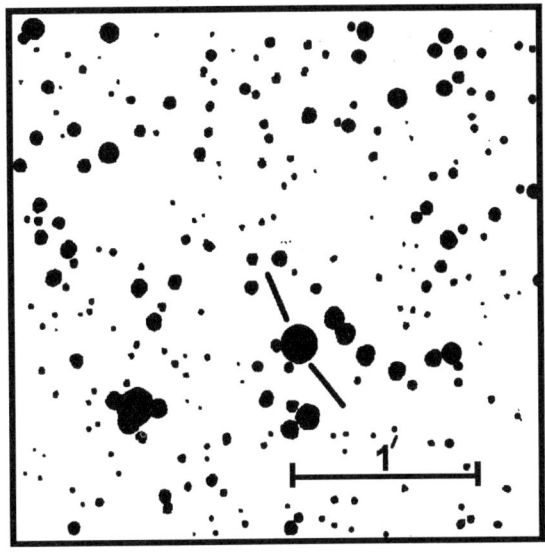

α (2000.0)= $00^h54^m34^s.8$
δ (2000.0)= $-73°40'43''.0$
   Type: X-ray transient
Spectrum: O9 III–V
   V=14.72–14.77

The comparison star used are:

   21 Pg=13.4
   29 Pg=13.3
   30 Pg=12.6

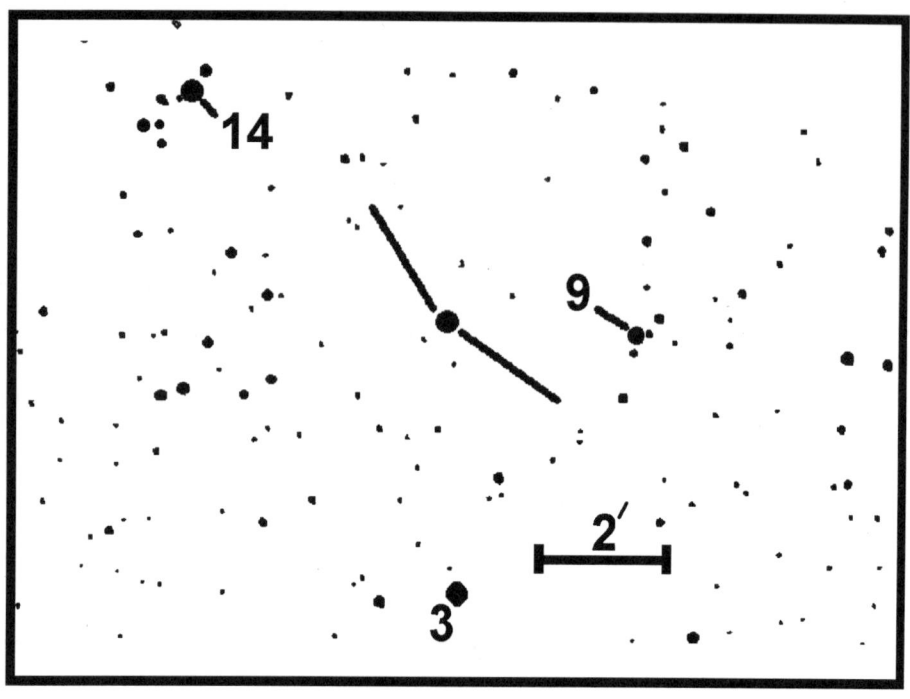

V 662 Cas/2S 0114+650/LSI+65°010 [33,101]

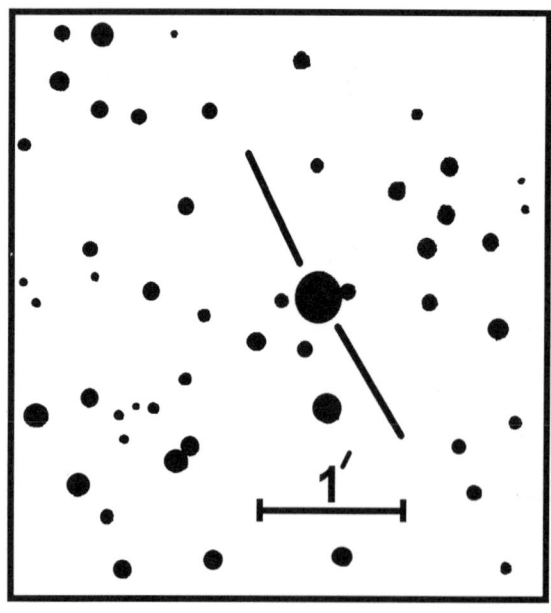

α (2000.0)=$01^h18^m02^s.8$
δ (2000.0)=$65°17'18''.6$
Type: X-ray transient
Spectrum: B0.5 Ib–IIIe
V=12.1–12.5
Max.VR=JD 2444134.3(2)
P=$11^d.588(3)$

Star	V
3	11.873(6)
9	12.328(6)
14	11.601(5)

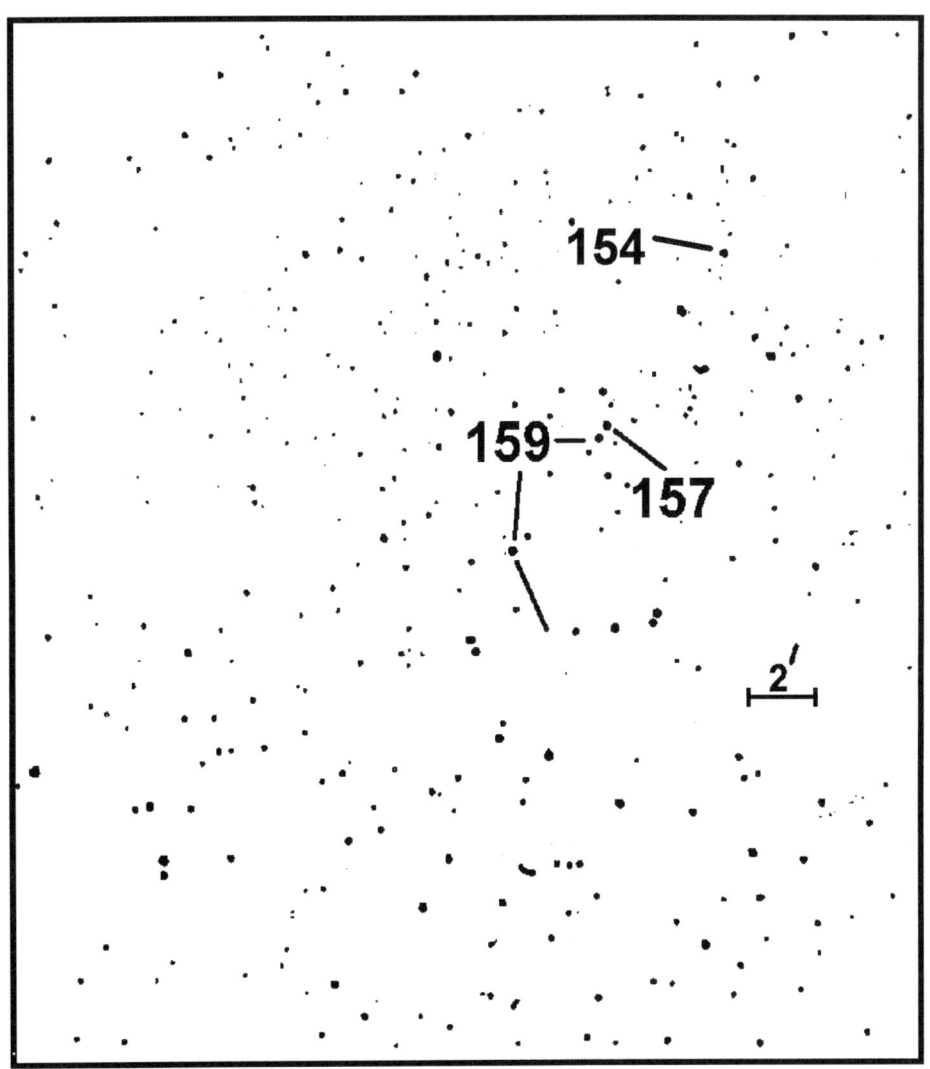

SMC X-1/Sk 160/4U 0115−737 [290]

α (2000.0)= $01^h17^m05^s.2$
δ (2000.0)=$-73°26'34''.8$
  Type: X-ray persistent binary
  Spectrum: B0.5 Ia
    V=13.12–13.34
  Min. V=JD 2447740.85906(3)
    P=$3^d.89238(2)$

Comparison stars:

Sk 159 (V=11.89, B=11.74)
Sk 157 (Pg=11.5)
Sk 154 (Pg=11.6)

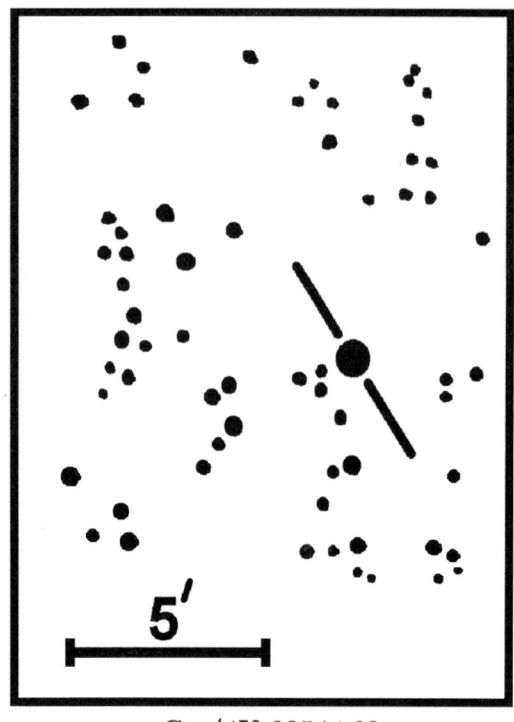

α (2000.0)= $00^h56^m43^s.3$
δ (2000.0)=$60°43'09''.8$
Type: X-ray transient
Spectrum: B0.5 IVe
V=1.65–2.8
$P\sim$4–7 yr

γ Cas/4U 0054+60

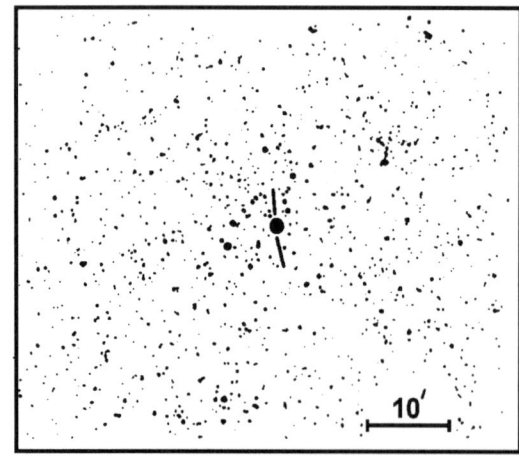

α (2000.0)= $09^h02^m06^s.7$
δ (2000.0)=$-40°33'16''.2$
Type: X-ray persistent binary
Spectrum: B0.5 Ibeq
V=6.76–6.99
Min.V=JD 2443821.860(6)
P=$8^d.9644(2)$

GP Vel/HD 77581/4U 0900–403/
/Vela X-1 [49, 95]

Star	α	δ	V	B−V	U−B
SAO 199881	$09^h00.7^m$	$-38°25'$	7.245	−0.100	−0.593
SAO 199829	08 57.8	−38 44	7.466	−0.039	−0.232

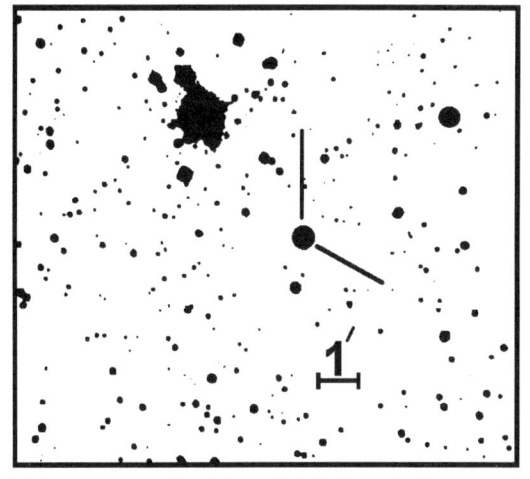

α (2000.0)=$02^h40^m31^s.6$
δ (2000.0)=$61°13'45''.4$
Type: X-ray transient
Spectrum: B0.5 IIIe
V=10.7–10.9
Min. of radio
  outbursts =JD 2443366.775
  P=$26^d.496(8)$

V 615 Cas/4U 0236+610/LSI+61°303 [312]

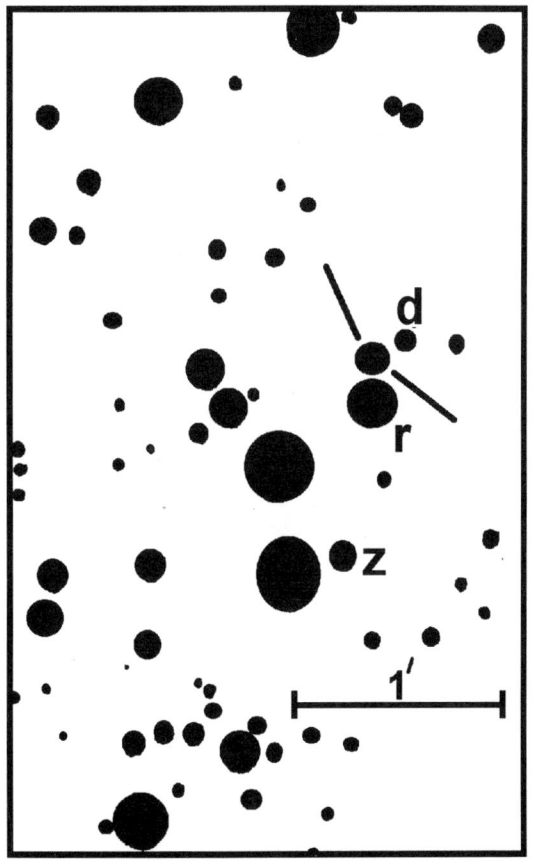

α (2000.0)= $05^h35^m40^s.5$
δ (2000.0)=$-66°51'52''.8$
Type: X-ray transient
Spectrum: B1–2 III
  V=12.5–15.3
  Max. V=JD 2445722.4(2)
  P=$16^d.6515(5)$

Star	*r*	*z*
Spectrum	B1–2 V	...
V	13.34(1)	14.85(1)
B	13.18(2)	14.62(2)
U	12.26(2)	13.69(1)

Star	*d*
Spectrum	B2–3 V
V	15.92(1)
B	15.71(2)
U	14.87(2)

The object is located in the Large Magellanic Cloud.

A 0538–66 [94, 240]

14   A.M.Cherepashchuk, N.A.Katysheva, T.S.Khruzina, S.Yu.Shugarov

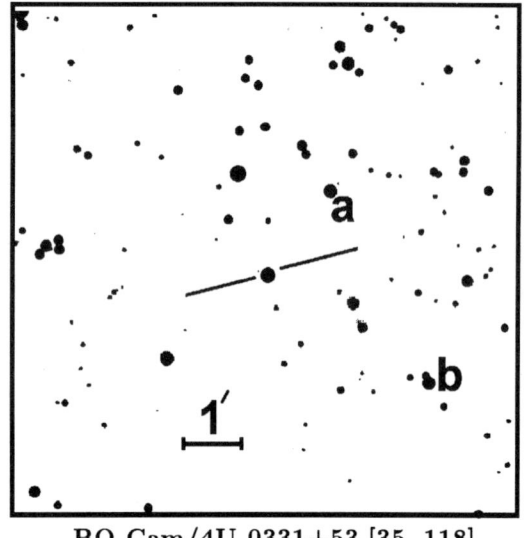

α (2000.0)=$03^h 34^m 59^s.9$
δ (2000.0)=$53°10'23''.0$
  Type: X-ray transient
Spectrum: B0–2 III–V
  V=15.1–15.4
  Max.V=JD 2445652(1)
  P=$34^d.25(10)$

Star	Pg
a	16.30
b	16.55

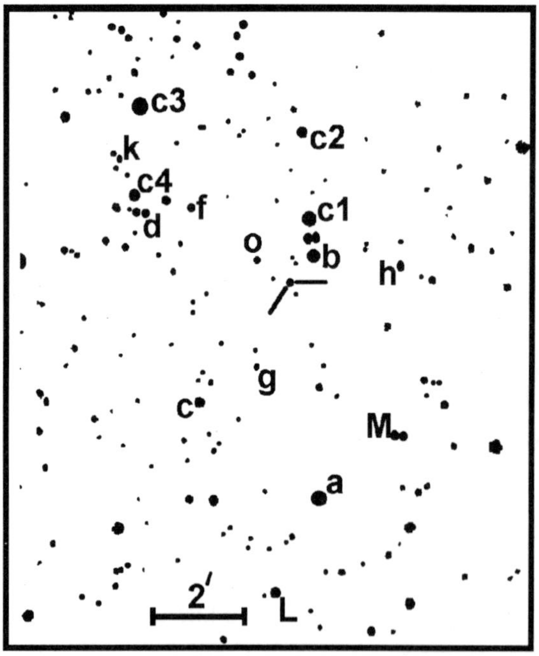

BQ Cam/4U 0331+53 [35, 118]

V 1343 Aql/SS 433/A 1909+04 [125]

Star	V	B	R
C1	11.51	12.91	10.06
C2	12.91	13.65	12.15
C3	10.92	11.21	...
C4	12.68	14.25	...
a	11.36	14.55	...
b	13.41	14.45	12.44
c	13.91	14.54	...
d	14.38	15.55	...
f	14.85	15.56	
g	15.04	15.49	
h	15.26	16.30	
k	15.30	...	
L	...	16.44	
M	15.50	16.74	
o	15.56	16.54	

α (2000.0)=$19^h 11^m 49^s.3$
δ (2000.0)=$04°58'57''.8$
  Type: X-ray persistent binary
   black hole candidate

Spectrum: OB
  V=13.7–15.5
  Min.V=JD 2446596.25(4)
  P=$13^d.077(4)$

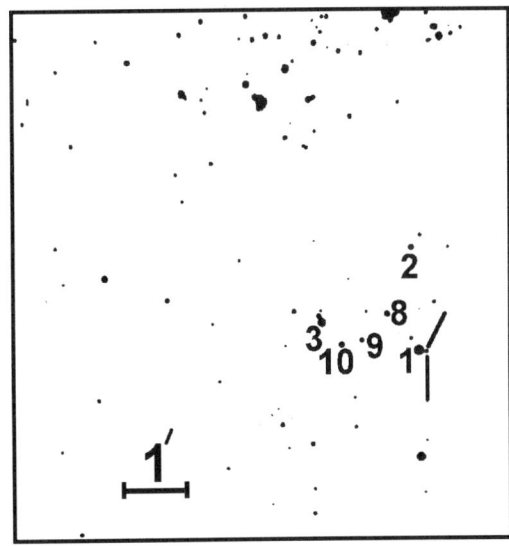

α (2000.0)= $05^h39^m37^s.2$
δ (2000.0)=$-69°44'46''.9$
 Type: X-ray persistent
  binary
Spectrum: O7–9 III
 V~14.37
 Min.V=JD 2445025.85(8)
 P=4^d.2288(6)

Star	V	B–V	U–B
1	12.05	0.26	-0.56
2	12.55	0.59	0.08
3	12.49	-0.08	-0.92
8	12.23	-0.06	-0.94
9	14.54	0.23	-0.79
10	14.27	-0.02	-0.86

LMC X-1/4U 0540–697 [108]

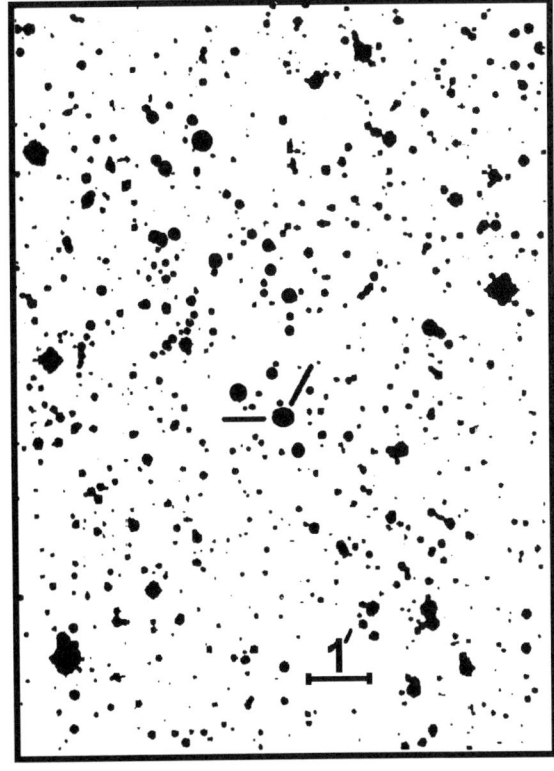

α (2000.0)= $07^h28^m53^s.5$
δ (2000.0)=$-26°06'28''.1$
 Type: X-ray transient
Spectrum: B0–1 III–V
 V=11.52–11.6

4U 0728–25 [325]

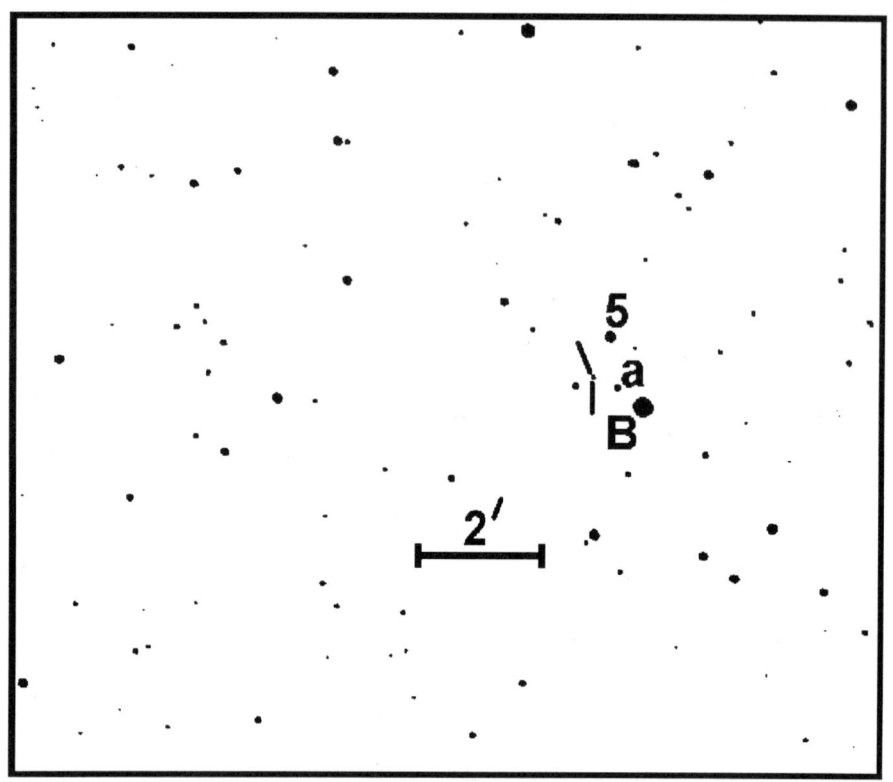

LMC X-3/4U 0538−641 [356, 389]

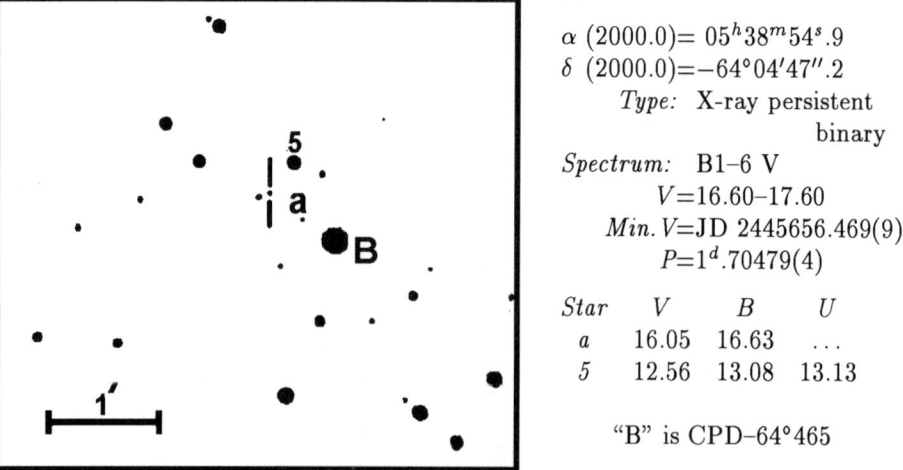

α (2000.0)= $05^h 38^m 54^s.9$
δ (2000.0)= $-64°04'47''.2$
Type: X-ray persistent binary
Spectrum: B1-6 V
V=16.60−17.60
Min.V=JD 2445656.469(9)
P=$1^d.70479(4)$

Star	V	B	U
a	16.05	16.63	...
5	12.56	13.08	13.13

"B" is CPD−64°465

Both the X-ray and optical flux of LMC X-3 show a pronounced, periodic variation on ∼198 (or possibly ∼99) day cycle, with alternating strong and weak intensity peaks. The ephemeris is $Max(X) = $ JD $2443733 + 197^d.8 \cdot E$.

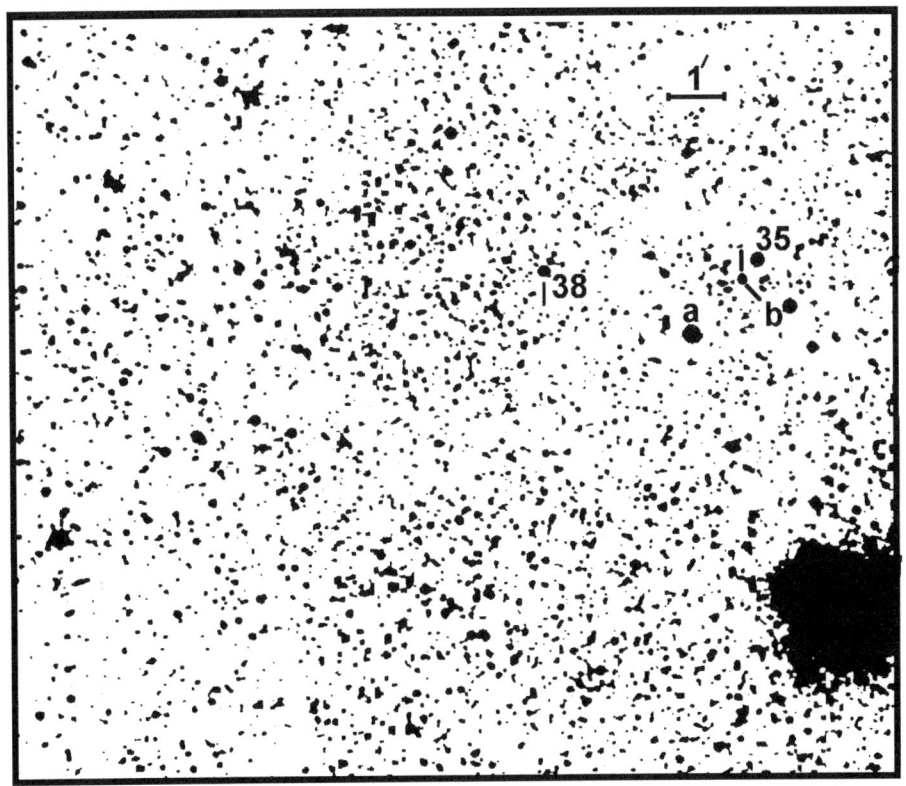

LMC X-4/4U 0532−664 [67, 399]

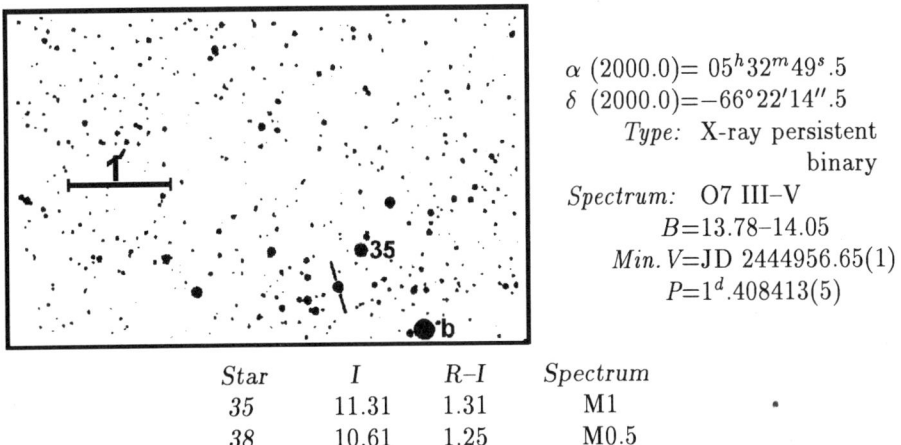

α (2000.0)= $05^h32^m49^s.5$
δ (2000.0)=$-66°22'14''.5$
Type: X-ray persistent binary
Spectrum: O7 III–V
B=13.78–14.05
Min.V=JD 2444956.65(1)
P=$1^d.408413(5)$

Star	I	R–I	Spectrum
35	11.31	1.31	M1
38	10.61	1.25	M0.5

Two brightest stars in the error box do not belong to LMC: the southern star marked "a" is HD 269743 ($V = 10.1$, Sp K0); the western star marked "b" is HD 279734 ($V = 12.0$, Sp K5).

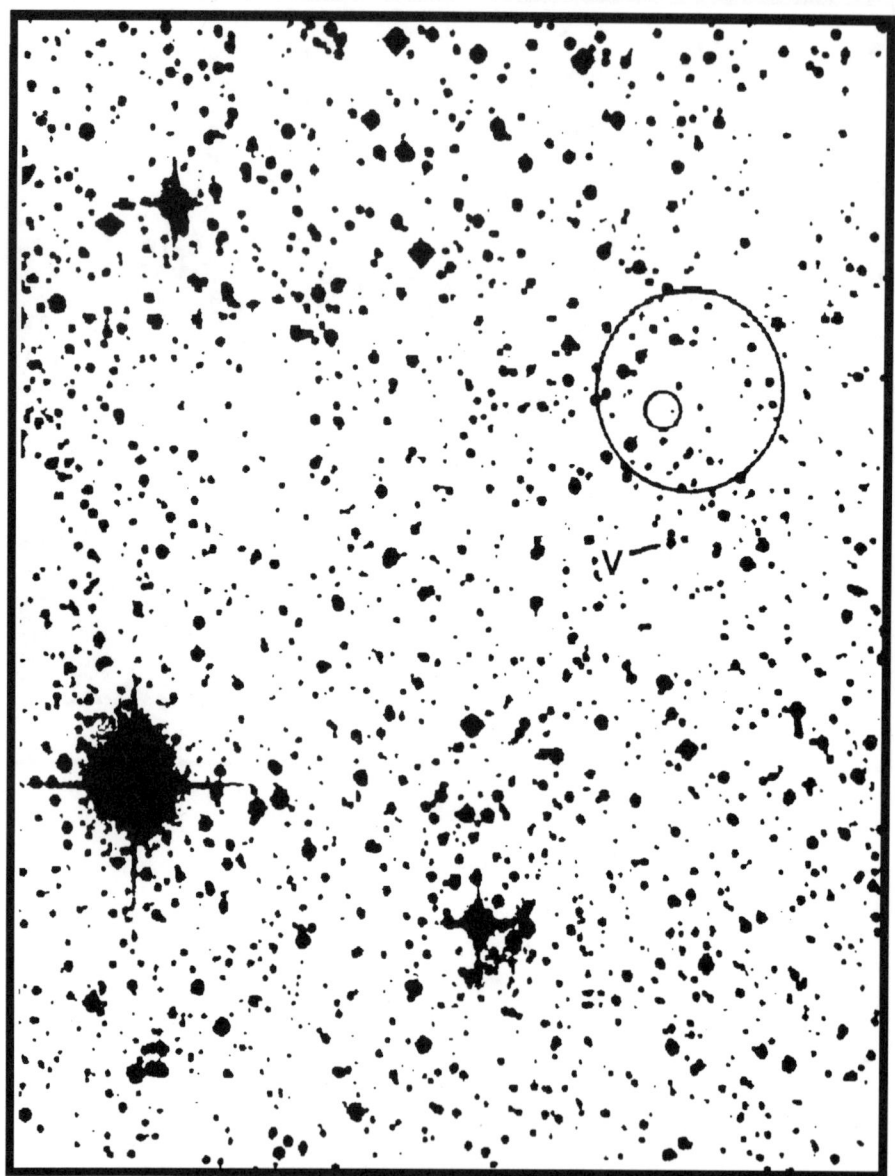

1E 1048.1−5937 [306]

α (2000.0)=$10^h 50^m 08^s.9$
δ (2000.0)=$-59°53'19''.8$
Type: X-ray transient

The probable optical counterpart with $V \geq 19$ mag lies in the smaller circle of errors. Star "V" is variable.

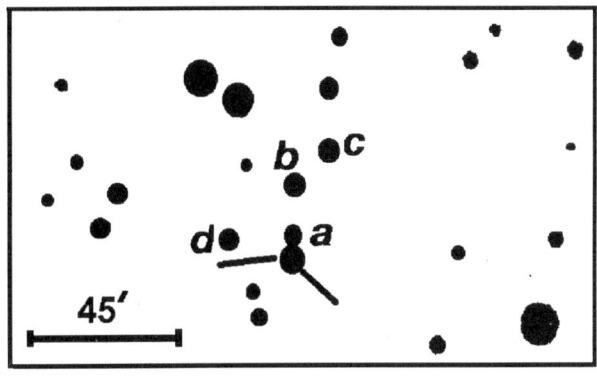

α (2000.0)=$19^h 58^m 21^s.8$
δ (2000.0)=$35°12'06''.2$
Type: X-ray persistent binary black hole candidate
Spectrum: O9.7 Iab
B=9.26–9.41
Min. V=JD 1163.631(1)
P=$5^d.59985(1)$

**4U 1956+35/V 1357 Cyg/HDE 226868
/BD+34°3815/Cyg X-1 [200]**

Star	a	b	c	d
V	9.976(10)	10.609(4)	9.053(20)	9.99
B	10.566(7)	10.973(6)	9.060(7)	10.14
U	10.630(7)	11.100(6)	9.044(10)	10.36

Star designated "a" is a visual binary, with a companion $V \sim 13^m$ at the distance $\sim 6''$. The magnitude of its combined brightness is given in the table.

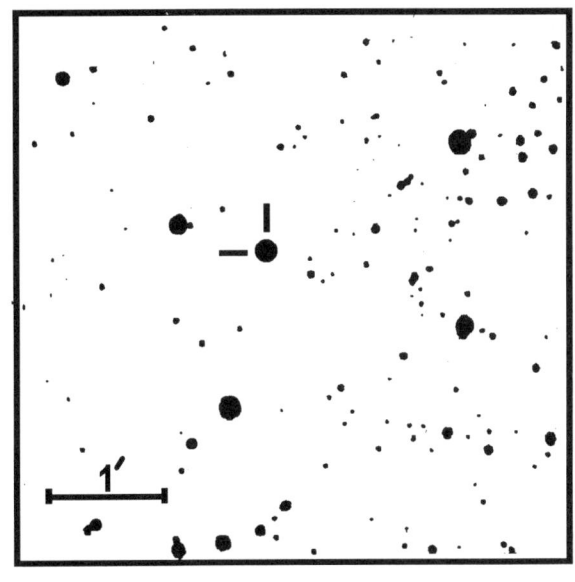

α (2000.0)=$13^h 01^m 17^s.1$
δ (2000.0)=$-61°36'05''.5$
Type: X-ray transient
Spectrum: B2 Vne
V=13.4–14.72
Min. V=JD 2443120(2)
P=$132^d.75(4)$

V 850 Cen/4U 1258–61/GX 304–1 [46]

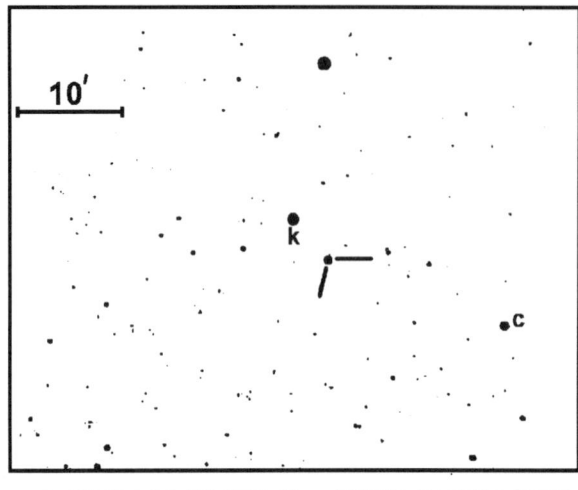

α (2000.0)=$05^h38^m54^s.6$
δ (2000.0)=$26°18'56''.8$
Type: X-ray transient
Spectrum: B0 IIIe
V=8.85–10.1
Max. V=JD 2441728(1)
P=$111^d.158(7)$

A 0535+26/V 725 Tau/HDE 245770 [202]

Identification chart with comparison stars "k" and "c" is given in [202], the other comparison stars are:

Star	HD 37170	HD 37696	BD+26° 887	BD+26° 876
V	7.931	7.962	10.54	7.96
B	8.228	8.002	10.99	8.24
U	8.507	7.230	11.18	8.34

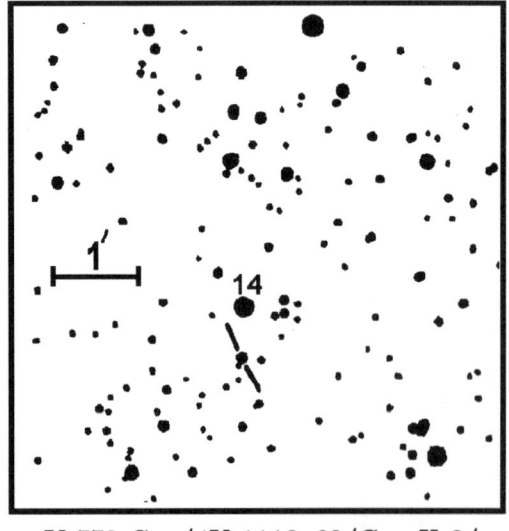

α (2000.0)= $11^h21^m16^s.3$
δ (2000.0)=$-60°37'21''.2$
Type: X-ray persistent
binary
Spectrum: O6–9 II–III
V=13.30–13.47
Min. V=JD 2447608.3688(8)
P=$2^d.087124(5)$

Star "14":
V=11.465(5)
$B-V$=0.24(1)
$U-B$=0.39(1)

**V 779 Cen/4U 1119–60/Cen X-3/
Krzeminski's star [180]**

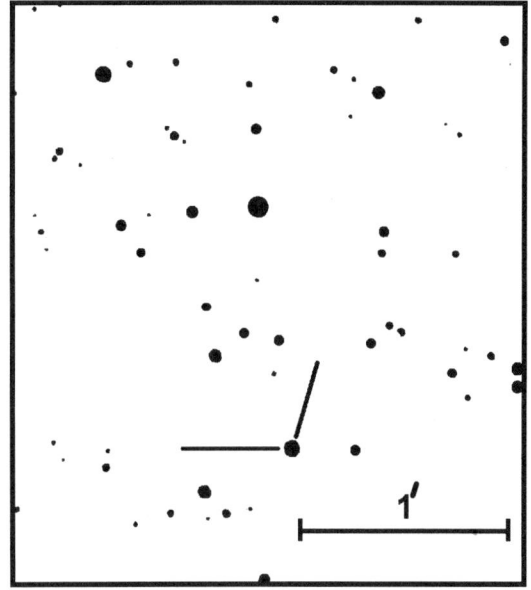

α (2000.0)=$10^h25^m56^s.6$
δ (2000.0)=$-57°48'41''.4$
Type: X-ray persistent binary
Spectrum: early O
$V\sim12.7$

1E 1024.0–5732/Wack 2134/TH 3542 [60]

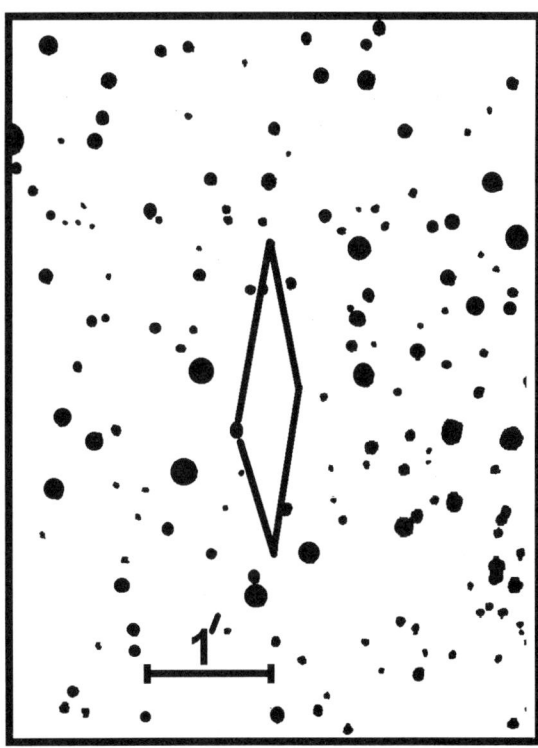

α (2000.0)=$17^h00^m47^s.9$
δ (2000.0)=$-41°40'22''.5$
Type: X-ray persistent binary
$P=10^d.444(4)$
T^*=JD 2448516.49(5)
(this is the epoch when the periastron longitude was equal to 90°)

No optical counterpart has yet been identified. On the finding chart [14] a parallelogram containing the object is shown.

OAO 1657–415 [14]

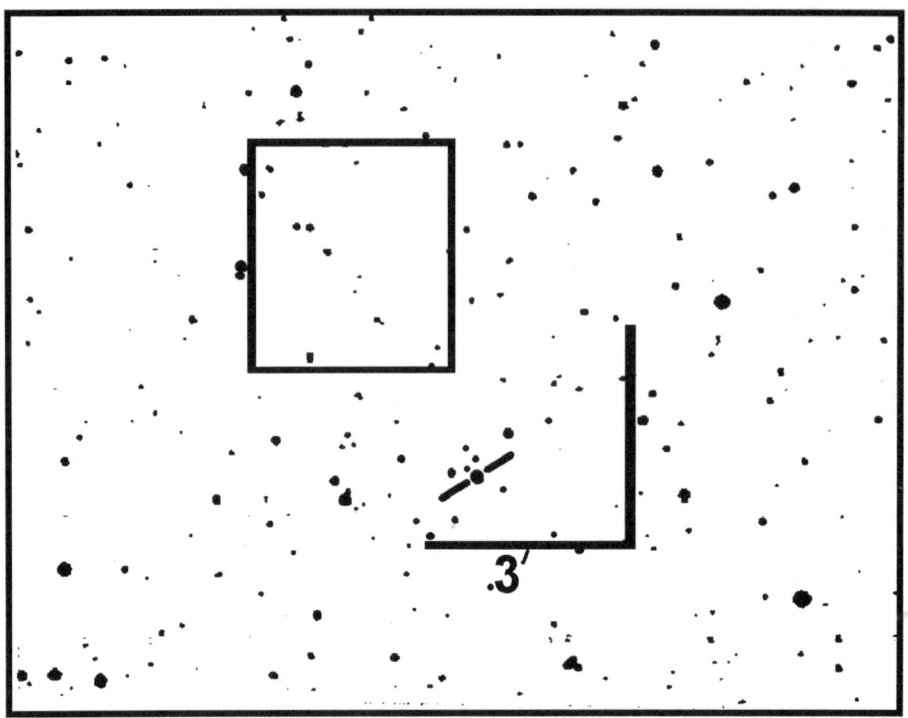

A 1118−615/WRA 793/He 3−640 [168]

α (2000.0)=$11^h20^m57^s.0$ *Spectrum:* O9.5 III–Ve
δ (2000.0)=$-61°54'58''.0$ B=13.07–13.11
Type: X-ray transient

The error box from *Ariel 5* data is wrong (see finding chart), the correct coordinates are about 3′SW [168].

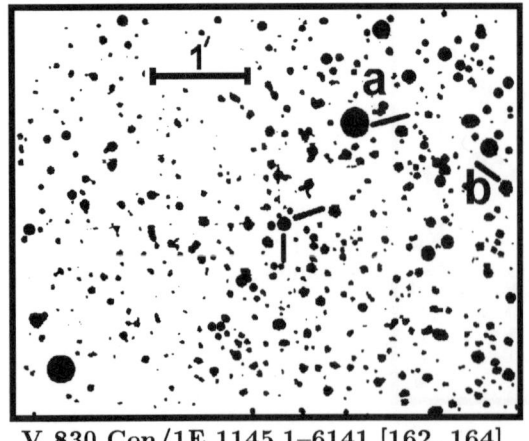

V 830 Cen/1E 1145.1−6141 [162, 164]

α (2000.0)= $11^h47^m28^s.6$
δ (2000.0)=$-61°57'13''.5$
Type: X-ray persistent binary
Spectrum: B2 Iae
  Min. V=JD 2445030.501(7)
    V=13.07–13.17
    P=$5^d.648(2)$

	a	b
V	9.33	12.31
B	9.93	13.48
U	10.02	14.18
R	8.99	11.71

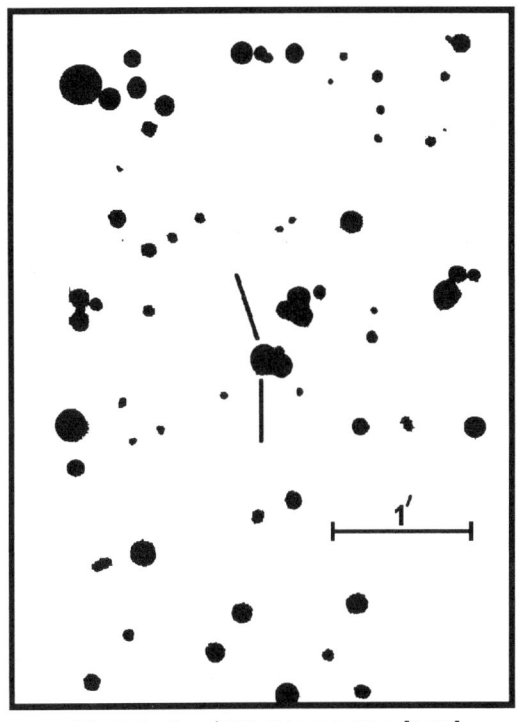

α (2000.0)=$01^h 18^m 31^s.9$
δ (2000.0)=$63°44'23''.9$
Type: X-ray transient
Spectrum: O–B3 V
V=14.5–16.3
Max.V=JD 2447942.030(6)
P=$24^d.31535(5)$

V 635 Cas/4U 0115+634 [172]

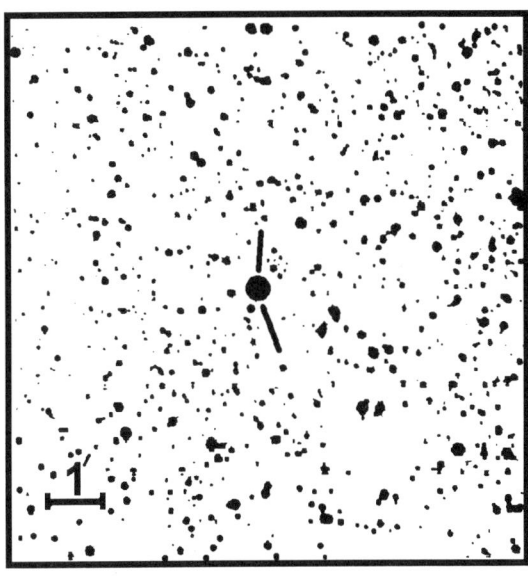

α (2000.0)= $11^h 48^m 00^s.1$
δ (2000.0)=$-62°12'24''.6$
Type: X-ray transient
Spectrum: B0.5 III–V
V=8.93–9.39
Max.V=JD 2442525.5(1)
P=$187^d.5 \pm 1^d.0$

V 801 Cen/4U 1145–619/Hen 715
/HD 102567 [46]

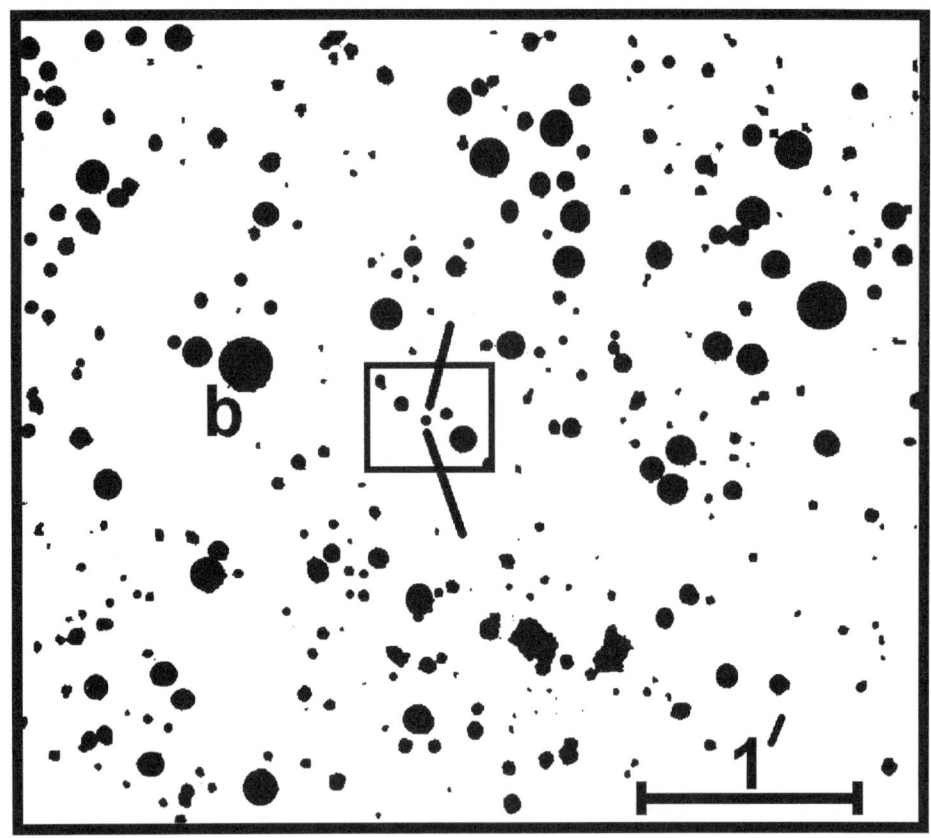

BR Cir/Cir X-1/4U 1516–56 [227]

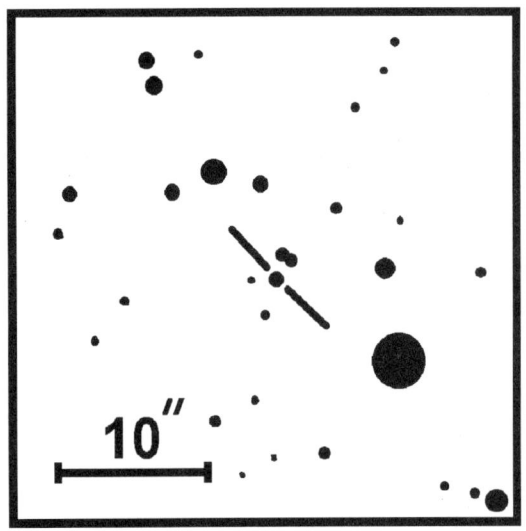

α (2000.0)=$15^h20^m40^s.9$
δ (2000.0)=$-57°09'59''.9$
  Type: X-ray transient
  Spectrum: G5 or later
    K=10.95–11.38
    P=$16^d.5696(1))$
Min. of radio
  outbursts =JD 2444618.23(2)
  Star "b" [445] has:
    V=12.16
    B=13.43
    U=14.33
Cir X-1/BR Cir is located in a highly obscured region of the galactic plane. It is surrounded with a diffuse radio nebula extending over an area of 5 arcmin^2.

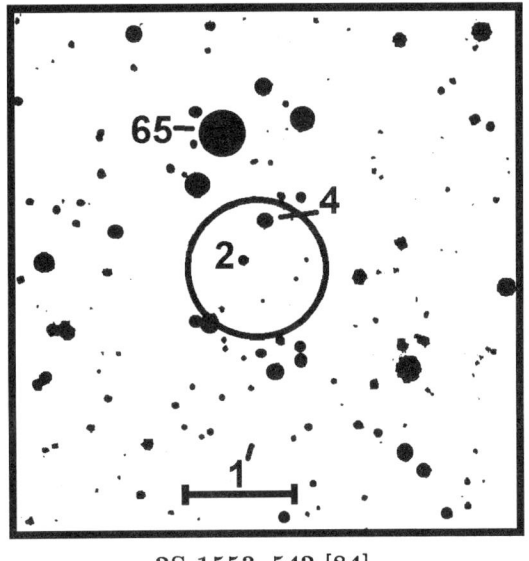

α (2000.0)=$15^h 57^m 49^s.2$
δ (2000.0)=$-54°24'51''.7$
  Type: X-ray transient
Spectrum: Be?
   $B>17$
 Max.X=JD 2442597.3(6)
   $P=30^d.6 \pm 2^d.2$

Star	V
2	15.0
4	13.8

2S 1553–542 [84]

No optical counterpart has yet been identified. The X-ray coordinates with the probable error of 30'' are given. The star "65" is HD 142565.

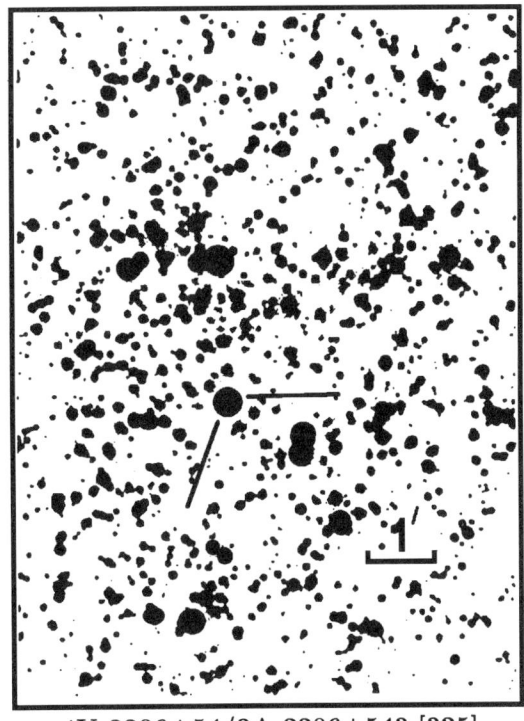

α (2000.0)=$22^h 07^m 57^s.0$
δ (2000.0)=$54°31'05''.8$
  Type: X-ray transient
Spectrum: B0–2 III–V
   $V=9.80-9.90$

Massive X-ray transient? The object shows an average luminosity 100–1000 times lower than for similar X-ray objects. It is possible that low outburst intensity of this star may be associated with unusually low envelope density.

4U 2206+54/3A 2206+543 [325]

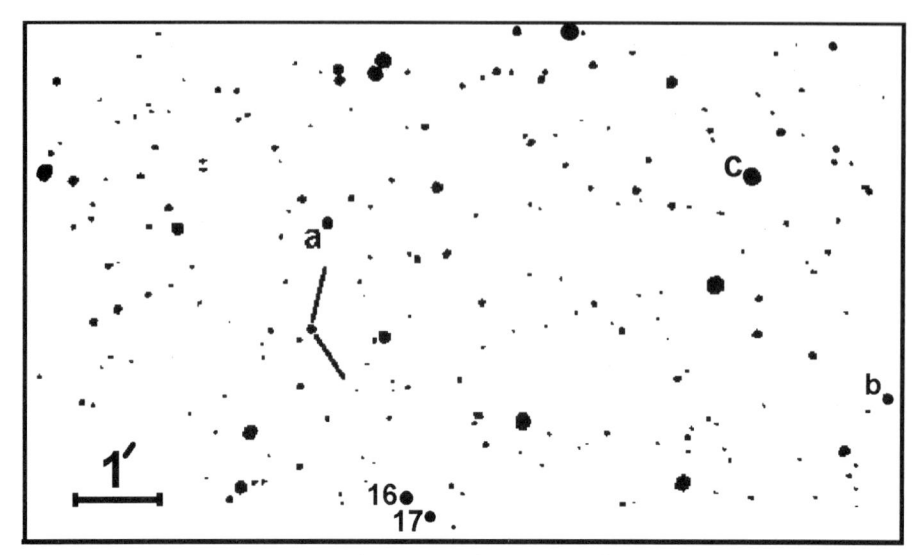

4U 1538–52/QV Nor/Nor X-2 [83, 163]

α (2000.0)= $15^h42^m23^s.3$
δ (2000.0)=$-52°23'10''.1$
  Type: X-ray persistent binary
  Spectrum: B0 Iabe
    B=16.19–16.31
    Min. V=JD 2445514.18(5)
    P=$3^d.72854(2)$

Star	V	B–V	U–B
a	13.60(2)	0.80(2)	0.25
b	13.1	0.2	...
c	11.6	1.0	...

Star "c" is CPD–51°8390. Stars "16" and "17" are variable.

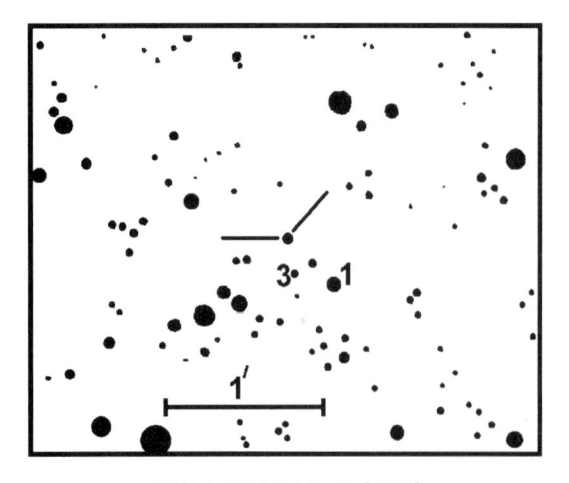

EXO 2030+375 [230]

α (2000.0)=$20^h32^m15^s.4$
δ (2000.0)=$37°38'18''.6$
  Type: X-ray transient
  Spectrum: Be
    V=19.6–19.76
    Max. V=JD 2446268(13)
    P=$45^d.9^{+1.8}_{-1.3}$

Star	Spectrum	V
1	A2–4 V	18.48(4)
3	G–K V	19.72(8)

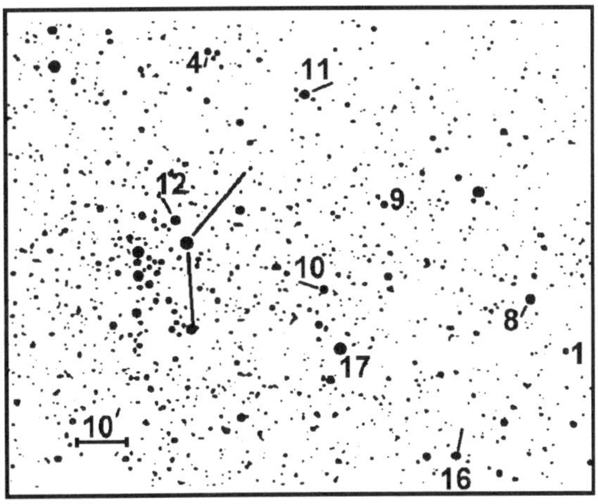

α (2000.0)= $17^h03^m57^s.0$
δ (2000.0)=$-37°50'39''.9$
Spectrum: O5.5–6.5 Iaf
V=6.51–6.60
Min. V=JD 2446161.341(3)
$P=3^d.411723(5)$
Type: X-ray persistent binary
Comparison stars used are:

	SAO 208304	HD 153767
V	7.465	7.430
B	7.605	7.435
U	6.838	...

$\alpha = 17^h01^m.1$ Sp A0
$\delta = -38°13'$

V 884 Sco/4U 1700–377/HD 153919 [269]

In the chart the stars are numbered:

Star	1	4	8	9	10	11	12	16	17
V	9.419	8.368	7.620	9.627	9.285	7.665	8.717	7.424	7.39

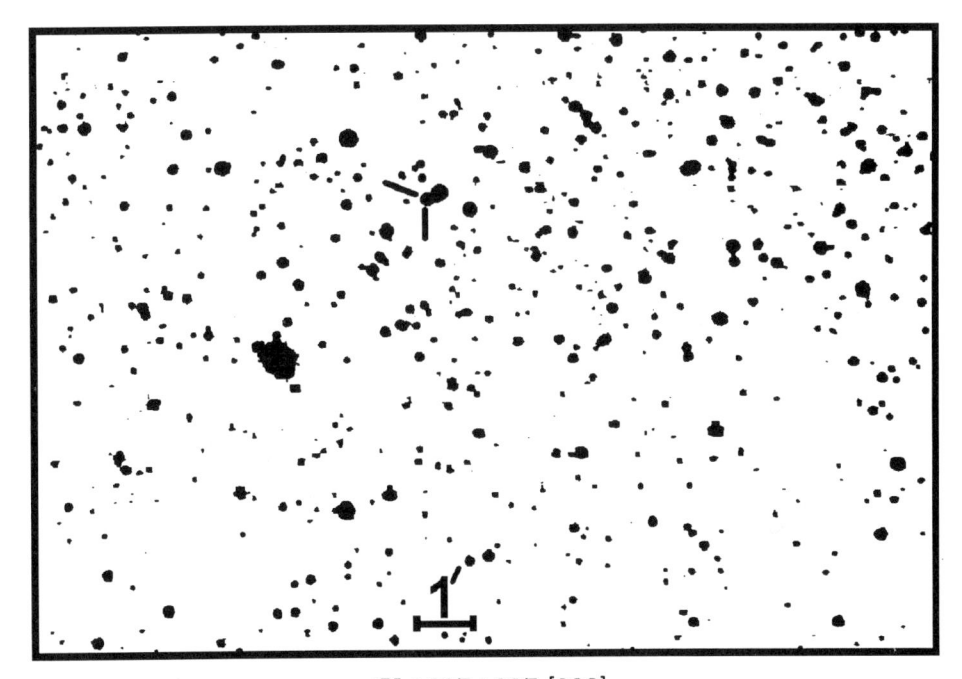

4U 1907+097 [302]

α (2000.0)=$19^h09^m37^s.9$ *Spectrum:* early B
δ (2000.0)=$09°49'48''.3$ *Max. V*=JD 2445576.316(1)
V=16.3–16.4 $P=8^d.375(4)$
Type: X-ray transient

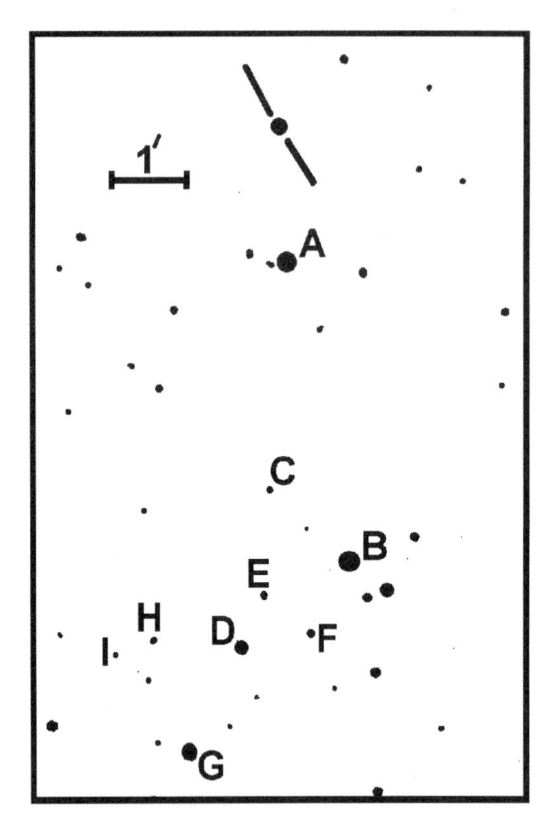

α (2000.0)= $12^h 26^m 36^s.9$
δ (2000.0)=$-62°46'13''.3$
  Type: X-ray transient
Spectrum: B1.5–2 Iae
  V=10.83–10.90
  Max.V=JD 2443906.56(16)
  $P=41^d.508(7)$

Star	V	B–V	U–B
A	9.903	0.180	-0.260
B	9.710	1.355	1.190
C	13.53:	0.56	-0.10
D	11.26	0.33	0.24
E	13.98	0.50	-0.14
F	13.48	1.04	0.12
G	10.54	0.20	-0.30
H	15.13	0.04	-0.55:
I	14.58	1.01	...

BP Cru/GX 301–2/4U 1223–62/WRA 977 [268]

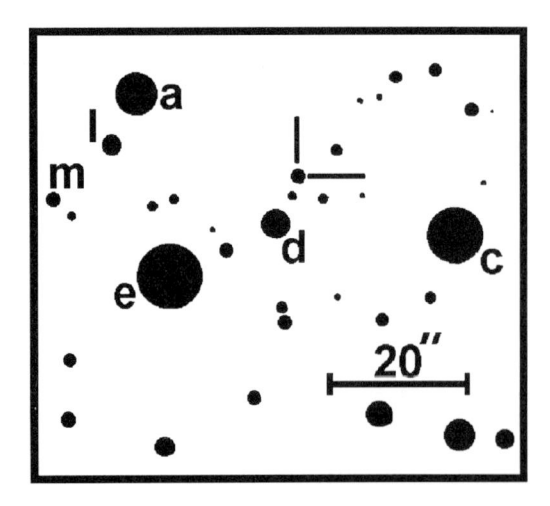

α (2000.0)=$20^h 32^m 25^s.8$
δ (2000.0)=$40°57'27''.9$
  Type: True WR$_2$ + c
         X-ray source
Spectrum: WN7
  K=11.4–12.0
  Min.V=JD 2440949.894(3)
  $P=0^d.19968396(2)$

Star	I
a	12.96(2)
c	12.30(2)
d	14.24(2)
e	11.95(2)
l	17.6(1)
m	18.6(1)

V 1521 Cyg/Cyg X-3/4U 2030+407
[379, 393]

LOW-MASS X-RAY CLOSE BINARIES

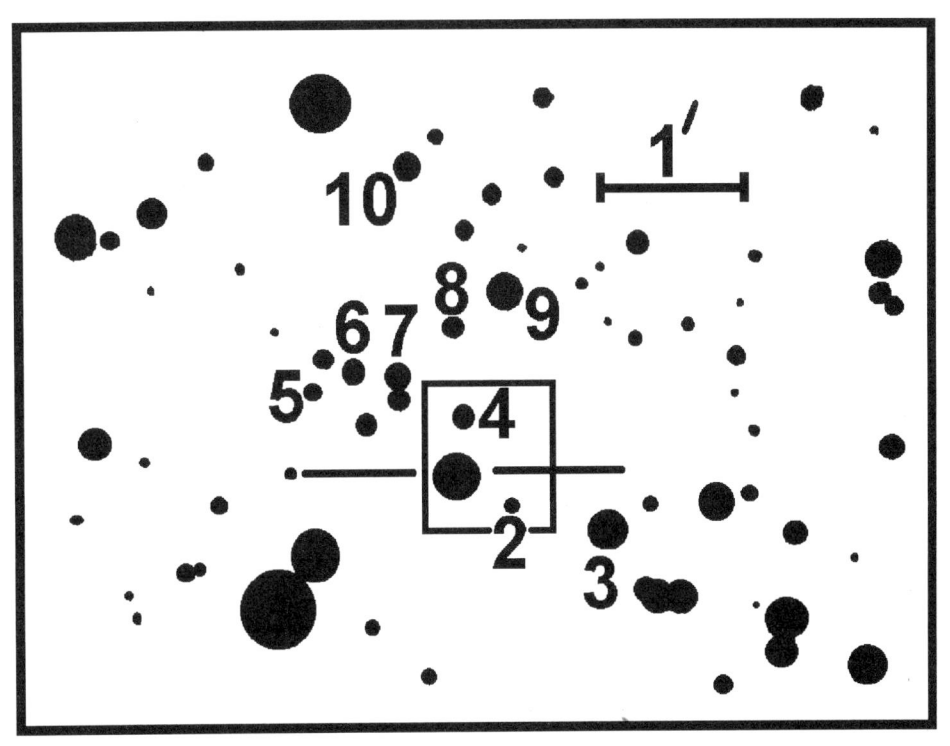

V 518 Per/GRO J 0422+32/XN Per 1992 [63, 73]

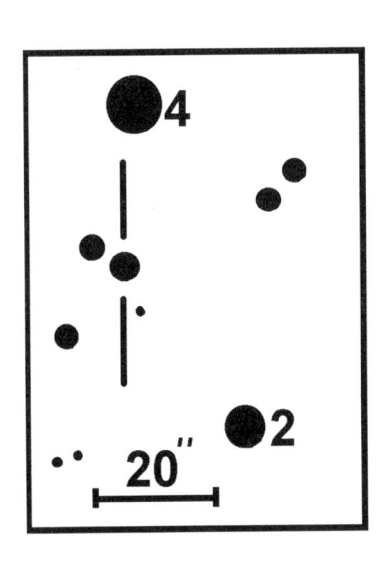

α (2000.0)=$04^h24^m56^s.1$
δ (2000.0)=$33°01'17''.3$
Type: X-ray Nova
Spectrum: $\sim$ M0 V
V=13.25–20.67
Min.V=JD 2449338.31
$P=0^d.21213(2)$

Star	V	B − V	V − I
2	17.01	0.94	1.08
3	14.14	0.82	1.00
4	16.62	1.04	1.18
5	17.74	1.02	1.17
6	16.92	0.88	1.02
7	15.74	0.98	1.14
8	15.67	1.09	1.20
9	14.52	1.30	1.40
10	14.35	1.30	1.39

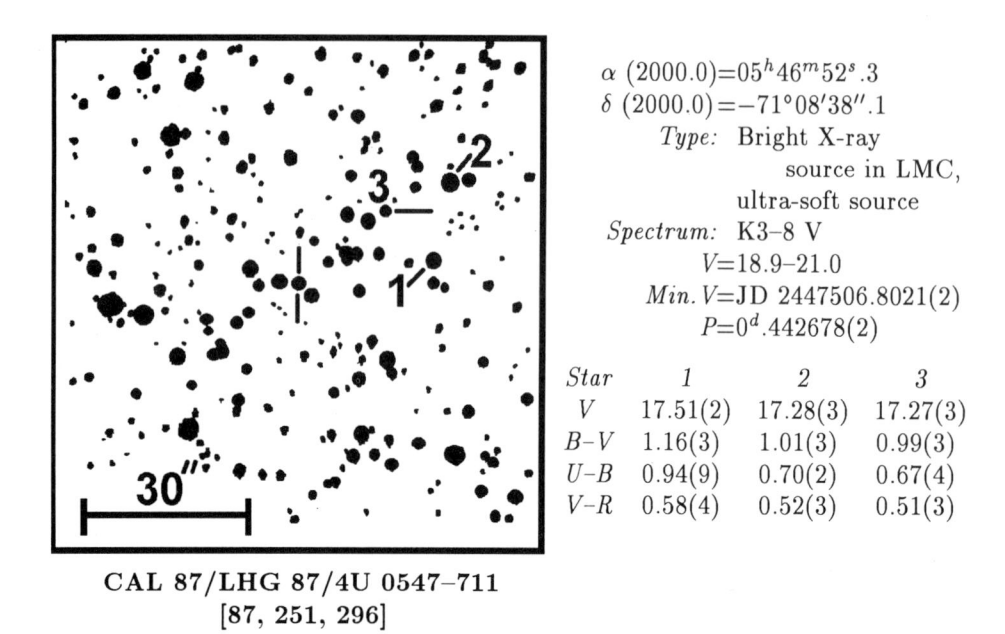

α (2000.0)=$05^h46^m52^s.3$
δ (2000.0)=$-71°08'38''.1$
Type: Bright X-ray
source in LMC,
ultra-soft source
Spectrum: K3–8 V
V=18.9–21.0
Min. V=JD 2447506.8021(2)
P=$0^d.442678(2)$

Star	1	2	3
V	17.51(2)	17.28(3)	17.27(3)
B-V	1.16(3)	1.01(3)	0.99(3)
U-B	0.94(9)	0.70(2)	0.67(4)
V-R	0.58(4)	0.52(3)	0.51(3)

CAL 87/LHG 87/4U 0547–711
[87, 251, 296]

The system has a visual companion with $V \sim 20.8$ mag located $0''.9$ away at position angle $340°$.

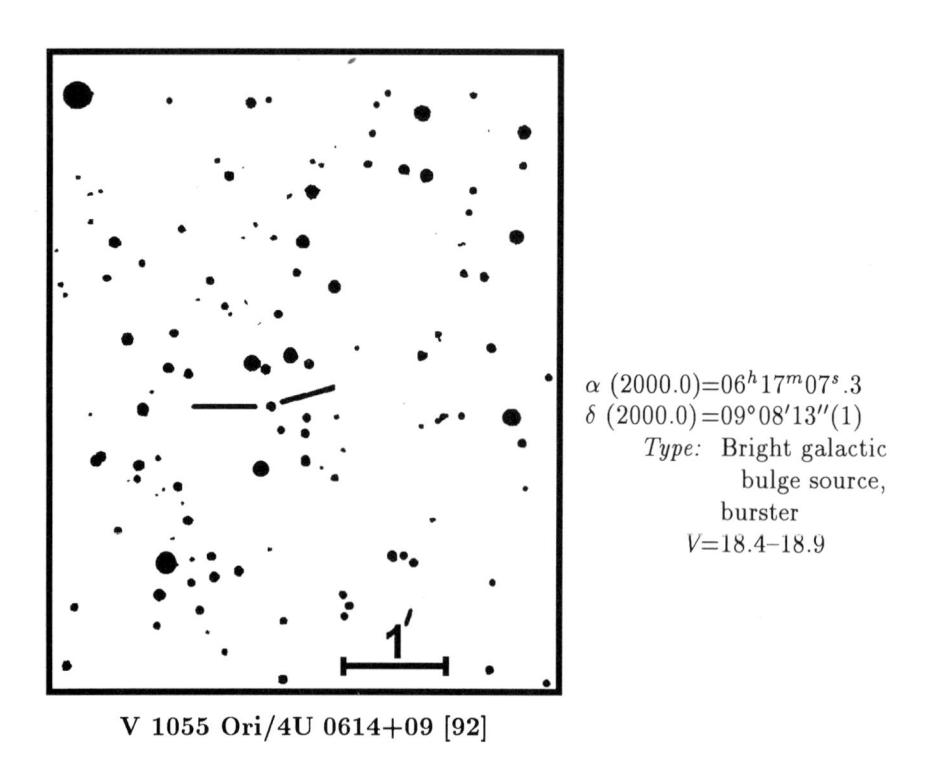

α (2000.0)=$06^h17^m07^s.3$
δ (2000.0)=$09°08'13''(1)$
Type: Bright galactic
bulge source,
burster
V=18.4–18.9

V 1055 Ori/4U 0614+09 [92]

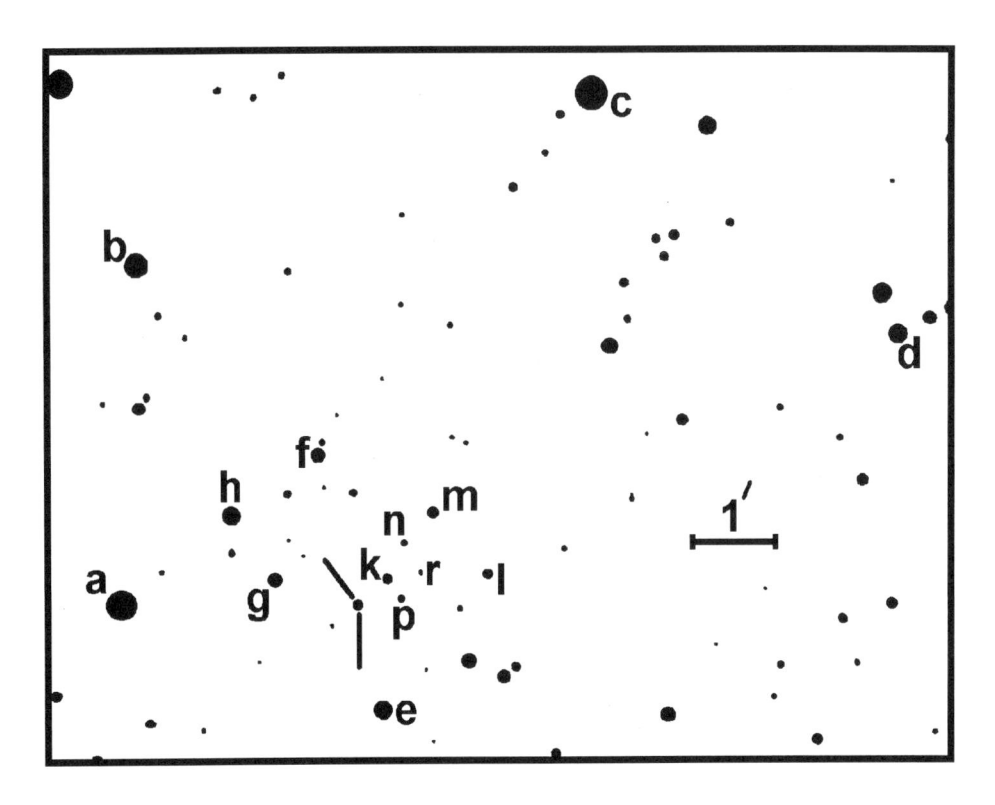

V 616 Mon/A 0620−00

α (2000.0)=$06^h22^m44^s.5$ B=11.26−20.2
δ (2000.0)=$-00°20'45''.7$ Min.V=JD 2445477.989(5)
Spectrum: K3–4 V $P=0^d.323014(4)$
          Type: X-ray transient, black hole candidate

Comparison stars [318]:

Star	a	b	c	d	e	f	g
B	9.79	10.04	10.98	11.78	12.56	13.56	13.57
B–V	0.12	0.15	1.90	0.31	0.47	0.62	1.30

Star	h	k	l	m	n	p	r
B	14.50	15.72	15.96	16.46	16.75	17.05	17.5:
B–V	1.39	0.80	...	...	...	...	...

The X-ray nova of 1917 and 1975. The mass function of the compact object in V 616 Mon, $f(M) = 2.91(8)M_\odot$ [213], established it as a strong **black hole candidate**. In the low state the optical orbital light curve of the system shows ellipsoidal variability.

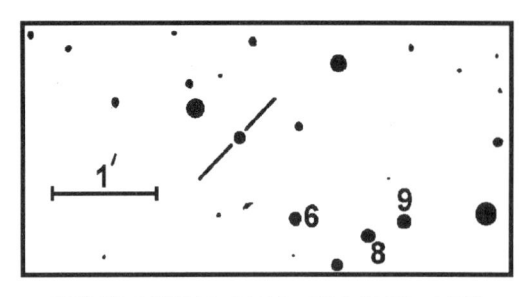

UY Vol/EXO 0748−676 [295, 377]

α (2000.0)=$07^h 48^m 33^s.7$
δ (2000.0)=$-67°45'08''.6$
Type: Low-mass transient
Spectrum: later than M0 V:
   B=15.3−23.0
   Min.V=JD 2446111.574549(5)
   $P=0^d.159337794(9)$

Star	V	B−V
6	15.252(8)	0.90(1)
8	15.364(9)	0.718(8)
9	15.24(1)	0.75(2)

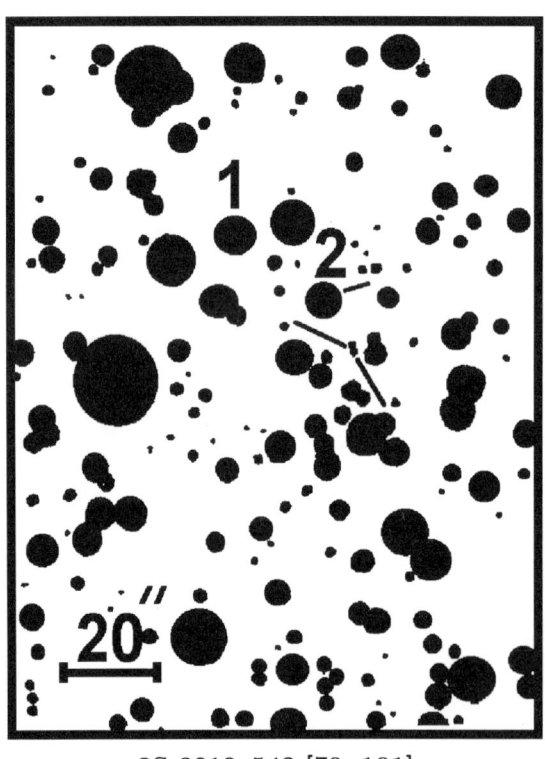

2S 0918−549 [70, 101]

α (2000.0)=$09^h 20^m 26^s.8$
δ (2000.0)=$-55°12'24''(3)$
Type: Bright galactic
   bulge source,
   V=20.87−21.08

Star	1	2
V	15.80(2)	16.30(2)
B−V	1.27(5)	0.75(2)
U−B	1.2(1)	0.16(7)

The optical counterpart — a blue star — forms a close visual pair with the nearest neighbour star ($V = 20.7(1)$, $B = 21.8(1)$, $U > 22.3$) located 2″NE of the 2S 0918−549.

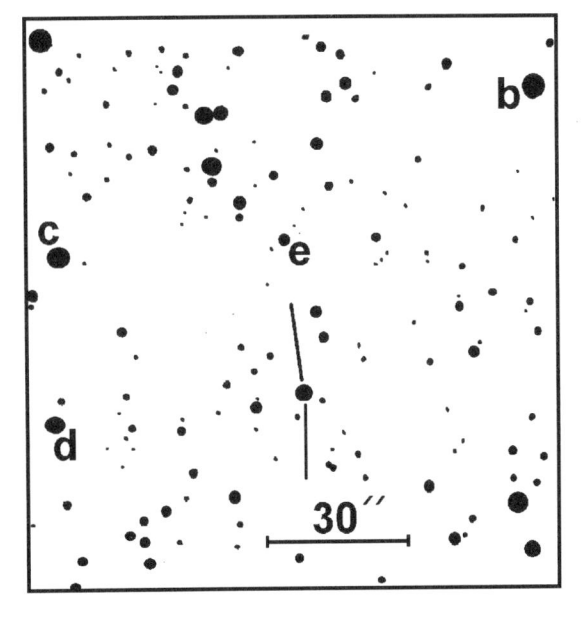

α (2000.0)=$05^h43^m06^s$
δ (2000.0)=$-68°22'38''$(1)
Type: Ultrasoft source
in LMC
V=16.2–17.1
Min.V=JD 2446051.58(2)
$P=1^d.0436 \pm 0^d.0044$

Star	V
a	12.7(1)
b	14.6(1)
c	14.8(1)
d	15.6(1)
e	17.0(1)

CAL 83/4U 0543−682/LHG 83 [323]

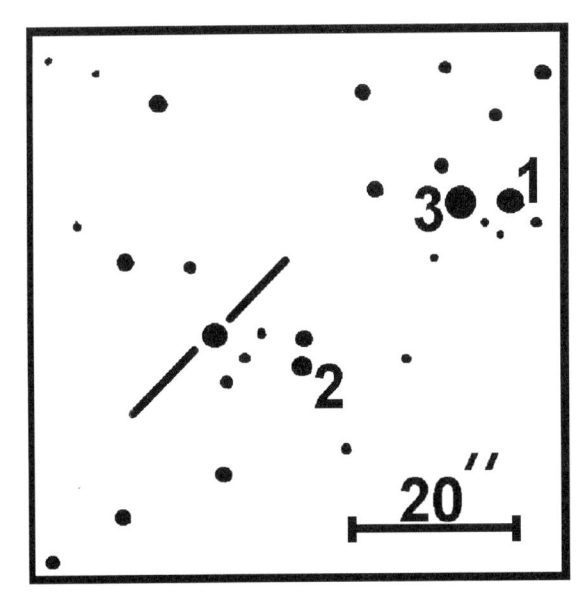

α (2000.0)=$11^h26^m26^s.7$
δ (2000.0)=$-68°40'32''.6$
Type: X-ray transient,
ultrasoft source,
black hole
candidate
Spectrum: K0–4 V
V=13.5–21.0
Min.V=JD 2448716.018(2)
$P=0^d.4333(6)$

Star	V	B−V
1	13.78	0.82
2	15.23	1.24
3	15.26	1.37

GU Mus/GRS 1124−684/XN Mus 1991 [19]

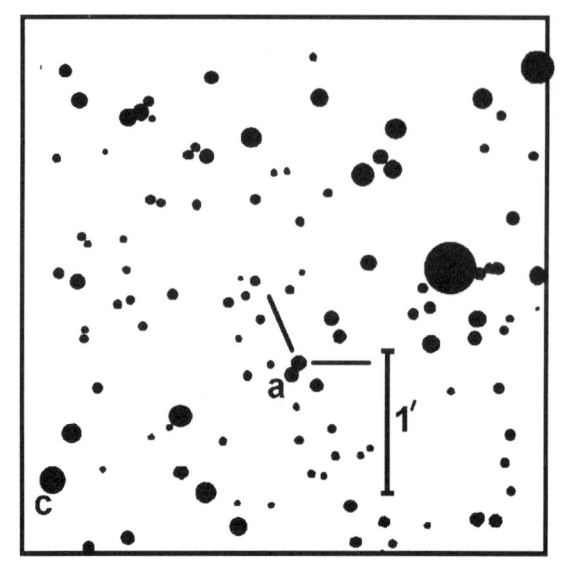

α (2000.0)=$09^h22^m34^s.8$
δ (2000.0)=$-63°17'41''(1)$
Type: Bright galactic
bulge source,
a coronal source
Spectrum: F8–G3 III
B=15.5–18.0
Min. V=JD 2446249.18(1)
P=$9^d.0115(5)$

Star	a	c
V	16.62	13.66
B–V	1.0	0.70
U–B	0.45–0.6	0.05

V 395 Car/2S 0921–630 [68, 195]

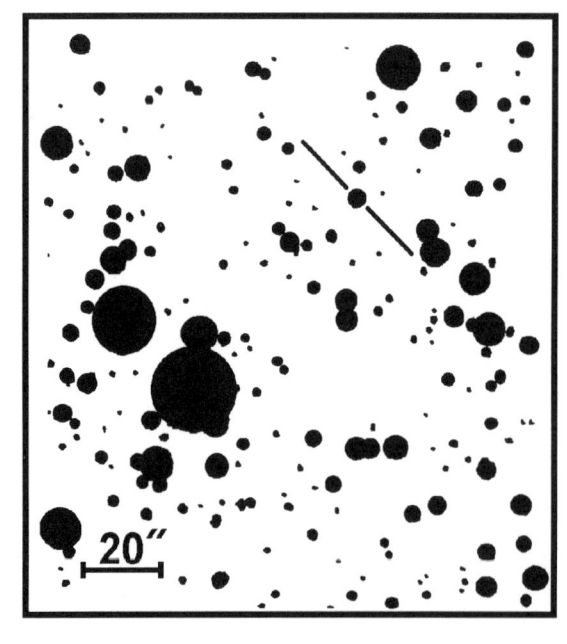

α (2000.0)=$13^h58^m09^s.7$
δ (2000.0)=$-64°44'05''.0$
Type: Soft X-ray
transient
V=16.8–17.0:

The source demonsrates optical variations of $>0^m.1$ in 8^h. There are two X-ray outbursts reported in this region: MX 1353–64 and Cen X-2. However, these two sources showed characteristics different from those of BW Cir.

**BW Cir/GS 1354–64/MX 1353–64?
/XN Cir 1987 [177, 267]**

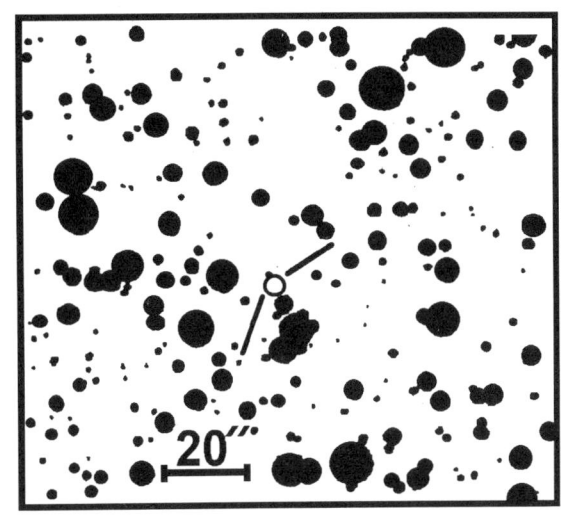

α (2000.0)=$13^h 26^m 36^s.1$
δ (2000.0)=$-62°08'10''.0$
 Type: X-ray burster
 $P=0^d.1221(2)$

The circle on a red-sensitive finding chart [254] denotes the *HRI* uncertainty region for an optical counterpart.

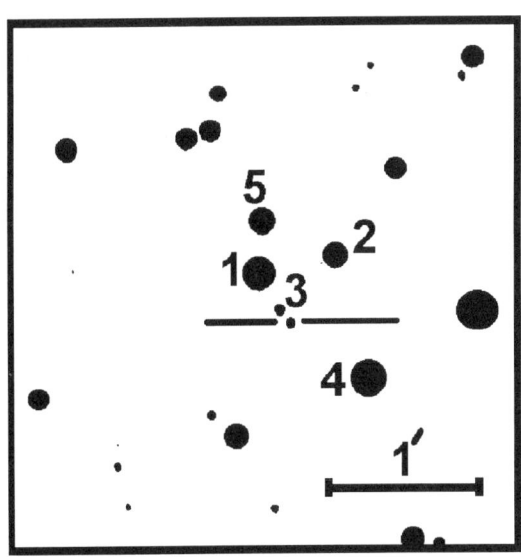

X 1323-619 [254]

α (2000.0)=$14^h 58^m 22^s.0$
δ (2000.0)=$-31°40'07''.4$
 Type: X-ray transient
 Spectrum: K7 V
 V=12.8-19.0
 Min. V=JD 2445778.342(4)
 $P=0^d.629063(5)$

The X-ray nova Cen X-4 underwent two outbursts, in 1969 and 1979. During the decay of these outbursts, Cen X-4 showed strong X-ray type I **bursts**.

V 822 Cen/A 1455-31/Cen X-4
[39, 72, 85, 313]

Star	J	H	K	V	B	U
1	...	...	...	16.537(7)	17.278(8)	...
3	15.74(2)	15.267(8)	15.07(9)	17.824(9)	18.90(2)	...
2	15.043(9)	14.587(6)	14.429(5)	17.02	17.99	18.20
4	12.833(7)	12.505(6)	12.440(9)	...	...	...
5	...	...	...	16.57	17.28	17.49

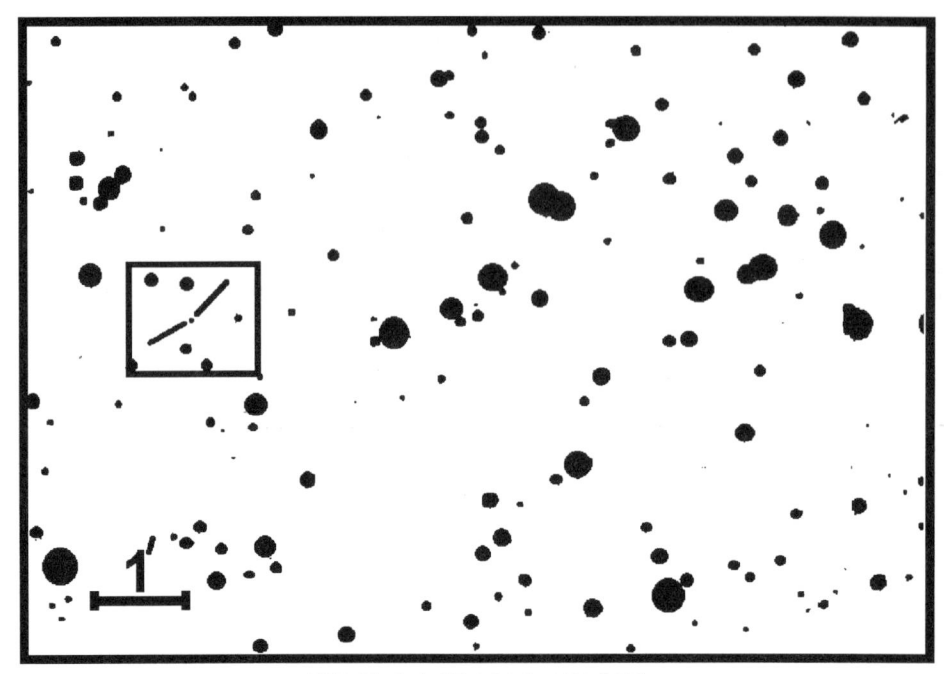

KZ TrA/4U 1626−673 [47]

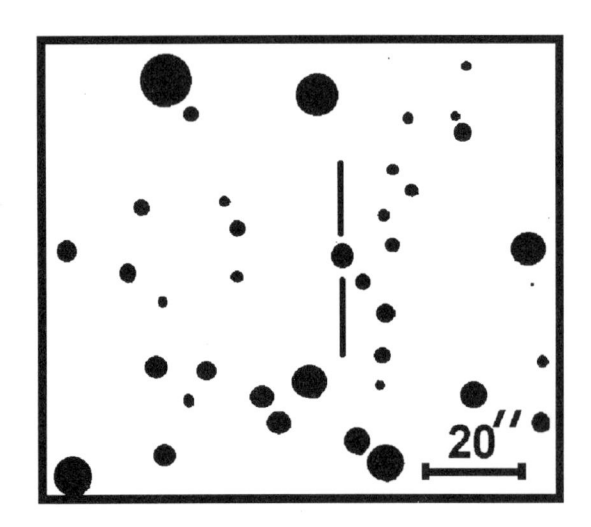

α (2000.0) $=16^h32^m16^s.8$
δ (2000.0) $=-67°27'43''(1)$
Type: Bright galactic
bulge source
X-ray pulsar
$V=18.23-19.10$
$P=0^d.028762:$

$P_s=7^s.6806273(5)$
at JD 2443227
$<\dot{P}_s/P_s>=-1.8\times 10^{-4}$ yr^{-1}

A relatively large fraction ($\sim$2.4%) of the optical light from the system is pulsed at the 7.7 s period of the X-ray source. In addition to periodic pulsations, the X-ray and optical intensity of this object shows quasi-periodic flaring behaviour with a characteristic time scale of $\sim 1 \times 10^3$ s. The oscillations are considerably stronger in the 1.5–3 keV X-ray energy band than in the 6–12 keV band.

HECBSs. Finding charts

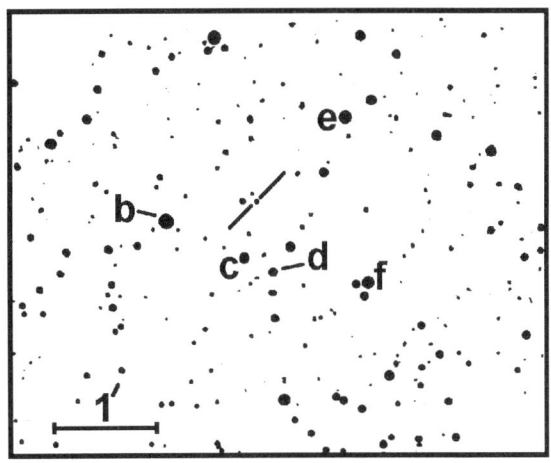

GR Mus/4U/XB 1254–690 [136, 268]

α (2000.0)=$12^h 57^m 37^s.2$
δ (2000.0)=$-69°17'21''(1)$
Type: Bright galactic bulge source, type I burster
Spectrum: M–K:
V=18.1–19.4
Min. V=JD 2445735.693(4)
$P=0^d.163890(9)$

Star	V	B–V	U–B
b	13.26	0.64	0.08
c	13.82	1.26	1.14
d	13.80	1.60	...
e	12.94	0.69	0.29
f	13.04	1.05	0.95

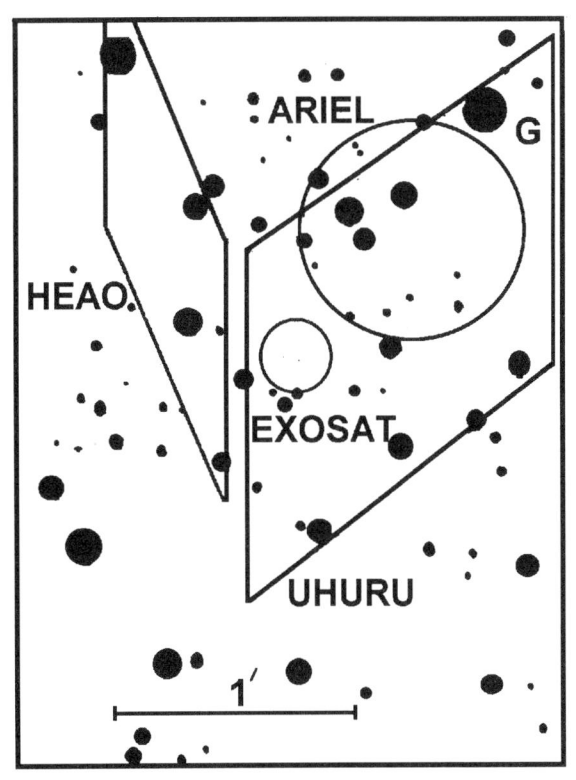

4U 1630–47 [137, 253]

α (2000.0)=$16^h 34^m 00^s.0$
δ (2000.0)=$-47°23'38''.9$
Type: ultrasoft recurrent X-ray transient

The *EXOSAT* error circle does not include the **G** candidate star of Grindlay [137] which is 83'' from the center. The finding chart for 4U 1630–47 (*V* band) is given in [253]. "*EXOSAT*" indicates the *EXOSAT* position, "*ARIEL*", "*HEAO*" and "*UHURU*" indicate the *Ariel 5*, *HEAO 1*, and *Uhuru* position, respectively.

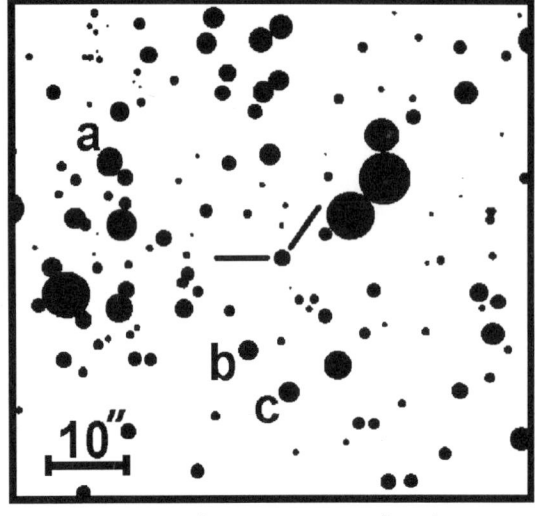

α (2000.0)=$16^h01^m02^s.3$
δ (2000.0)=$-60°44'17''.4$
Type: Bright galactic
bulge source
V=18.6–19.5
P=$0^d.0.198(1)$

Star	V	B–V	V–R
a	16.13(3)	0.83(3)	0.47(3)
b	17.21(3)	1.08(4)	0.63(4)
c	17.30(3)	1.15(6)	

LU Tra/4U 1556–605 [231]

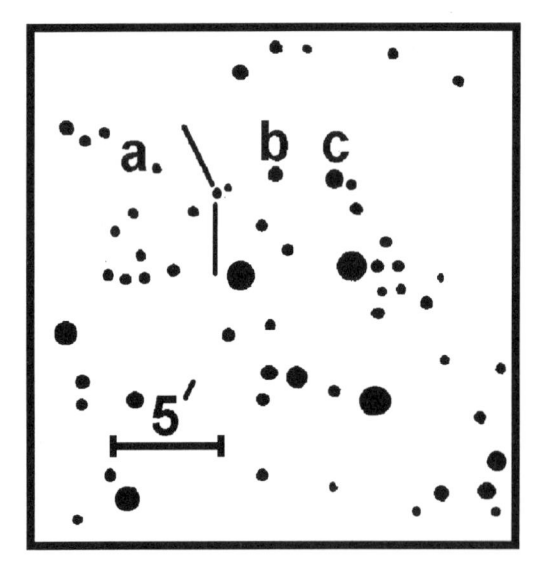

α (2000.0)=$16^h57^m50^s.1$
δ (2000.0)=$35°20'33''.8$
Type: Bright galactic
bulge source
X-ray pulsar
Spectrum: B0–F5 Ve
B=12.8–15.2
Min.V=JD 2447664.40043(2)
P=$1^d.70016772(1)$
P_x=$1^s.23775776(1)$
at JD 2447664

Star	V	B	U
a	12.56(2)	13.12(1)	13.10(2)
b	13.16(2)	14.28(2)	15.29(6)
c	13.47(2)	14.10(3)	14.19(7)

HZ Her/Her X-1/3U 1653+35 [186, 199]

The system lies **intermediate between high and low-mass X-ray binaries**. The ephemeris of long-term variability is: $Min.long$ = JD 2444784.73 + 34.875 · E. The instability of the long-term period is about $\pm 2P_{orb}$ on 600–900 day time scale.

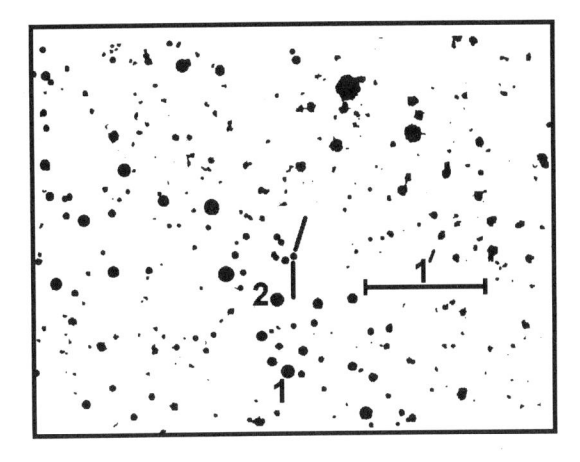

α (2000.0)=$05^h20^m29^s.3$
δ (2000.0)=$-71°57'36''(8)$
Type: Bright X-ray
source in LMC
V=17.8–19.1
Min. V=JD 2447506.36
P=$12^d.539$

Star	1	2
V	15.84(4)	16.61(3)
B	16.74(6)	17.36(5)
U	17.29(7)	17.64(7)
R	15.32(5)	16.17(9)

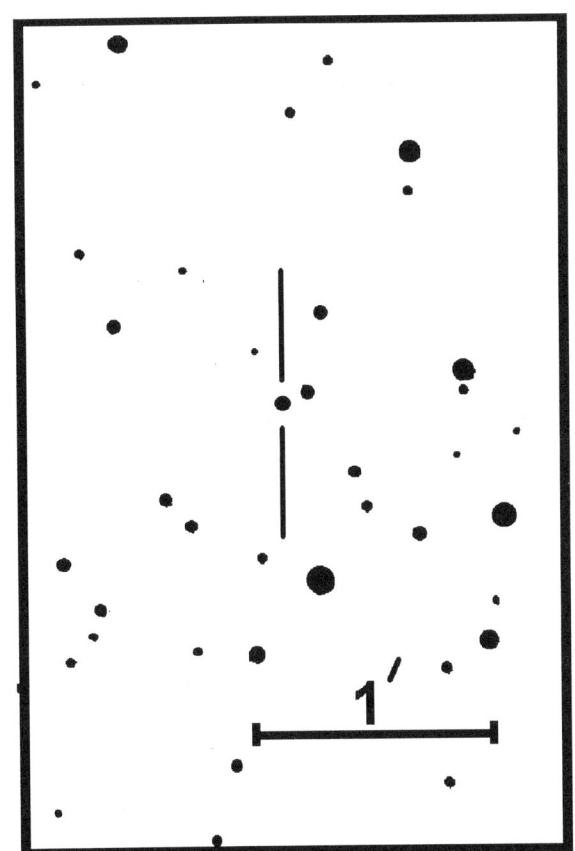

LMC X-2/4U 0521–720 [88, 250]

α (2000.0)=$17^h02^m06^s.3$
δ (2000.0)=$-29°56'45''(1)$
Type: soft X-tay transient
eclipsing binary
burster
Spectrum: G5–K0 V
V=18.3–23
P=$0^d.29650474(7)$

The system exhibits high ($\sim(1.5-3) \times 10^{37}$ erg/s) and low ($< 10^{36}$ erg/s) levels of quiescent state.

V 2134 Oph/MXB 1658–298 [106]

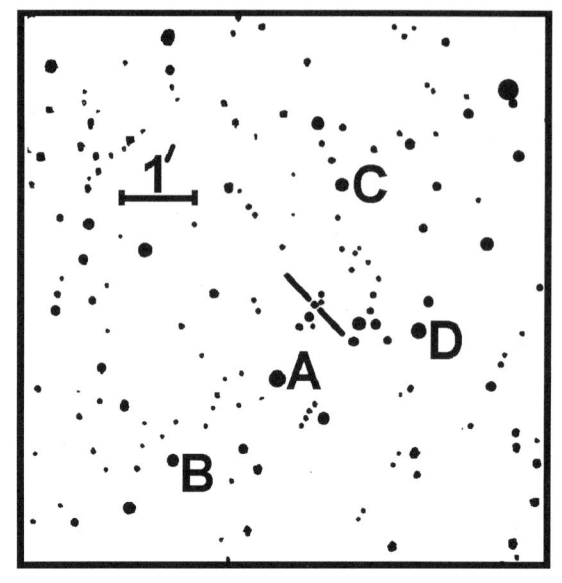

α (2000.0)=$15^h47^m08^s.6$
δ (2000.0)=$-47°40'09''.6$
Type: recurrent X-ray
slow transient
Spectrum: M2 III
V=14.94–16.0

Star	V	B–V	U–B
A	13.90	0.87	0.20
B	14.79	0.76	0.47
C	13.92	0.87	...
D	12.06	1.69	2.3:

HL Lup/4U 1543–475 [117, 268]

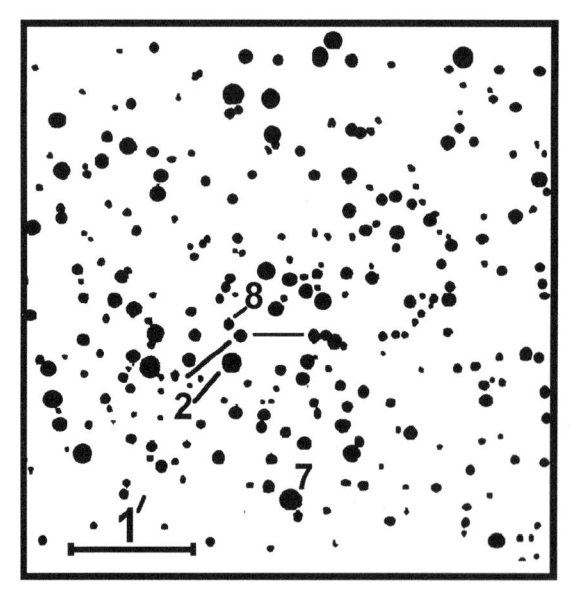

α (2000.0)=$17^h02^m49^s.4$
δ (2000.0)=$-48°47'21''.5$
Type: bright galactic
bulge source
QPO source
Z source
Spectrum: F2–G5 V
B=15.4–21:
P=$0^d.618(4)$:

Star	V	B	Sp
2	14.0	16.0	K0 IV
7	12.6	14.1	G5 III
8	17.5	19.3	K2 IV-V

V 821 Ara/GX 339–4/4U 1659–487[106,138]

V 821 Ara is a close visual binary system with $\Delta\rho = 1''.13(3)$, p.a. $\sim 125°$, the southern component is V 821 Ara, the northern component has $V = 20.32(5)$, $B - V = 1.5(1)$, $V - I = 1.84(8)$.

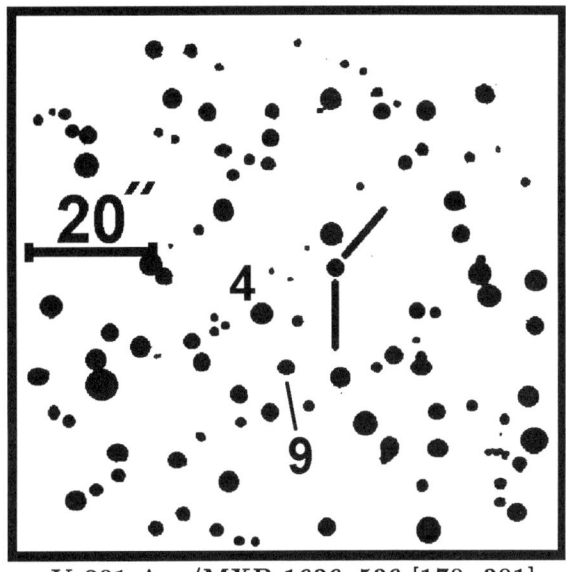

α (2000.0)=$16^h40^m55^s.5$
δ (2000.0)=$-53°45'05''(1)$
Type: optical and X-ray
burster
Spectrum: ~M0 V
V=17.12–18.55
Max.V=JD 2444431.093(5)
$P=0^d.1584949(9)$

Star	V
4	16.2(1)
9	17.8(1)

V 801 Ara/MXB 1636–536 [170, 301]

Others comparison stars [301]:

Star	$\Delta\alpha,''$	$\Delta\delta,''$	U	B	V	R	I
1	-21.6	93.6	17.3(2)	15.5(1)	14.1(1)	13.1(1)	12.3(2)
2	151.2	-15.8	15.8(2)	15.5(1)	14.5(1)	14.0(1)	13.5(2)

$\Delta\alpha$, $\Delta\delta$ designate the distances (in arc seconds) between the comparison star and V 801 Ara.

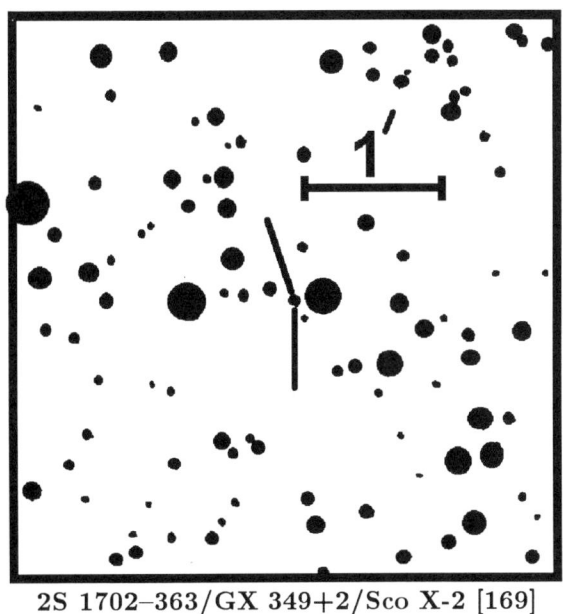

α (2000.0)=$17^h05^m44^s.9$
δ (2000.0)=$-36°25'24''(1)$
Type: Bright galactic
bulge source
Z source
V=18.3–18.7

2S 1702–363/GX 349+2/Sco X-2 [169]

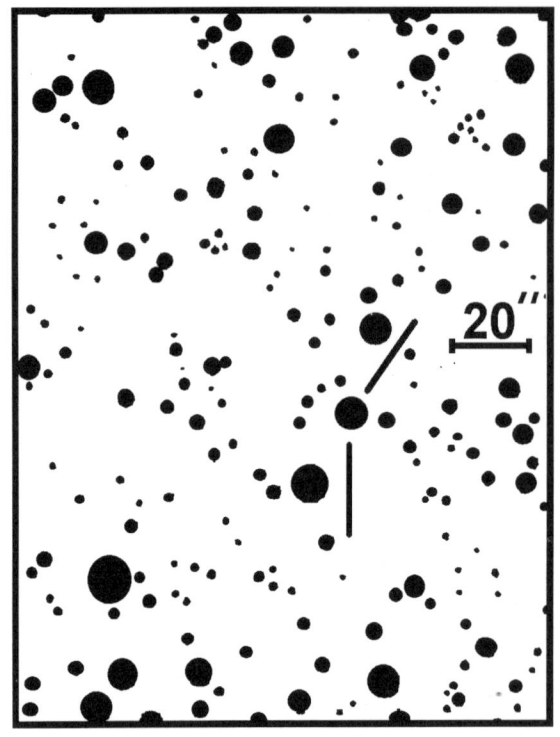

α (2000.0)=$17^h08^m14^s.6(2)$
δ (2000.0)=$-25°05'29''.0$
Type: X-ray transient
$V=16.5-\geq 21$

The variable star II Oph is 2'NW of the Nova.

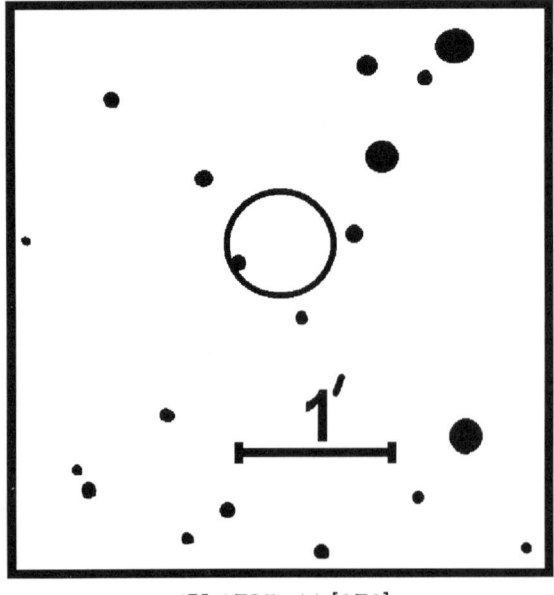

V 2107 Oph/H 1705–250 [135]

α (2000.0)=$17^h08^m55^s.0$
δ (2000.0)=$-44°05'59''.3$
Type: burster
Atoll source
dipped source
$P=0^d.054$:

The source has not yet been optically identified.

4U 1705–44 [171]

HECBSs. Finding charts 43

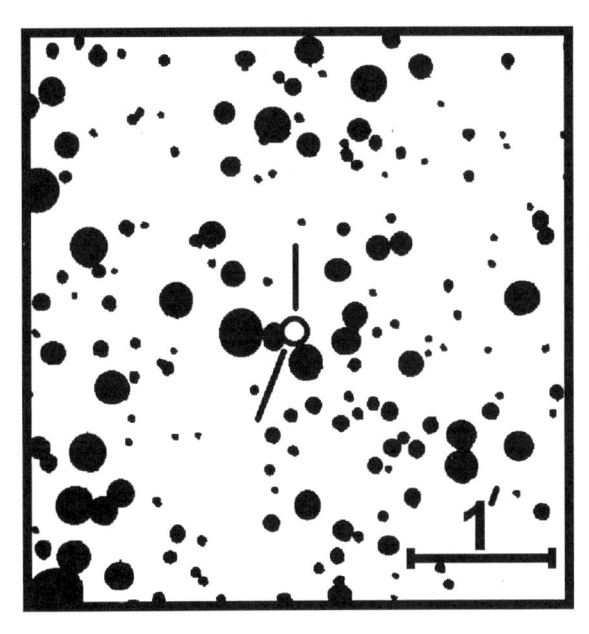

α (2000.0)=$18^h01^m13^s.2(6)$
δ (2000.0)=$-25°44'29''(10)$
Type: bright galactic
bulge source
Z source

GRS 1758–258/GX 5–1 [247]

There is a very red star ("x") within the radio error box, with a faint companion to the NE ("c"). The combination of these stars does not vary significantly ($< 0^m.066$), but star "x" is amongst the reddest objects in the field [247].

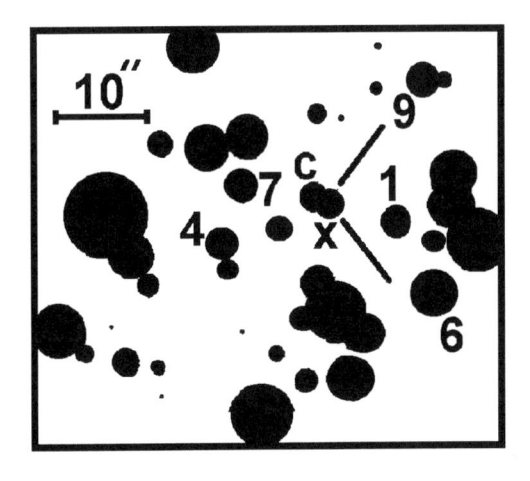

Star	J	H	K
x	16.2(2)	14.48(7)	13.67(4)
c	17.0(3)	15.9(3)	14.8(1)
1	14.76(4)	14.00(5)	13.39(3)
4	14.25(3)	13.88(4)	13.66(4)
6	14.26(3)	13.04(2)	12.35(2)
7	14.13(2)	13.81(4)	13.52(4)
9	15.16(6)	13.85(4)	13.18(5)

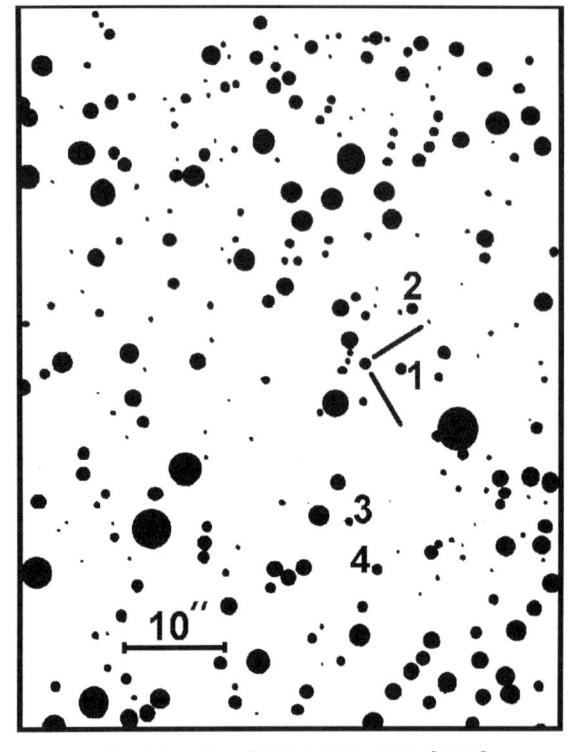

α (2000.0)=$17^h58^m40^s.0$
δ (2000.0)=$-33°48'27''(1)$
Type: bright galactic
  bulge source
  ultrasoft source
  dipped source
V=18.2–18.85
Max.V=JD 2445900.0
$P=0^d.1858$

Star	V
1	17.97(1)
2	18.24(1)
3	18.34(1)
4	17.98(1)

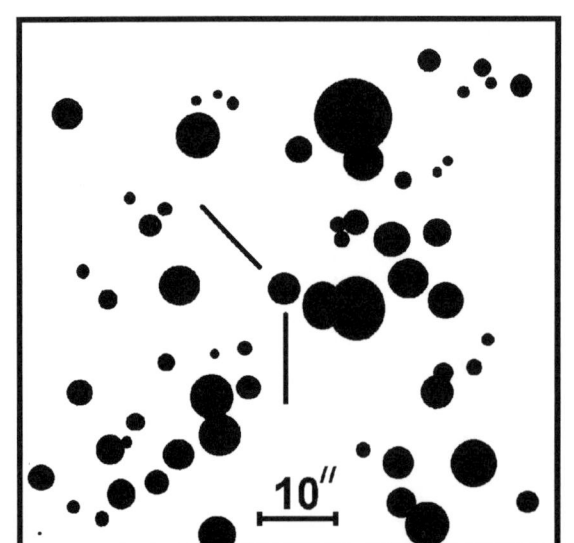

V 4134 Sgr/4U 1755−338 [210]

α (2000.0)=$18^h14^m31^s.0$
δ (2000.0)=$-17°09'28''.2$
Type: bright galactic
  bulge source
  Atoll source
K=11.90–12.30

The finding chart in J band is given in [243].

GX 13+1/4U 1811−17 [64, 243]

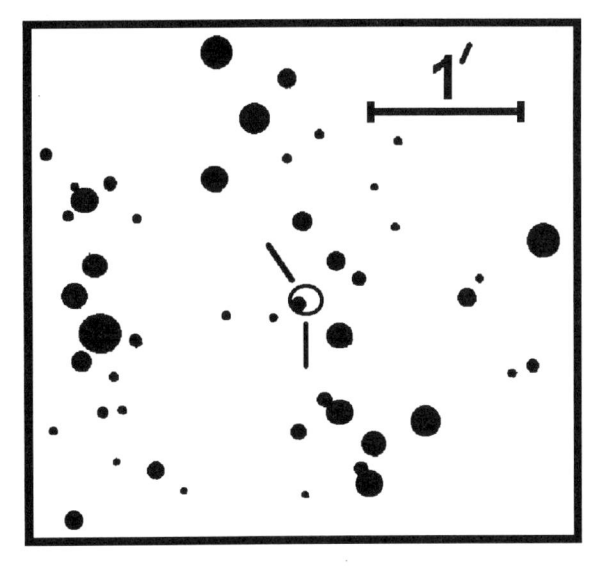

α (2000.0)=$18^h16^m01^s.4$
δ (2000.0)=$-14°02'11''(1)$
Type: bright galactic
       bulge source
       Z source
Spectrum: G–K V(?)
       K=13.63–13.71

NP Ser/4U 1813–14/GX 17+2/Ser X-2 [243]

Accurate coordinates of the rapidly varying radio source are coincident with the X-ray error box and the position of the very red TR star (Sp ~ G8) with a faint companion to the NE (c). In X-rays the object exhibits periods of 31.8 min [400], 1.4 hr [189], 19.5(2) hr [150] and 6.44 days [271]: Bailyn et al. [20] suggest that GX 17+2/NP Ser is a triple system, with a giant star ($M \approx 0.8 M_\odot$, $R \approx 3 R_\odot$) in a 6-day orbit around an LMXB ($M_x \sim 1.6 M_\odot$, $M_c \sim 0.2 M_\odot$) with an orbital period of 1.4 hr. The finding chart with comparison stars is given for K band.

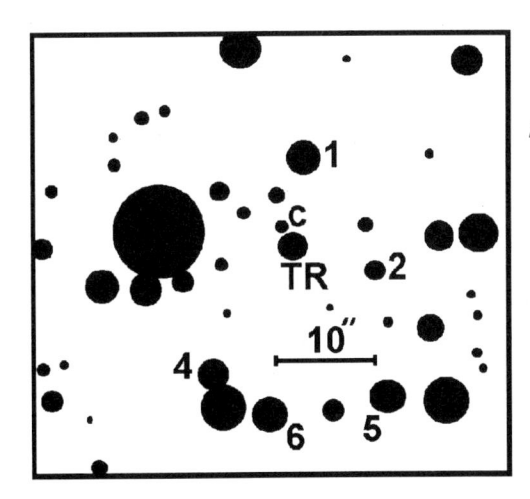

Star	J	H	K
TR	16.2(2)	14.48(7)	13.67(4)
c	17.0(3)	15.9(3)	14.8(1)
1	15.39(9)	14.20(7)	13.56(5)
2	16.2(2)	14.6(1)	14.14(8)
4	15.15(7)	13.95(5)	13.44(4)
5	15.48(9)	13.74(4)	13.11(3)
6	15.14(7)	13.56(4)	12.52(2)

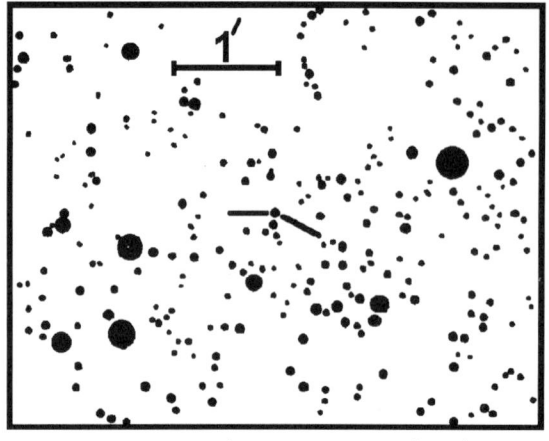

α (2000.0)=$18^h25^m46^s.8$
δ (2000.0)=$-37°06'19''$(1)
Type: bright galactic
   bulge source
   coronal source
Spectrum: K0–M2 V
   B=14.8–16.6
   Min.V=JD 2445615.3095(2)
   P=$0^d.23210898$(4)

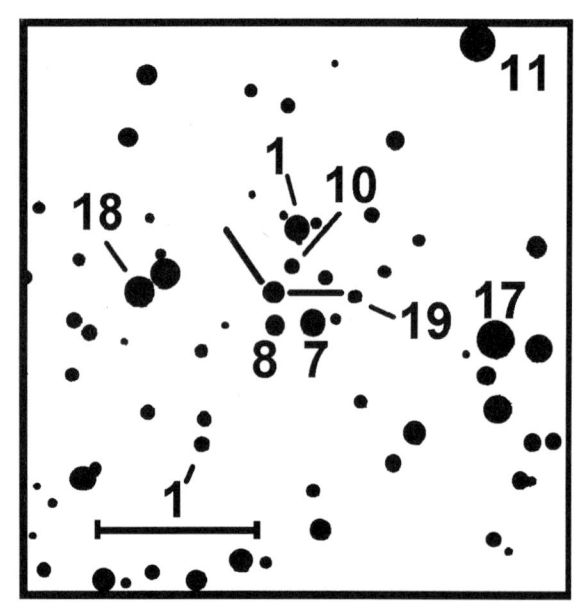

V 691 CrA/4U 1822−371 [136]

α (2000.0)=$18^h39^m57^s.5$
δ (2000.0)=$05°02'08''.5$
Type: burster
   B=18.9–19.2
   P=$3^d.40252$:

MM Ser is a visual binary system. The optical counterpart is the southern, weakest in *V*, component of a close pair with narrow He II emission lines in its spectrum.

MM Ser/Ser X-1/4U 1837+05 [2, 275]

Star	V	B	U	R	Spectrum
1	14.73	15.75	16.57	13.92	...
7	15.09	16.31	16.98	14.01	F4–7
8	16.54	17.79	18.40	15.62	...
10	16.86	17.94	18.44	15.87	...
11	14.39	15.38	15.76	13.59	F6–G2
17	13.66	15.20	...	12.41	G0–4
18	15.15	16.01	...	...	F4–8
19	...	18.91	...	16.45	...

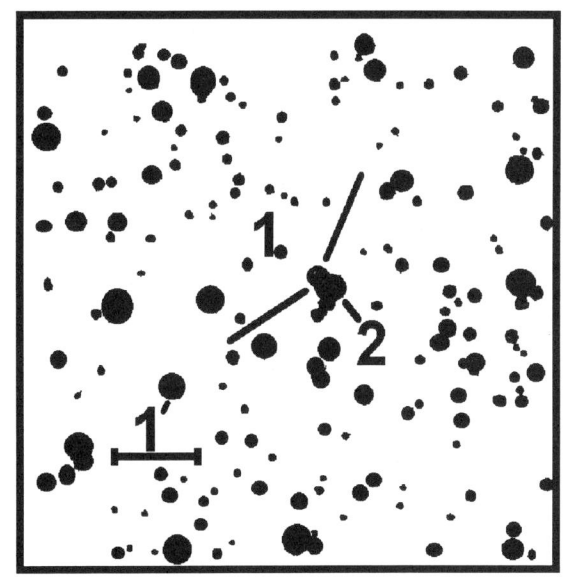

α (2000.0)=$17^h31^m44^s.2$
δ (2000.0)=$-16°57'42''(1)$
Type: bright galactic
        bulge source
Spectrum: M0–5 V:
    B=16.8–17.35
    Max. V=JD 2446963.746(4)
    P=$0^d.1749(1)$

Star	A	F
U	18.78(3)	
B	18.19(3)	
V	17.33(3)	17.73(3)
R	16.74(3)	16.90(3)

V 2216 Oph/4U 1728–169/GX 9+9 [105, 293]

The optical star is a visual triple system [293]: a bright companion A lies $2''.7N$, $1''.4E$ from V 2216 Oph; a faint companion F lies $2''.3N$, $0''.3W$ from V 2216 Oph; the nearby "bright"star "2" lies at $2''.9S$, $4''.9W$ from the object. Star "1" is a variable star with a small amplitude, $\Delta B \sim 0.3$, and a period of $1^d.61$.

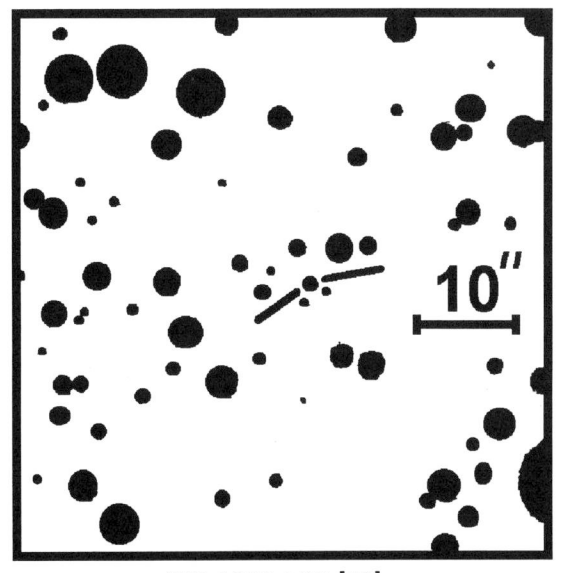

α (2000.0)=$19^h08^m27^s.08$
δ (2000.0) =$00°10'07''.68$
Type: burster
    V=20.46–20.54

Two fainter red stars lie close to the burster.

XB 1905+00 [69]

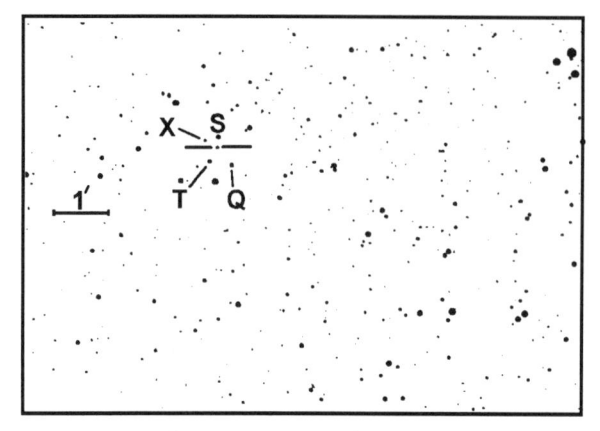

α (2000.0)=$15^h28^m16^s.6$
δ (2000.0)=$-61°52'58''.2$
  Type: ultrasoft X-ray
           transient
  B=17.5–20.5

Star	V	B
S	14.33	15.15
T	16.40	17.23
X	17.6	18.6
Q	17.9	19.6

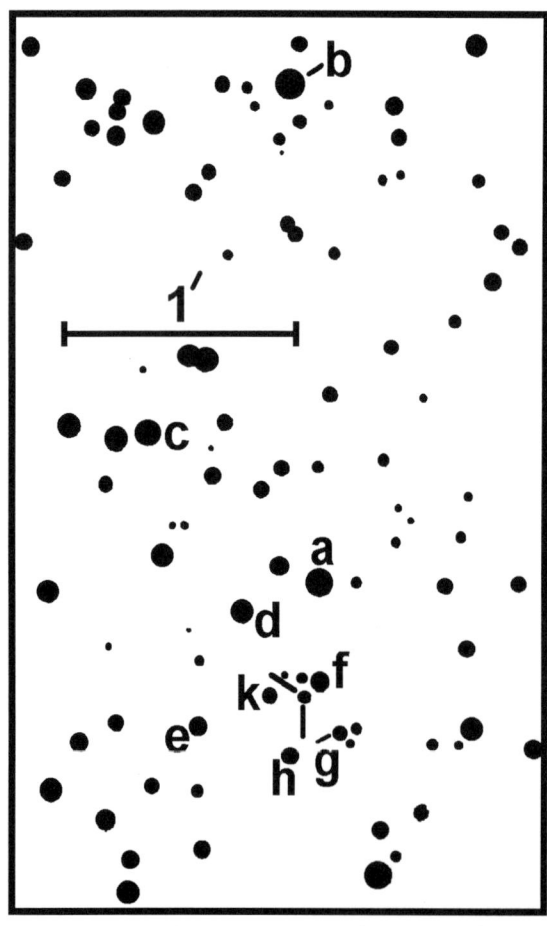

KY TrA/4U 1524–617/TrA X-1 [241]

α (2000.0)=$19^h11^m16^s.1$
δ (2000.0)=$00°35'06''(2)$
  Type: X-ray transient
 Spectrum: K0 V
   V=14.60–19.4
  Min. V=JD 2442593.6(2):
  P=$1^d.30(4)$:

Star	V	B
a	12.44(3)	13.09(3)
b	12.55(2)	13.09(1)
c	12.98(3)	13.80(2)
d	14.05(3)	15.99(5)
e	14.58(2)	15.24(3)
f	15.08(2)	17.31(8)
g	15.41(3)	16.90(5)
h	15.91(4)	18.20(9)
k	16.19(5)	18.02(7)

Star	U
a	13.64(4)
b	13.53(2)
c	14.17(4)
d	17.5(3)
e	15.36(5)
f	18.7(2)
g	17.7(4)
h	>18.8
k	18.5(3)

V 1333 Aql/4U 1908+005/Aql X-1 [198]

α (2000.0) = $19^h18^m47^s.6$
δ (2000.0) = $-05°14'10''(6)$
Type: bright galactic
bulge source
burster
dipped source
Spectrum: G:
V=20.5–21.0
Min.V=JD 2446900.01012(3)
P=0^d.0350393(4)

Star	V	B–V
1	18.621	1.380
4	18.636	1.294
5	18.638	1.597
6	20.533	0.639

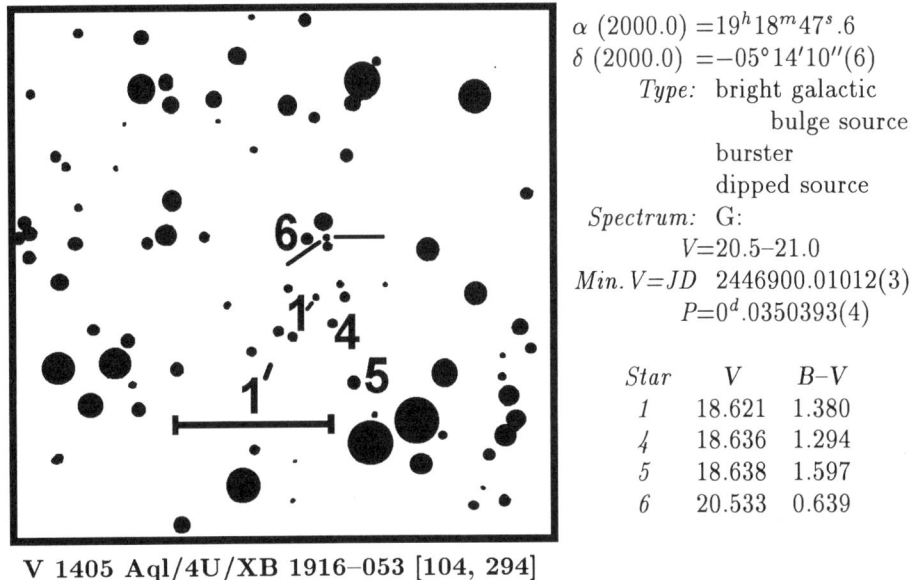

V 1405 Aql/4U/XB 1916–053 [104, 294]

The periods obtained from optical and X-ray observations show different values: $P_{opt} = 50.458(3)$ min and $P_x = 50.00(8)$ min.

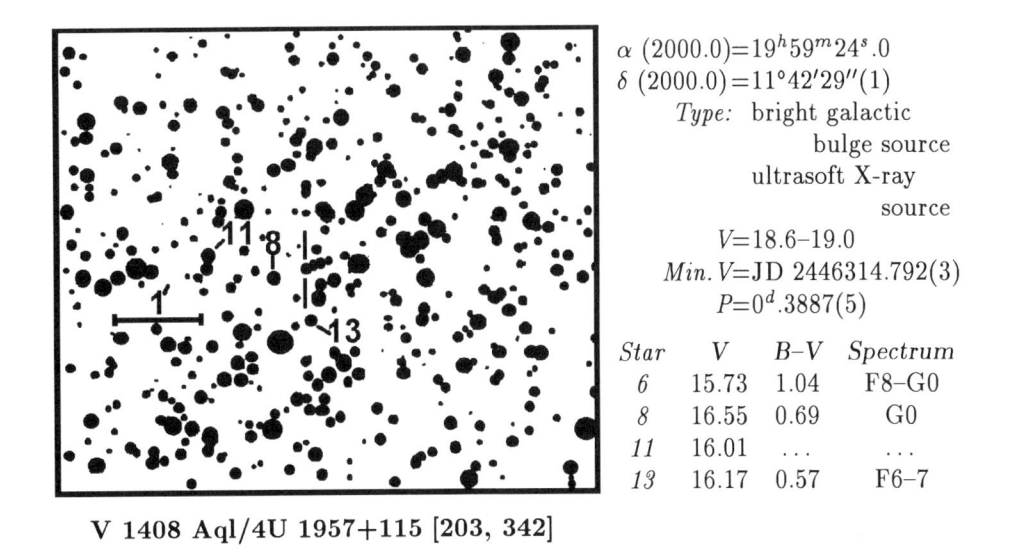

α (2000.0) = $19^h59^m24^s.0$
δ (2000.0) = $11°42'29''(1)$
Type: bright galactic
bulge source
ultrasoft X-ray
source
V=18.6–19.0
Min.V=JD 2446314.792(3)
P=0^d.3887(5)

Star	V	B–V	Spectrum
6	15.73	1.04	F8–G0
8	16.55	0.69	G0
11	16.01	...	...
13	16.17	0.57	F6–7

V 1408 Aql/4U 1957+115 [203, 342]

V 1408 Aql has a visual companion at a separation of only $6''.2$.

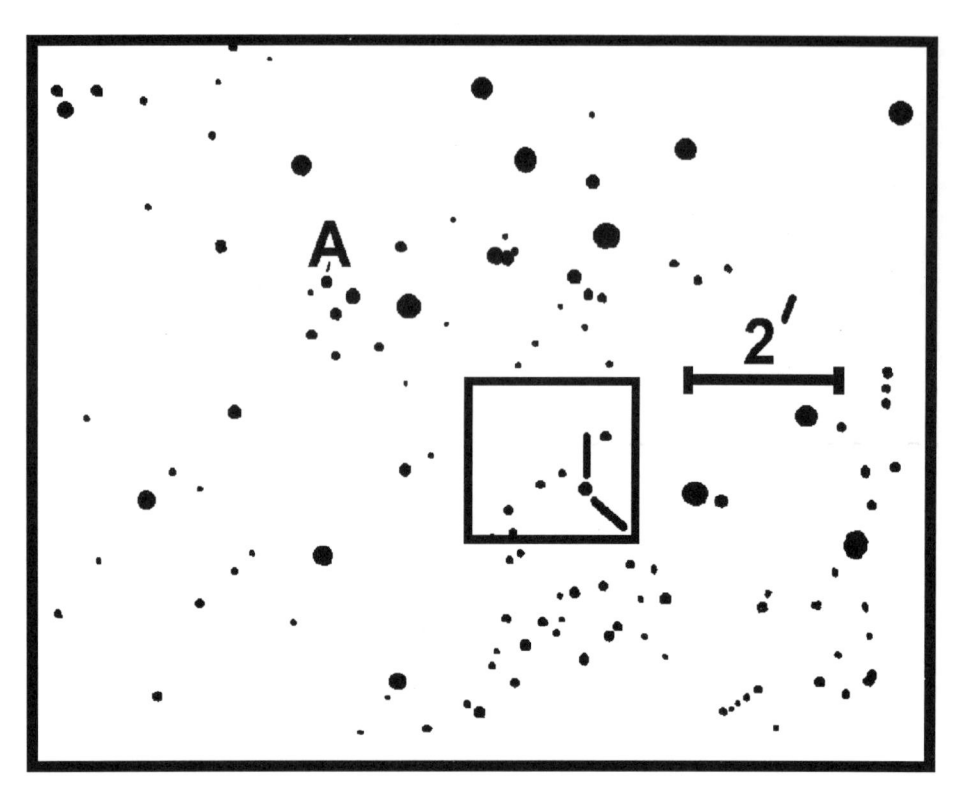

QZ Vul/GS 2000+25/XN Vul 1988 [264, 345]

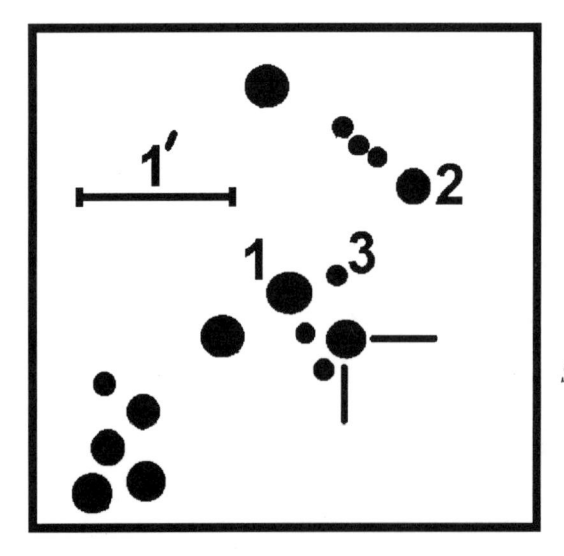

α (2000.0)=$20^h02^m49^s.6$
δ (2000.0)=$25°14'11''.2(7)$
Type: ultrasoft X-ray transient
Spectrum: K2 V
B=17.5->21
Min.V=JD 2447371.462(6)
$P=0^d.344098(5)$

Star	V	B
1	16.97 ± 0.20	18.37 ± 0.10
2	18.06 ± 0.20	18.56 ± 0.15
3	...	19.75 ± 0.20

The star A is a red variable object $(B - V = +4^m)$ with $B \sim$ 17.5–18.5.

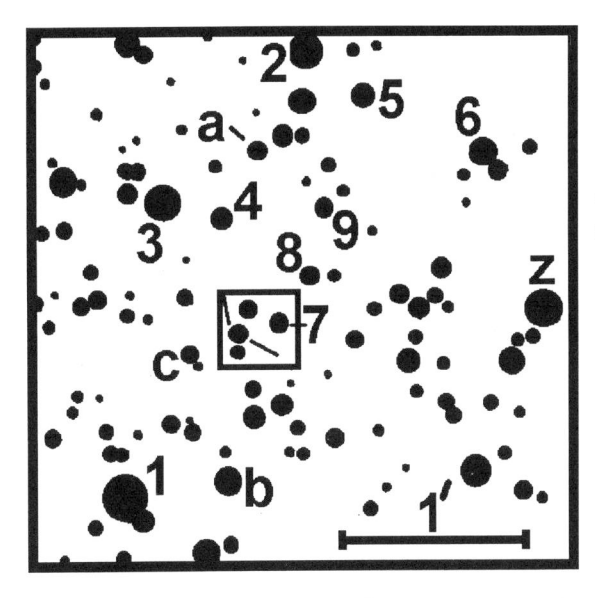

α (2000.0)=$20^h24^m03^s.9$
δ (2000.0) =$33°52'02''.2(3)$
Type: ultrasoft X-ray
transient
black hole
candidate
Spectrum: K0($\pm$1) IV
V=11.64–19.4
Min.V=JD 2448477.35(3)
P=$6^d.4726(9)$

V 404 Cyg/GS 2023+338/XN Cyg 1989 [13, 353]

The comparison star magnitudes have been determined by S.Yu. Shugarov. CCD R image of V 404 Cyg shows a faint companion star d, $1''.43(3)$ to the north.

This system is one of the best candidates for a **black hole**, the mass function of the compact star is $6.26 \pm 0.31 M_\odot$ [61]. Casares *et al.* [62] found evidence for the 6-hr modulation in both the H$_\alpha$ emission radial velocity and the photometry with the ephemeris: $Min.phot. = $ HJD $2448095.58 + 0^d.23585 (or\ 0^d.24511) \cdot E$.

The outburst recurrence time is 20–30 yr: the Nova had outbursts in 1938, 1956, and possibly 1979.

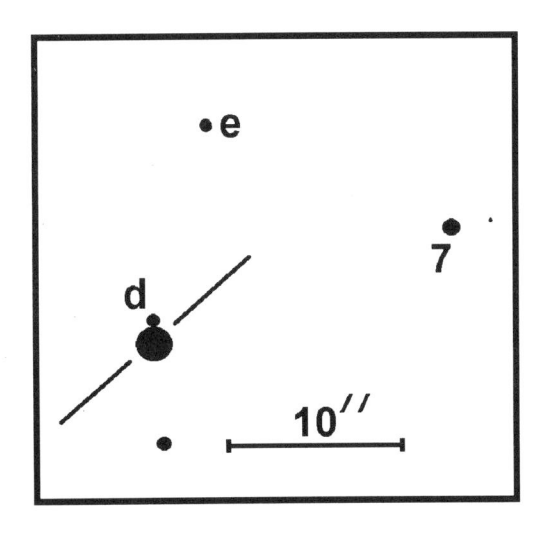

Star	B	V	R
1	13.523	12.817	12.29
2	...	15.73	15.70
3	18.410	15.940	14.47
4	17.877	16.715	15.92
5	...	17.24	15.72
6	...	15.32	14.51
7	20.439	18.014	16.07
8	...	18.34	17.25:
9	...	18.43:	17.0:
a	...	17.94	16.90:
b	...	16.72	15.75
c	...	17.87	16.8:
d	20.59	18.90	17.52
e	21.64	19.02	16.70

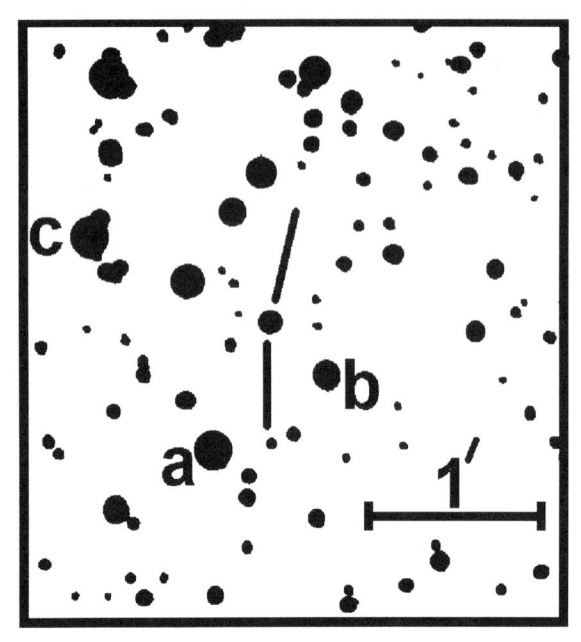

α (2000.0) = $17^h 32^m 02^s.3$
δ (2000.0) = $-24°44'44''.6$
Type: bright galactic
bulge source
X-ray pulsar
Spectrum: M6 III
V=18.66–19.36
Epoch of
periastron =JD 2441574.5:
P =304^d :
P_s=$113^s.626(2)$
at JD 2447766
$\dot{P}_s/P_s$=-2.7×10^{-2} yr^{-1}

Star	V	B–V	U–B
a	13.37	1.20	0.39
b	15.90	1.46	0.9:
c	14.58	1.69	...

V 2116 Oph/4U 1728–247/GX 1+4
[91, 105, 268]

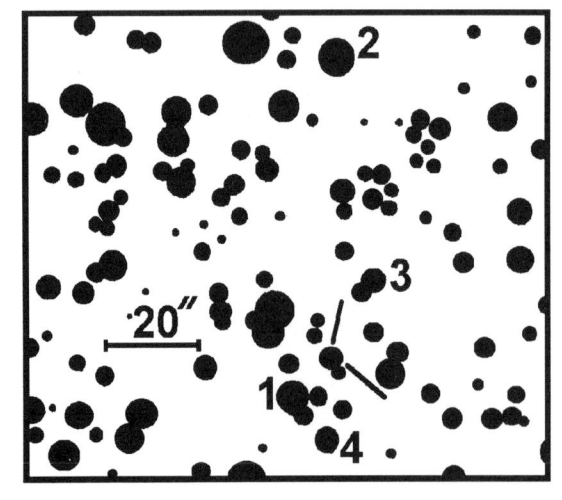

α (2000.0)=$21^h 31^m 26^s.2$
δ (2000.0) =$47°17'24''(1)$
Type: bright galactic
bulge source
coronal source
Spectrum: K5V–F8 IV
V=15.6–18.5
Min.V=JD 2444403.743(2)
P=$0^d.2182579(8)$

Star	V	B	U
1	15.99	16.48	16.76
2	15.40	16.24	...
3	17.35	18.74	...
4	17.56	18.46	...

V 1727 Cyg/4U 2129+470 [71, 86, 340]

High-resolution imaging obtained by Chevalier *et al.*[71] shows five very close objects at the position of V 1727 Cyg, hence one cannot rule out a superposition of the F8 IV star at the approximate position of the X-ray binary.

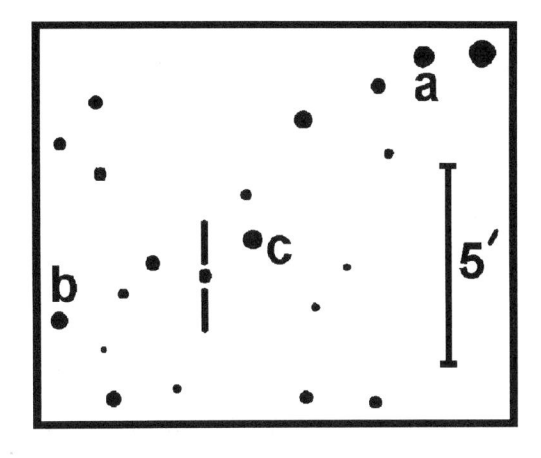

α (2000.0)=$16^h19^m55^s.1$
δ (2000.0)=$-15°38'24''.24(7)$
Type: Bright galactic
bulge source
Z source
QPO source
B=11.1–14.1
Min.V=JD 2440081.13
$P=0^d.787313$

Star	V	B	U
a	9.88(1)	10.21(1)	10.48(1)
b	11.36(3)	12.05(1)	12.00(1)
c	11.40(1)	12.68(1)	13.98(2)

V 818 Sco/Sco X-1/4U 1617–155 [199]

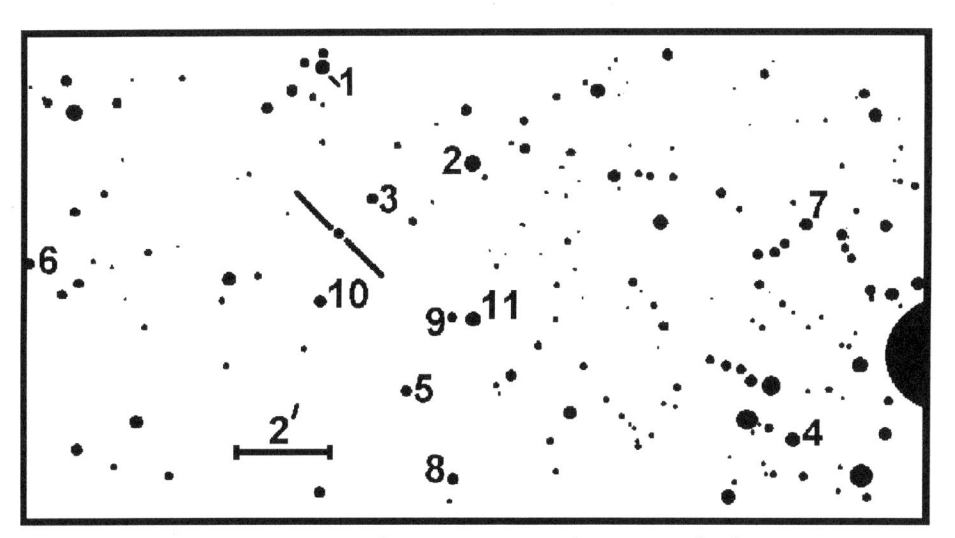

V 1341 Cyg/4U 2142+380/Cyg X-2 [25]

α (2000.0)=$21^h44^m41^s.2$
δ (2000.0)=$38°19'18''(1)$
Type: bright galactic
bulge source
Z source
QPO source
Spectrum: A5–F2III–IV
B=14.41–15.7
Min.V=JD 2443166.7(5)
$P=9^d.844(1)$

Star	V	B–V	U–B
1	12.82	0.93	0.41
2	12.99	0.62	0.00
3	13.49	1.31	0.86
4	13.51	0.68	0.17
5	14.47	0.78	0.05
7	13.86	0.85	0.24
8	14.95	0.56	0.00
10	13.54	1.43	1.18
11	12.09	1.84	1.45

The system shows both high and low intensity states, possibly bursts, long-term variability and irregular intensity dips which are seen mainly during the high state.

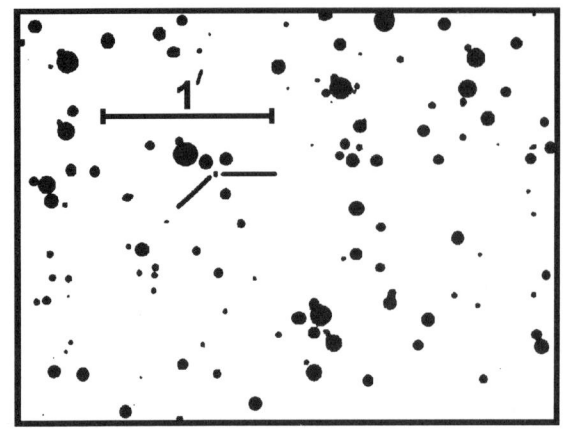

QX Nor/4U 1608–522 [139]

α (2000.0)=$16^h12^m43^s.0$
δ (2000.0)=$-52°25'22''.8$
Type: soft X-ray transient
a probable **Atoll**
source
I=18.2–23:

The finding chart [139] corresponds to a burst stage.

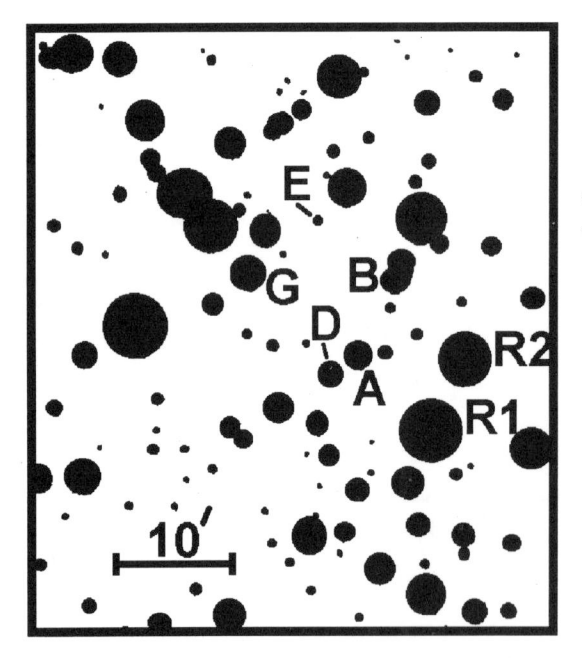

V 665 Cas(?)/1E 2259+586 [77, 93]

α (2000.0)=$23^h01^m08^s.1$
δ (2000.0)=$58°52'45''(3)$
Type: bright galactic
bulge source
X-ray pulsar
Spectrum: K–M:
V>23
P_s=$6^s.978725(8)$
at JD 2446035
$\dot{P}_s/P_s$=$3.3(2) \times 10^{-6}$ yr^{-1}

X-ray coordinates are given as the optical counterpart is currently unknown, it is certainly very faint ($V > 23$ mag). The searches for the optical counterpart found several possible candidates, one of which (star "D") (V 665 Cas) was tentatively identified with the X-ray source [223] on the basis of IR pulsations with $P_{IR} \approx P_x$. $P_x = 2300 \pm 90$ s was registered sometimes in the X-ray band [143]; however other groups of observers failed to find any orbital pulsations from the system. At the finding chart the previously known optical candidates for 1E 2259+586 are labeled.

α (2000.0)=$17^h38^m58^s.3$
δ (2000.0)=$-44°27'00''$(1)
Type: Burster
  Atoll source
Spectrum: K7–M0 V
  V=17.40–18.0
  Max.V=JD 2446290.354(3)
  $P=0^d.19382(1)$

Star "*3*" has $V = 17.60$.

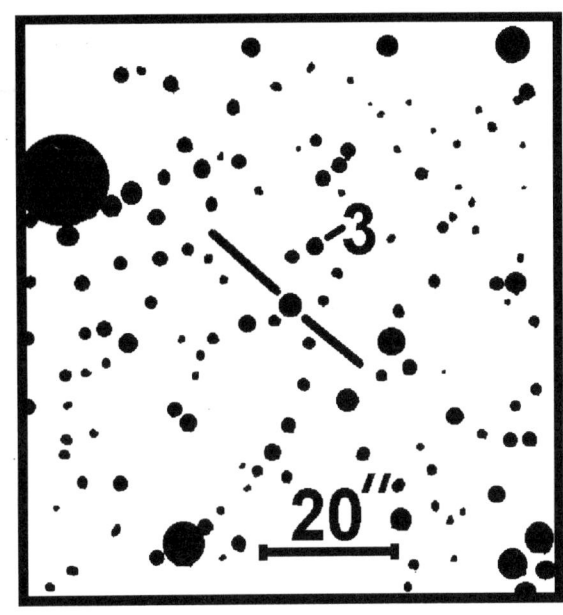

V 926 Sco/4U/X/MXB 1735−444 [170]

PRECATACLYSMIC CLOSE BINARY SYSTEMS

RR Cae/LFT 349/BPM 31582

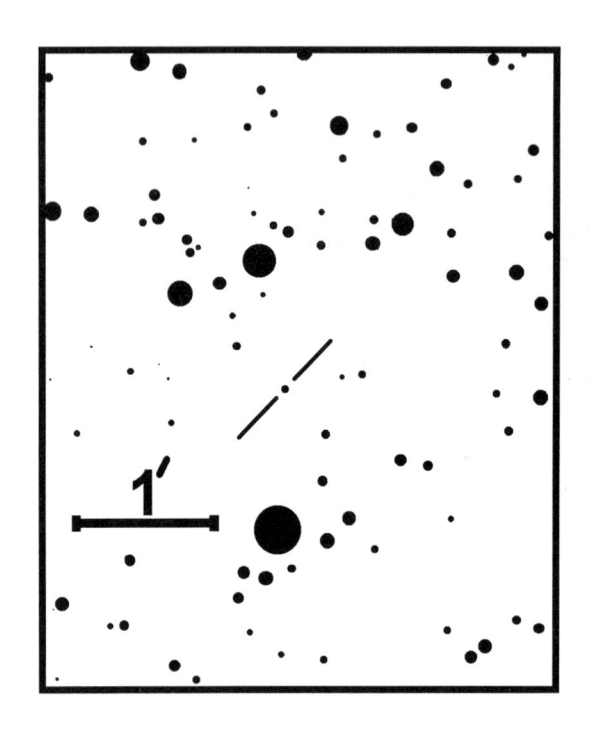

α (2000.0)= $04^h21^m01^s$
δ (2000.0)=$-48°39'04''$
Type: pre CV, E
Spectrum: DAwk + Me V
B=14.4–18.2
P=0^d.30371

The system RR Cae is not a nucleus of a planetary nebula. The spectrum contains contains several weak features which include the hydrogen Balmer lines and a shallow band near λ 5180 Å.

References: [394]

V 664 Cas

α (2000.0)= $03^h03^m47^s$.1
δ (2000.0)= $64°54'35''$.2
Type: pre CV
Spectrum: wd: + K–M
B=13.5–14.7
P=0^d.581647(1)

The system V 664 Cas is the central star of the extended planetary nebula HFG 1

References: [102]

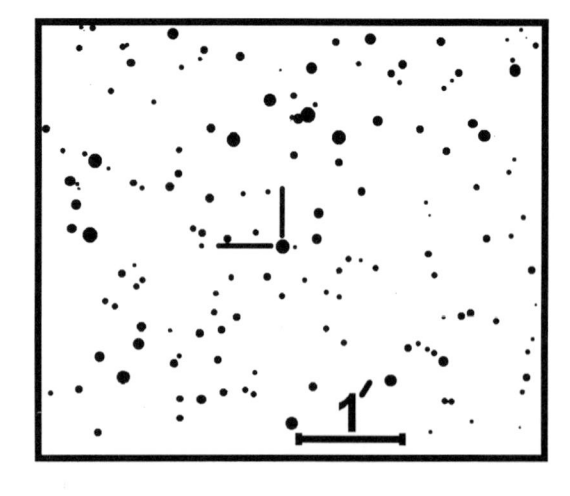

IN Com/(HD 112313/SAO 82570/BD+26°2405)

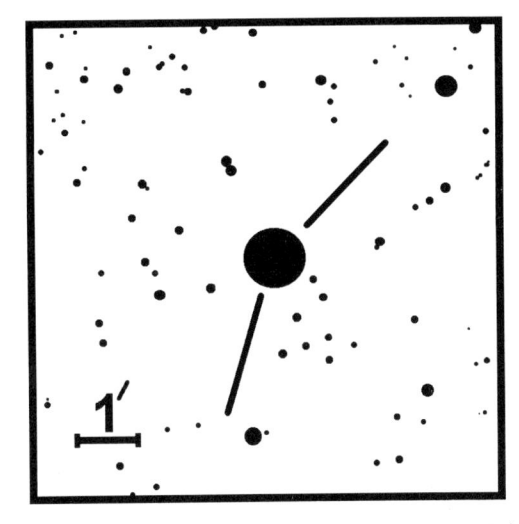

α (2000.0)=$12^h 55^m 33^s.8$
δ (2000.0) =25°53'30''
Type: pre CV
Spectrum: sdO + G5 III
V=8.7–9.4
P=$1^d.7545$:
Min.phot.=JD 2445805.93

The system IN Com (HD 112313 = SAO 82570 = BD+26°2405) is the nucleus of the planetary nebula LT5. The spectroscopic orbital period is given above. *Min.phot.* is the epoch of inferior conjunction of the emission line source. IN Com is a triple system. The orbital period of the outer system consisting of hot subdwarf ($M_3 \sim 1.1 M_\odot$) and a pre CV binary is $P_{orb} = 2000^d$, *Epoch* =JD 2446158.0, e=0.65, $\omega = 180°$, $f(M) = 0.20 M_\odot$. The existence of the period $5.^d9522$ is possible [245].

References: [196, 245]

V 477 Lyr

α (2000.0)=$18^h 31^m 18^s.5$
δ (2000.0) =26°56'11''
Type: pre CV, E
Spectrum: sdO
B=14.6–15.9
P=$0^d.4717291(5)$
Min.phot.=JD 2448135.50464(6)

V 477 Lyr is the nucleus of the planetary nebula Abell 46. It is an *eclipsing* system with large reflection effect. The depths of the minima are: $\Delta m_I = 1^m.42\ B$, $\Delta m_{II} \simeq 0^m.1 B$. The light curve resembles that of UU Sge.

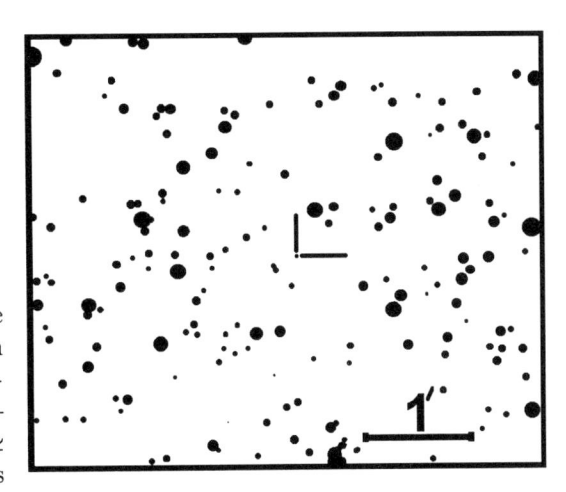

References: [187]

UU Sge/H 1938+16/Abell 63

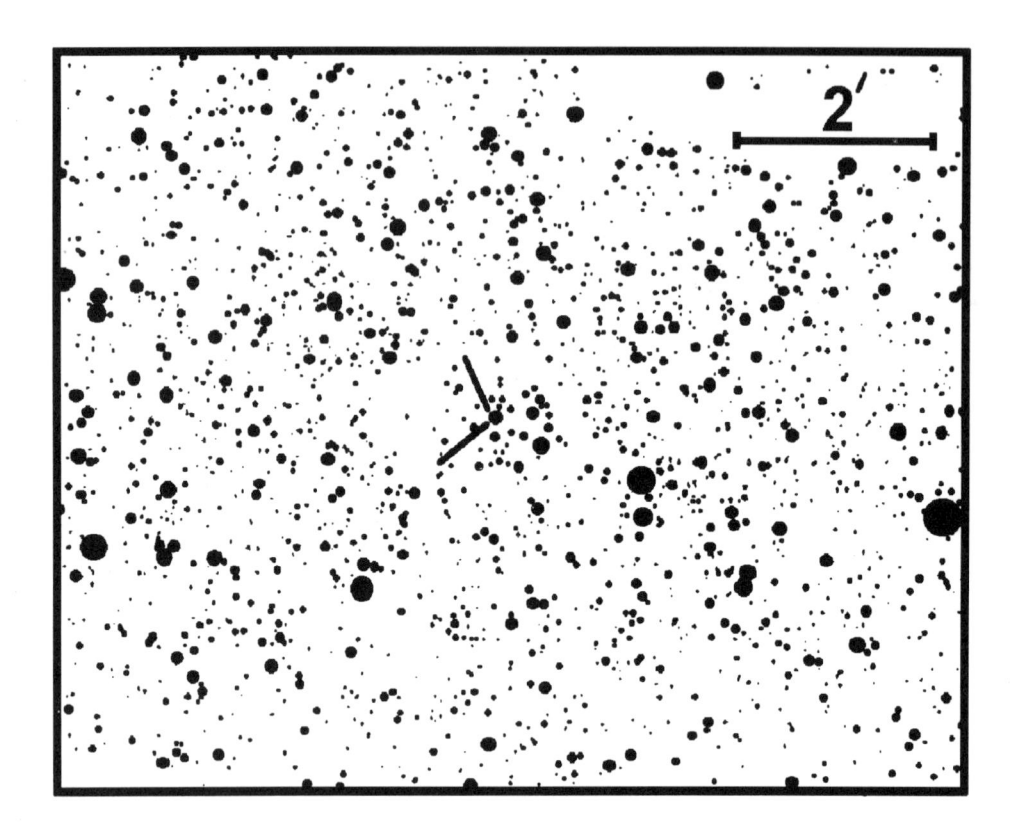

α (2000.0)=$19^h42^m10^s.2$
δ (2000.0)=$17°05'15''$
Type: pre CV, E
Spectrum: sdO + G7 V

V=14.18–15.59
P=$0^d.46506918(6)$
Min.phot.=JD 2442953.9328

This is the nucleus of Abell 63. It is an eclipsing system with large reflection effect. The primary eclipse is total and very deep, $\Delta m_I = 4^m.3$; $\Delta m_{II} \sim 0.2$. $L_X(0.18$–2.5 keV$) = (1 \pm 0.6) \times 10^{32}$ erg/s. The secondary has V=18.04.

References: [42, 354]

V 471 Tau/BD+16°516/H 0349+17/G 7–23/WD 0347+171

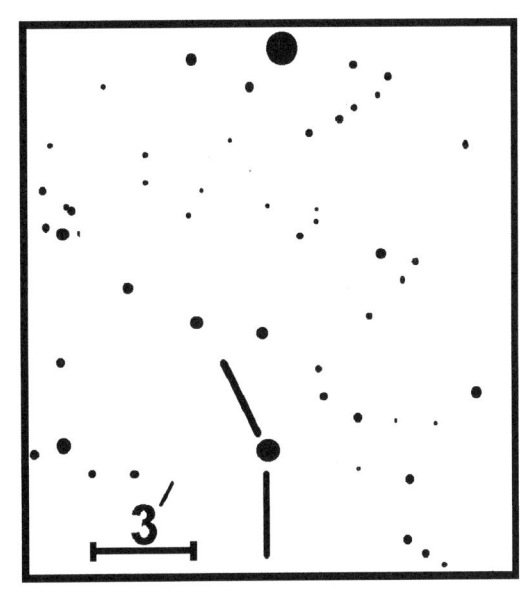

α (2000.0)=$03^h50^m02^s.4$
δ (2000.0)=$17°14'48''$
Type: pre CV, E
V=9.2–9.8
Spectrum: DA + K2 V
$P=0^d.5211831(2)$
Min.phot.=JD 2447471.962896(5)

V 471 Tau is an *eclipsing* binary system, member of Hyades cluster. It is often classified as an RS CVn star: a K star exhibits Ca II in emission, "starspots", orbital period changes, radio emission, coronal X-ray emission with $kT \sim 1$ keV. There happened an abrupt period change.
Beavers *et al.* [31] explained this period change by the third body with the following parameters: $P_{orb} = 24.6$ yr, $e = 0.36(1)$, $\omega = 65°.0(2)$. Ultraviolet short-period variations ($P \simeq 13$ min and $P \simeq 9.2$ min) with the amplitude $\simeq$4–5% are observed [16]. Soft X-flux pulsations with $P = 277, 554.7(3)$ s and 562 s were reported.

References: [16, 31, 122, 354]

Feige 24/WD 0232+035

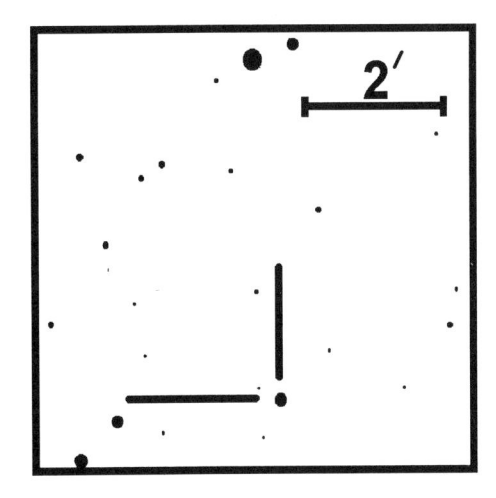

α (2000.0)=$02^h35^m07^s.0$
δ (2000.0)=$03°43'57''$
Type: pre CV
$V\sim$12.4
Spectrum: DA + M1.5 V
$P=4^d.23160$
Min.phot.=JD 2448578.409(33)

References: [114]

BE UMa/PG 1155+492

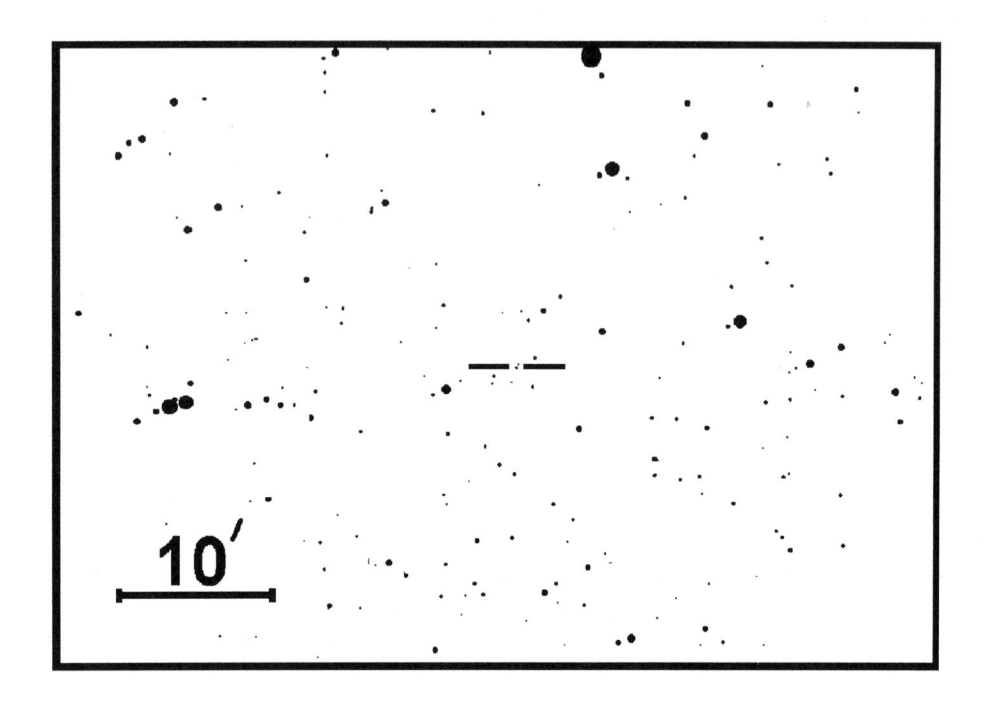

α (2000.0)=$11^h57^m44^s.7$ B=14.8–17.8
δ (2000.0)=$48°56'19''$ $P=2^d.2911667$
Type: pre CV, E   *Min.phot.*=JD 2447628.5381
Spectrum: wd + G-K V

BE UMa is an *eclipsing* binary system. It is a classical case of the reflection effect with the hot spot's temperature 1.5×10^4 K [116]. No planetary nebula is observed.

References: [116, 318]

GK Vir/PG 1413+015

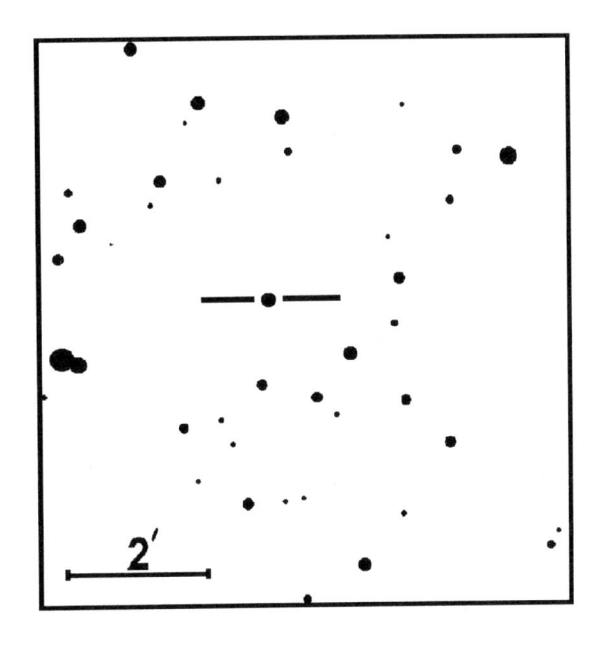

α (2000.0)=$14^h15^m36^s.4$
δ (2000.0)=$01°17'18''$
Type: pre CV, E
Spectrum: wd + M3–5 V
V=17.0–23
P=$0^d.344330809$
Min.phot.=JD 2442543.837148

GK Vir is an *eclipsing* precataclysmic binary. It has a high bolometric luminosity $L_{bol} \leq 10^{35} \mathrm{erg\,s^{-1}}$, L_x (0.18–2.5 keV) $\sim 5 \times 10^{30} \mathrm{erg\,s^{-1}}$ [354].

References: [134, 354]

EUVE 2013+40

α (2000.0)=$20^h13^m09^s.2$
δ (2000.0)=$40°02'25''$
Type: pre CV
Spectrum: wd + dM3–3.5 V
$V\sim$14:
P=$0^d.7059$
Min.phot.=JD 2449257.899(1)

The M3 star contributes about 15–25% of the total light near λ 6500 Å. The resulting H_α emission is twice as strong as in prototype Feige 24, with an equivalent width of $\sim$50 Å with respect to the red dwarf continuum.

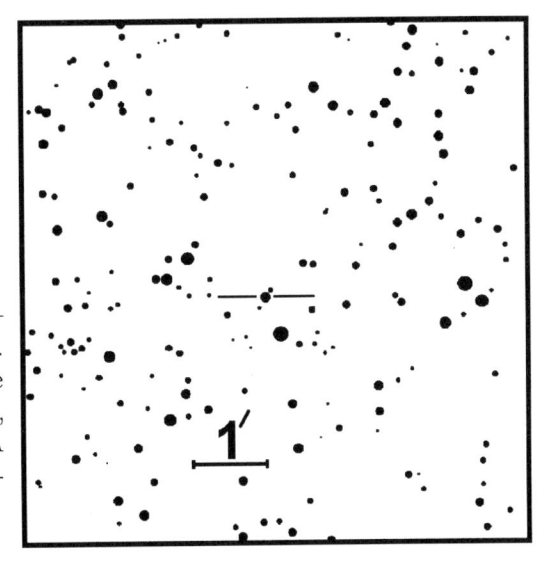

References: [344]

AM CANUS VENATICORUM TYPE STARS

V 803 Cen/AE 1

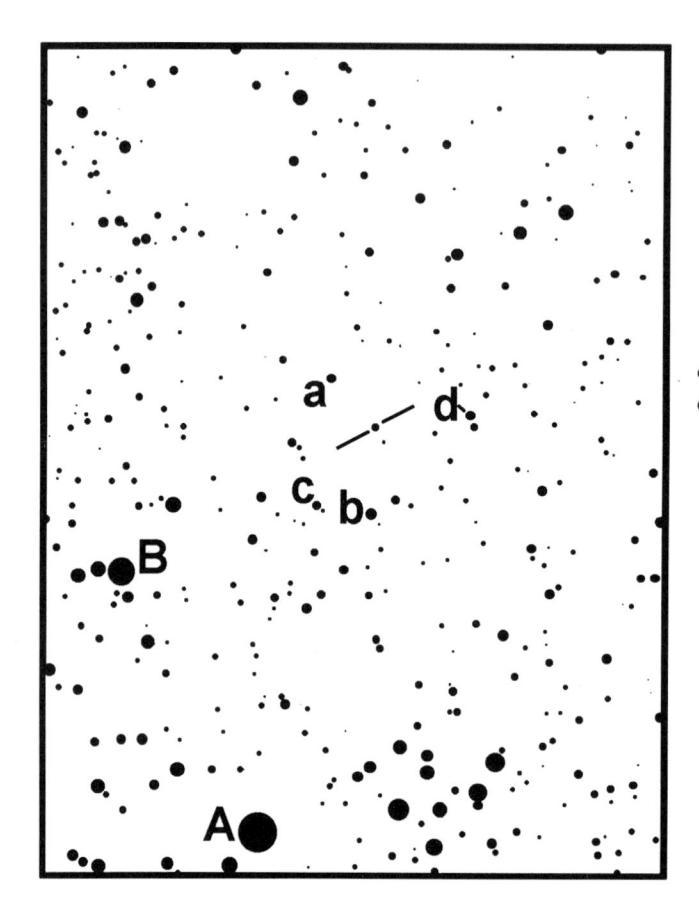

α (2000.0)=$13^h23^m45^s.0$
δ (2000.0)=$-41°44'34''$
Type: AM CVn,
       NL
V=13.2–16.8
P=$0^d.01865$:

The comparison stars are [112]:

Star	V	B–V	U–B
a	13.95	0.75	0.17
b	13.40	0.59	−0.01
c	14.20	0.83	0.42
d	14.14	0.73	0.16

The star A is HD 116385, the star B is HD 116431. V 803 Cen has He I lines in its spectrum. It is similar to AM CVn and CR Boo.

References: [112]

AM CVn/CBS 354/HZ 29/EG 91

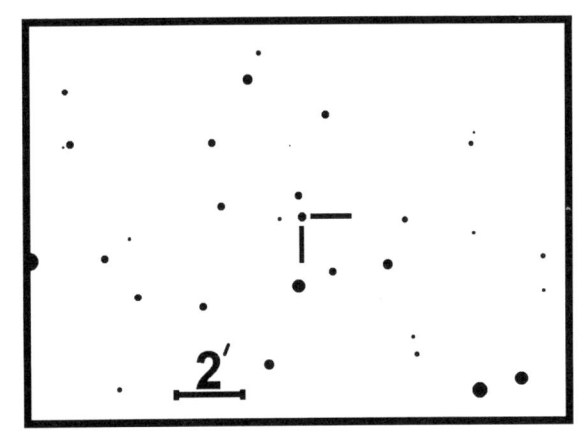

α (2000.0)=$12^h 34^m 54^s.4$
δ (2000.0) =$37°37'43''$
Type: AM CVn
   NL XS
V=13.10–14.18
P=$0^d.012165$

The orbital period during 1967–1990 was probably 1051.22(3) s. There are also such periods as 525 s, 1023 s (orbital?), 1051 s (superhump?), $10^h.8$ (disc precession?) and $13^h.38$ [260, 262]. It has asymmetric He I absorption lines, weak He II λ 4686 Å. The energy of the outburst in 1986 (7 February) was $E = 2.7 \times 10^{36}$ erg s^{-1}.

References: [260, 262, 263, 321]

CP Eri

α (2000.0)=$03^h 10^m 32^s.8$
δ (2000.0) =$-09°45'06''.2$
Type: AM CVn, NL
V=16.7–19.7
P=$0^d.023738$:

CP Eri has no hydrogen lines in its spectrum. In 1962 it underwent an outburst to 14^m. Howell et al. [160] found periodic modulations in B and V light, $\Delta B = 0.2 - 0.4$, $P_B = 28.6 \pm 1.3$ min, $\Delta V = 0.1 - 0.2$, $P_V = 29.5 \pm 1.5$ min.

References: [160, 197]

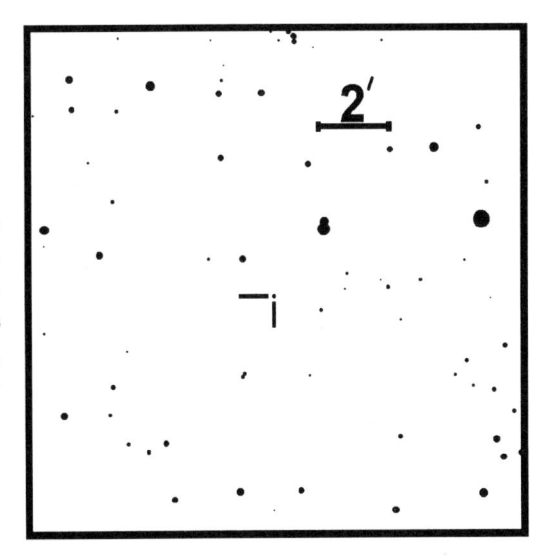

GP Com/G 61-29/GR 389/1ES 1303+1817

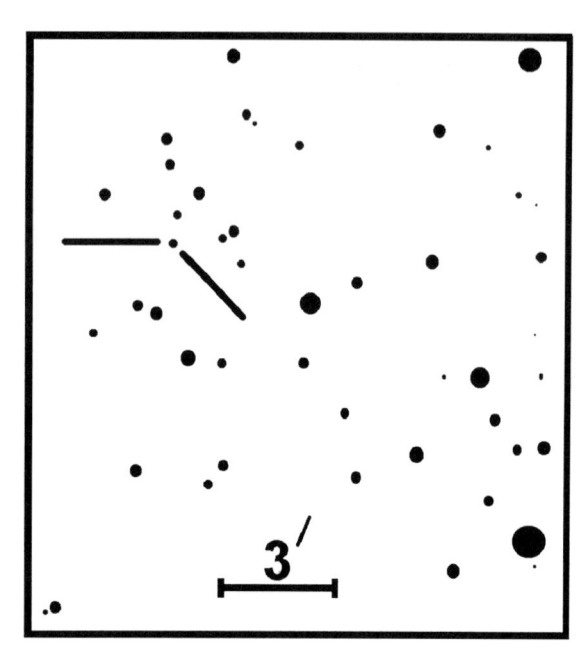

α (2000.0)=$13^h05^m43^s.3$
δ (2000.0)=$18°01'02''$
  Type: AM CVn, NL, XS
  V=15.7–16.3
  $P=0^d.03231$

GP Com has only helium lines in emission.

References: [114]

SYMBIOTIC BINARY STARS

Z And/HD 221650/IRAS 23312+4832

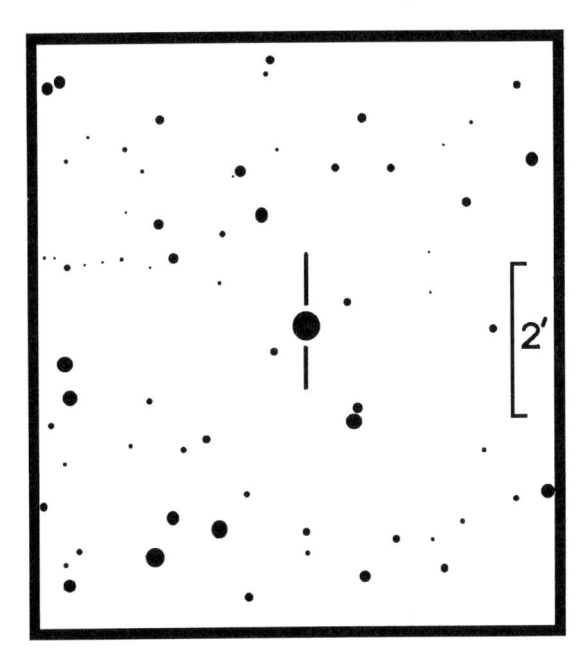

α (2000.0)=$23^h 33^m 39^s.97$
δ (2000.0)=$48°49'05''.59$
 Type: S
 Pg=8.0–12.4
 Spectrum: M3.5 III+B1eq
 P=$756^d.85$
Min.phot. JD 2421298

Z And shows quiescent and increased activity phases. The luminosity may increase by two magnitudes. The last period of quiescence continued for ten years (1974–1983). A new outburst began in spring of 1984 and then, in autumn of 1985, more powerful outburst followed ($\Delta U=3.4$, $\Delta B=2.3$, and $\Delta R=1.0$) [34].

References: [4, 34]

EG And/HD 4174/BD+39°167/IRAS 00415+4024/SAO 36618

α (2000.0)=$00^h 44^m 37^s.19$
δ (2000.0)=$40°40'46''.06$
 Type: S
 V=7.08–7.8
 Spectrum: M2-3 III+sd
 P=$481^d.1(4)$
Min.phot.=JD 2446336.7

This symbiotic star is a bright IR source. The following periods were also found: 492±4.3 and 470±10 days [121]. The peculiar velocity of the hydrogen absorption lines is –96 km/s, that of the emission lines, –92.8 km/s. No radio emission is observed.

References: [4, 121]

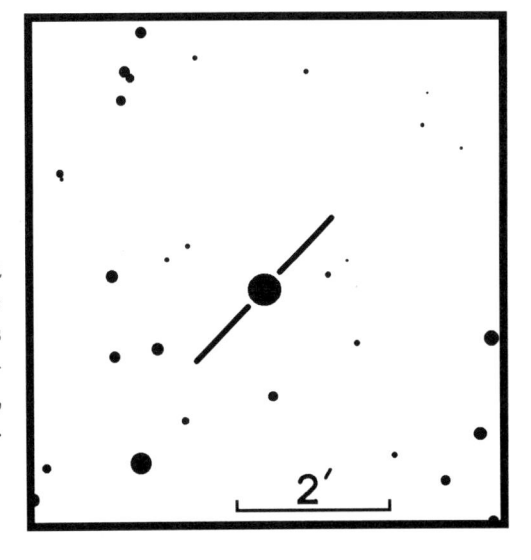

V 1413 Aql/AS 338/SS 428

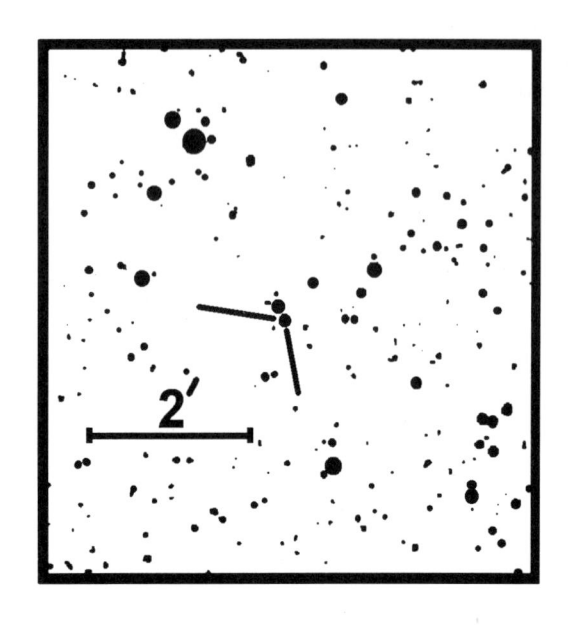

α (2000.0)=$19^h03^m46^s.86$
δ (2000.0)=$16°26'17''.57$
Type: S
V=10.6–15.1
Spectrum: M5 III+Be? wd?
P=$434^d.1(2)$
Min.phot.=JD 2446650±15

This is a symbiotic nova that entered the outburst phase in late 1981 (ΔV=3.4, ΔB=4.0)

References: [4, 238]

R Aqr/HD 222800

α (2000.0)=$23^h43^m49^s.48$
δ (2000.0)=$-15°17'02''.75$
Type: D, Mira
V=5.8–12.4
Spectrum: M7e III+sd? wd?
P=16060^d

R Aqr/IRAS 23412–1533 is the nearest symbiotic binary located in the center of a planetary nebula of a complex shape. The cool star has a pulsational period of $386^d.83$. The system underwent a violent outburst episode in 1928–1934. Jet-like structures are present.

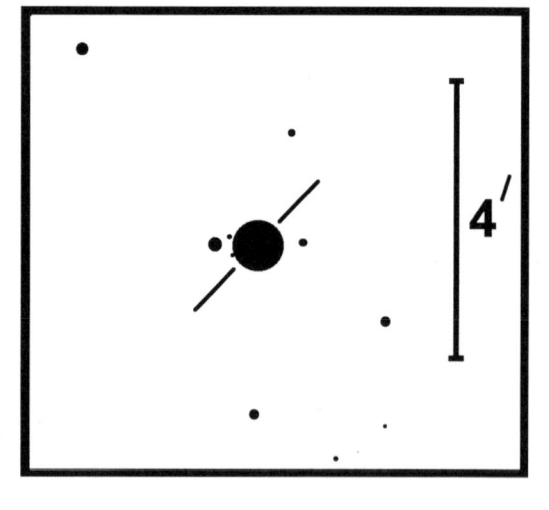

References: [4]

UV Aur/HD 34842

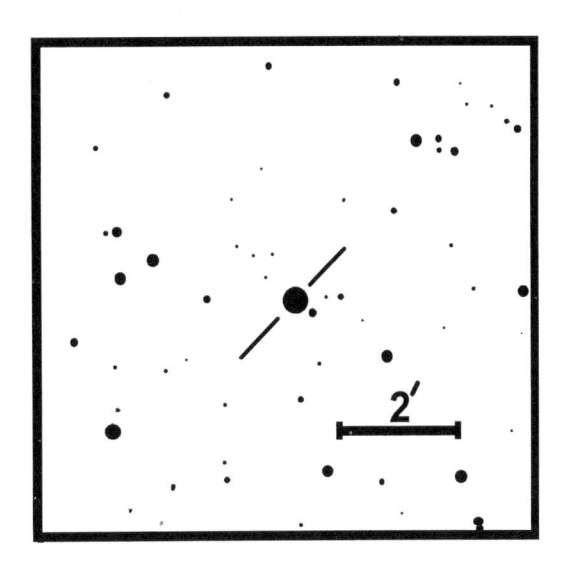

α (2000.0)=$05^h21^m48^s.97$
δ (2000.0)=$32°31'44''.17$
Type: D
V=7.4–10.4
Spectrum: G6.2–8.2p + B
P=$395^d.2$
Max.phot.=JD 2415016

The finding chart corresponds a moment before the outburst. The system is classified as a carbon variable.

References: [4]

T CrB/IRAS 15374+2603

α (2000.0)=$15^h59^m30^s.19$
δ (2000.0)=$25°55'11''.47$
Type: S
V=2.0–9.9
Spectrum: M4.1 III+wd?
P=$227^d.5$
Min.phot.=JD 2433687

T CrB/IRAS 15374+2603 is a recurrent nova (outbursts in 1866 and 1946). $V_{max} \sim 3^m$ in 1946. Expansion velocity near the 1946 maximum reached -5000 km s^{-1} [148, 292]. $L_{UV} \geq 10^{35}$ erg s^{-1}, L_x(2–10 keV)$\simeq 10^{31}$ erg s^{-1}.

References: [4, 141, 148, 292]

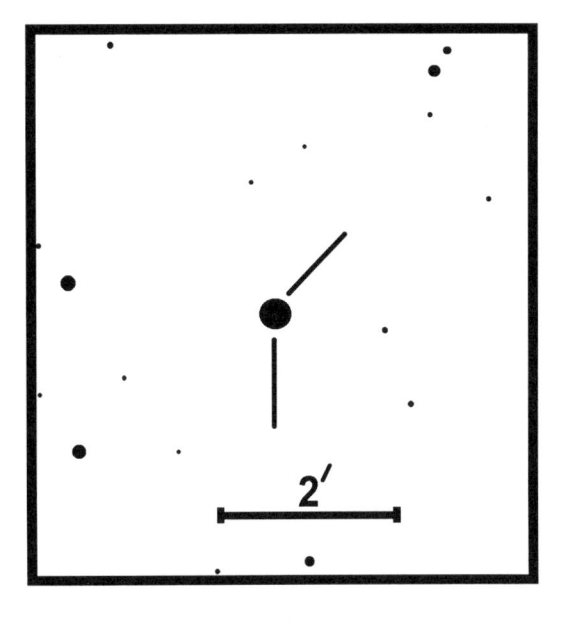

BF Cyg

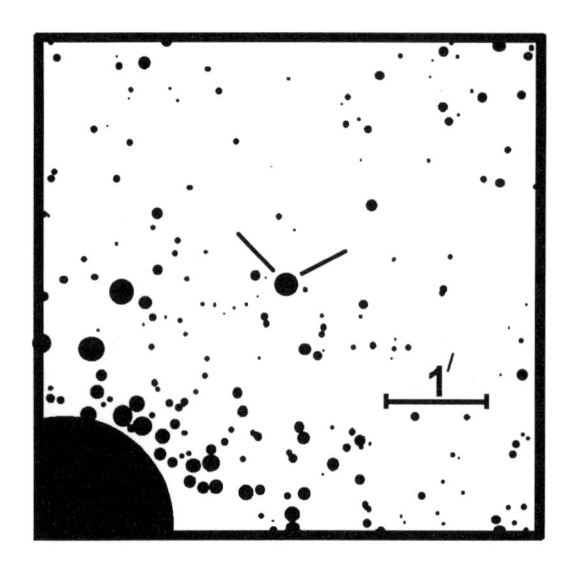

α (2000.0)=$19^h23^m53^s.46$
δ (2000.0)=$29°40'28''.63$
Type: S
Pg=9.3–13.4
Spectrum: M5 III+wd
P=$756^d.8(7)$
Min.phot.=JD 2415058±12

This is a symbiotic star with a strong C IV P Cyg profile. Eclipsing binary. Nuclear burning near WD surface.

References: [4]

CH Cyg/BD+49°299/HD 182917/SAO 031632/IRAS 19232+5008

α (2000.0)=$19^h24^m33^s.16$
δ (2000.0)=$50°14'29''.65$
Type: S
V=5.6–9.2
Spectrum: M6.5 III+wd
P=$751^d.5 \pm 2^d.3$
Min.phot.=JD 2447302±12

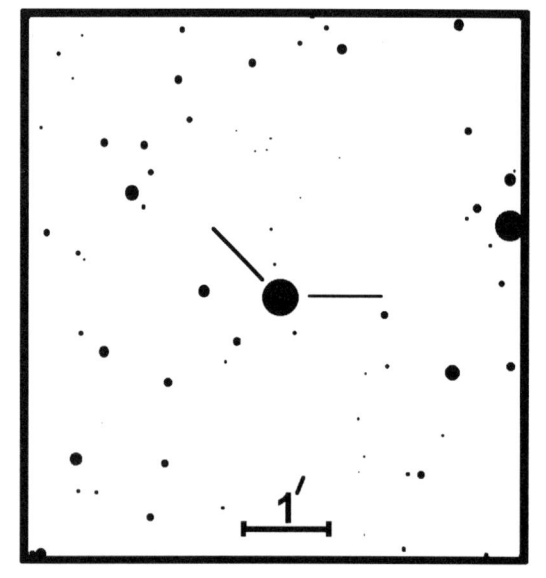

CH Cyg is the second brightest symbiotic star in K band ($K = -1$). The semiregular variability ($\sim 100^d$) is known. A prominent outburst was observed in 1984 July. The faint radiosource H 1926+503. The bipolar jet was formed in mid 1984. The system's light exhibits a 2600 s period, similar to intermediate polar periods [155]. CH Cyg is a triple system with a noninteracting G–K dwarf ($14^y.5$) and a symbiotic system (2 yr orbit) [154]. Its orbital elements: e=0.067±0.029, ω=207°±23°, T=JD 2445517±327, P=$5298^d±98^d$, $a\sin i$=$489.9 R_\odot$, $f(M)$=$0.06 M_\odot$.

References: [4, 154, 155]

CI Cyg/MWC 415/IRAS 19483+3533/He 3-1784

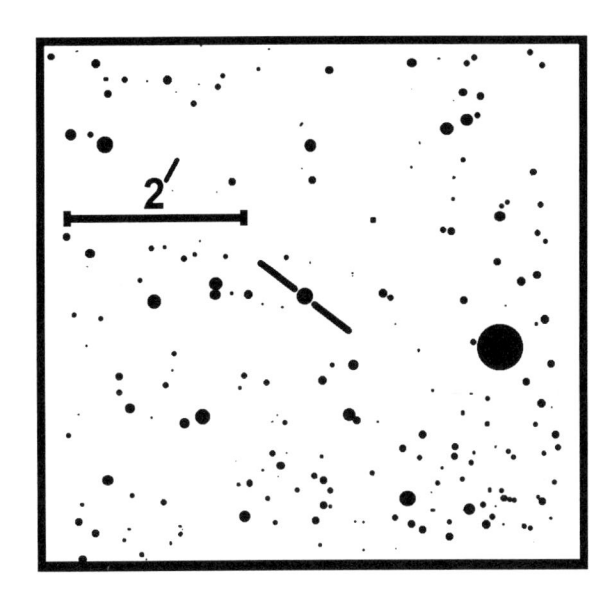

α (2000.0)=$19^h50^m11^s.87$
δ (2000.0)=$35°41'02''.83$
Type: S
Pg=9.9–13.1
Spectrum: M4.9 IIIep+B9ep
P=$855^d.25$
Min.phot.=JD 2411902

CI Cyg is an *eclipsing* system. The finding chart corresponds to an epoch before the outburst.

References: [4]

V 1016 Cyg/AS 373/IRAS 19553+3941

α (2000.0)=$19^h57^m05^s.01$
δ (2000.0)=$39°49'36''.55$
Type: D, Mira, N
V=10.5–17.5
Spectrum: M7 III+?
P=450^d IR

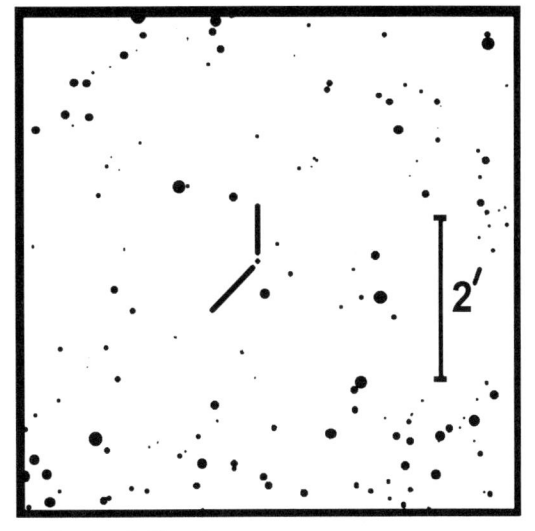

V 1016 Cyg is a symbiotic nova. The finding chart corresponds to an epoch before the outburst. The last nova-like outburst was in 1965, with $\Delta V \sim 4.5$. The preoutburst spectrum resembled a Mira [242], and then the spectrum of a high excitation planetary nebula [76]. Before the outburst it was a *Mira* with $P \sim 478^d$. The system exhibits strong IR-radiation [408] with amplitude $\Delta K \simeq 1$.

References: [4, 76, 242, 408]

V 1329 Cyg/HBV 475

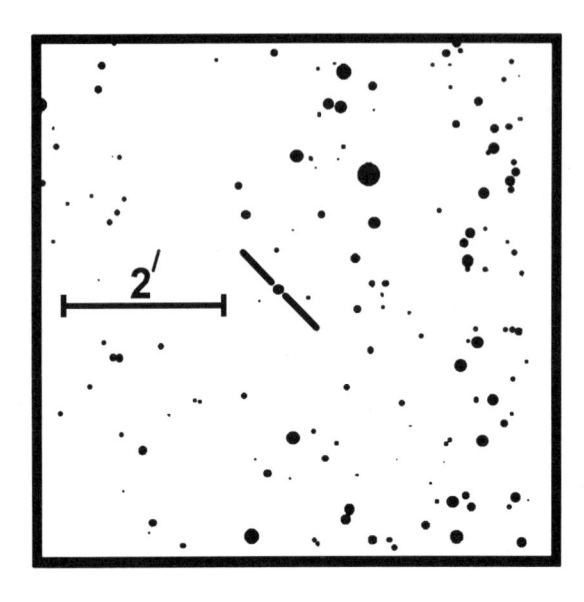

$\alpha\ (2000.0) = 20^h 51^m 01^s.28$
$\delta\ (2000.0) = 35°34'54''.00$
Type: S, N
$B = 12.1–18$
Spectrum: M4(1) III
$P = 964 \pm 10^d$
Min.phot. = JD 244487816±16

V 1329 Cyg/HBV 475 shows the pure nebular spectrum. The system has no X-rays, radio emisson is low [297].

References: [4, 297, 427]

AG Dra/BD+67°922/IRAS 16013+6656

$\alpha\ (2000.0) = 16^h 01^m 41^s.04$
$\delta\ (2000.0) = 66°48'10''.01$
Type: S, N
$Pg = 8.9–11.8$
Spectrum: K3 IIIep+sd?
$P = 552.4 \pm 2.2^d$
Max.phot. = JD 2442514±11

AG Dra is a symbiotic nova. During the recent 10 years there were three events: outbursts from the end of 1980 till the end of 1983, in spring 1985 and in the beginning of 1986. 11 outbursts were noted between 1890 and 1966.

References: [4, 434]

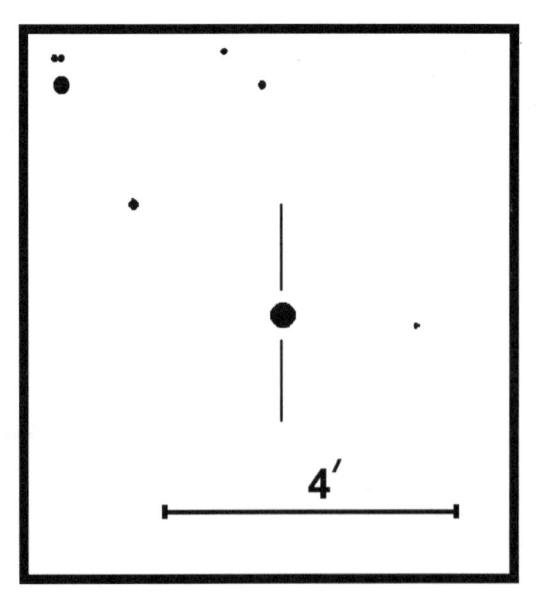

YY Her/AS 297

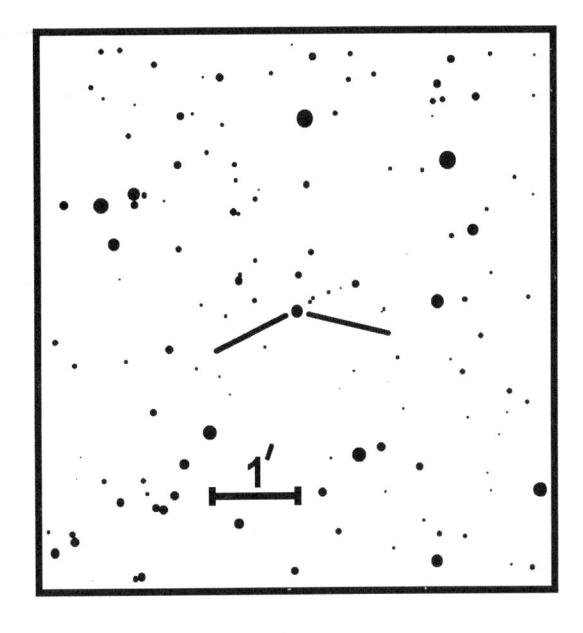

α (2000.0)=$18^h14^m34^s.12$
δ (2000.0)=$20°59'19''.03$
 Type: S
  B=11.1–13.6
Spectrum: M3 III

The finding chart corresponds to an epoch before the outburst.

References: [4]

V 443 Her/MWC 603

α (2000.0)=$18^h22^m07^s.85$
δ (2000.0)=$23°27'19''.12$
  Type: S
   V=11.42–11.72
Spectrum: M5.1±0.4
   P=597^d

The star was identified as a symbiotic one in 1958 [298].

References: [4, 298]

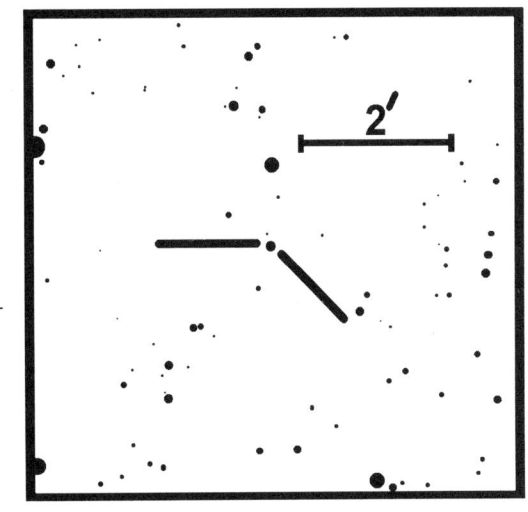

RW Hya/HD 117970

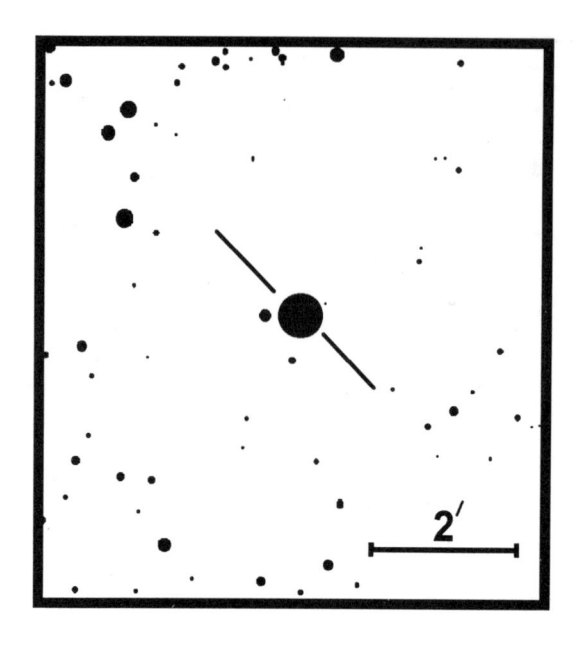

α (2000.0)=$13^h 34^m 18^s.10$
δ (2000.0)=$-25°22'49''.93$
Type: S
Pg=10.0–11.2
Spectrum: M2 III+wd
P=$370 \pm 0^d.9$
Max.phot.=JD 2445071.6±11.3

References: [4, 423]

BX Mon/AS 150

α (2000.0)=$07^h 25^m 22^s.78$
δ (2000.0)=$-03°35'50''.61$
Type: S, Mira
Pg=9.5–13.4
Spectrum: M4 III+F0 II–III
P=1374^d
Max.phot.=JD 2430345

References: [4]

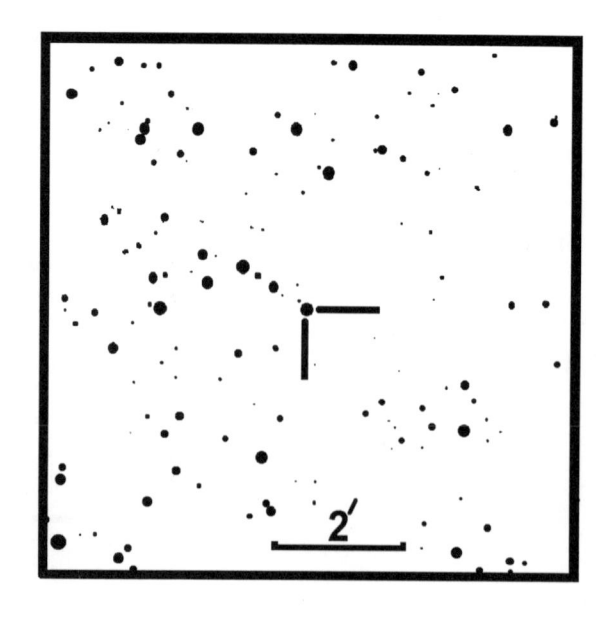

SY Mus/HD 100336

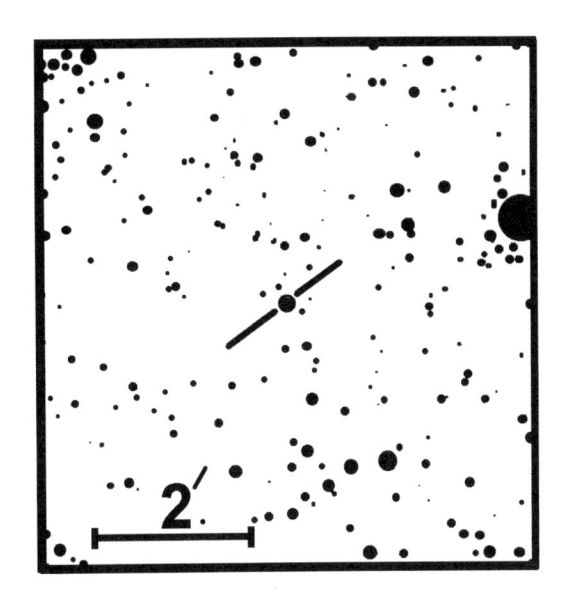

α (2000.0)=$11^h32^m10^s.47$
δ (2000.0)=$-65°25'10''.34$
Type: S
V=10.2–12.7
Spectrum: M4.5±0.5
P=$628^d.3$
Max.vr=JD 2449405

This is a well detached *eclipsing* system [298].

References: [4, 298]

AR Pav/MWC 600

α (2000.0)=$18^h20^m28^s.65$
δ (2000.0)=$-66°04'48''.65$
Type: S
B=7.4–13.62
Spectrum: M3-4 II–III
P=$604^d.6$
Min.phot.=JD 2420330

References: [4]

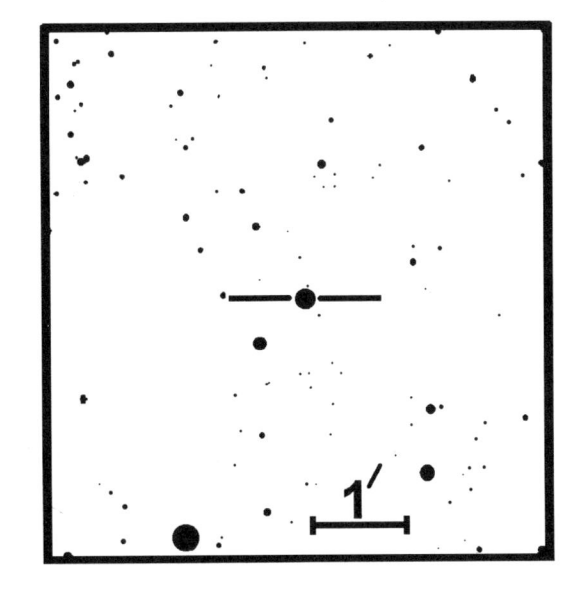

AG Peg/BD+11°4673/HD 207757/IRAS 21486+1223

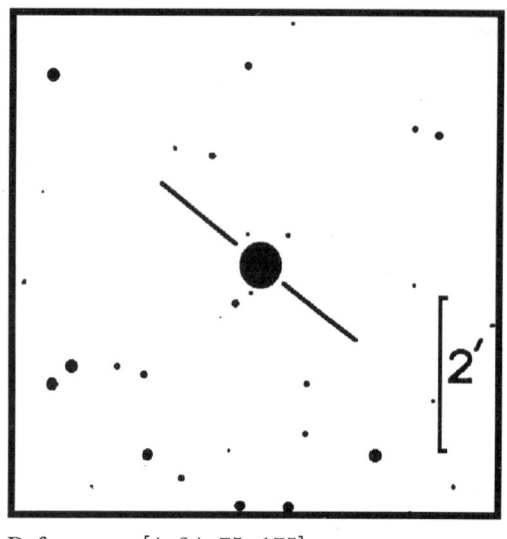

α (2000.0)=$21^h51^m02^s.01$
δ (2000.0)=$12°37'32''.16$
Type: S, N
V=6.0–9.4
Spectrum: M3.0 III+B3–5
P=$813 - 827^d$

AG Peg is one of the bright symbiotic novae. The last outburst was in the middle of the 19th century. The fading rate was $\dot{V} \simeq 0.604$ mag yr^{-1}. The luminosity was changing from ~$3000L_\odot$ (1870–1980) to ~ $700L_\odot$ (1985–1991).

References: [4, 34, 75, 175]

AX Per/MWC 411/MW 228

α (2000.0)=$01^h36^m16^s.56$
δ (2000.0)=$54°15'35''.48$
Type: S
Pg=9.4–13.6
Spectrum: M5.2 II–III+A0
P=$680.8 \pm 0^d.2$
Min.phot.=JD 2436667

The last eruption of AX Per was in 1986–1987. The accretion disc surrounding a main-sequence star produces the UV continuum, the radiation from the boundary layer ionizes a small nebula. The orbital period decreases systematically at the rate $\dot{P} \sim (4.6 \pm 1) \times 10^{-5}$ s^{-1}, (or ~2 days per 100 yr). The observed recurrence time is ~15 yr.

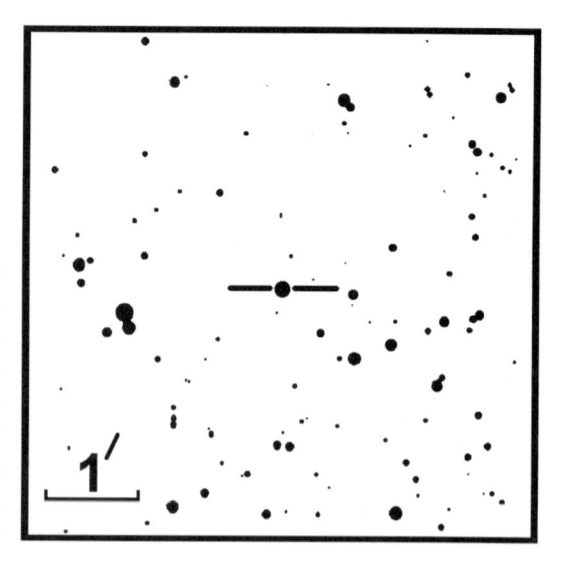

References: [4, 224]

RX Pup/HD 69190/IRAS 08124-4133

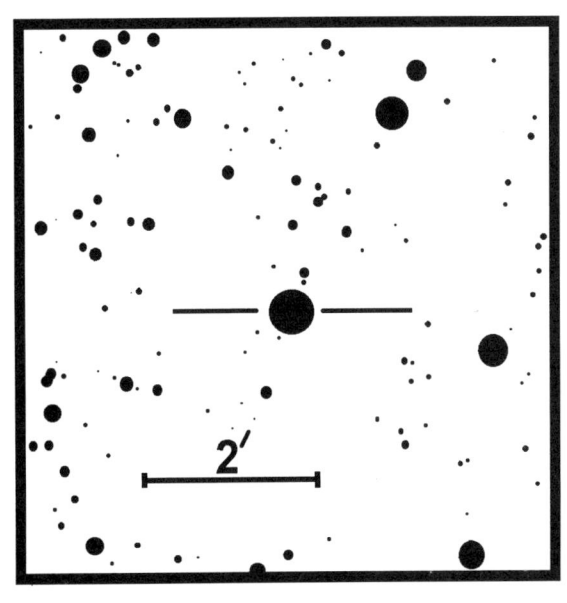

α (2000.0)=$08^h14^m12^s.33$
δ (2000.0)=$-41°42'28''.71$
Type: D, Mira
V=10.9-14.5
Spectrum: M5.3 III+A II?
$P=650^d \pm 10^d$
Min.phot.=JD 2448492

RX Pup is related to slow novae and contains a 580^d Mira. The low excitation stage was in 1960–1975, the high excitation stage, in 1905(?), 1940, 1981–1988 [5]. The radio emission between 100 and 40 GHz at the low stage is of the order of 20 mJy, in the high state, ≥ 500 mJy.

References: [4, 5]

CL Sco/AS 213/Hen 1286/MHα71-5

α (2000.0)=$16^h54^m51^s.91$
δ (2000.0)=$-30°37'17''.72$
Type: S
V=11.2-13.9
Spectrum: K5?+pec
$P=624.7 \pm 1^d.5$
Min.phot.=JD 2427020

References: [4]

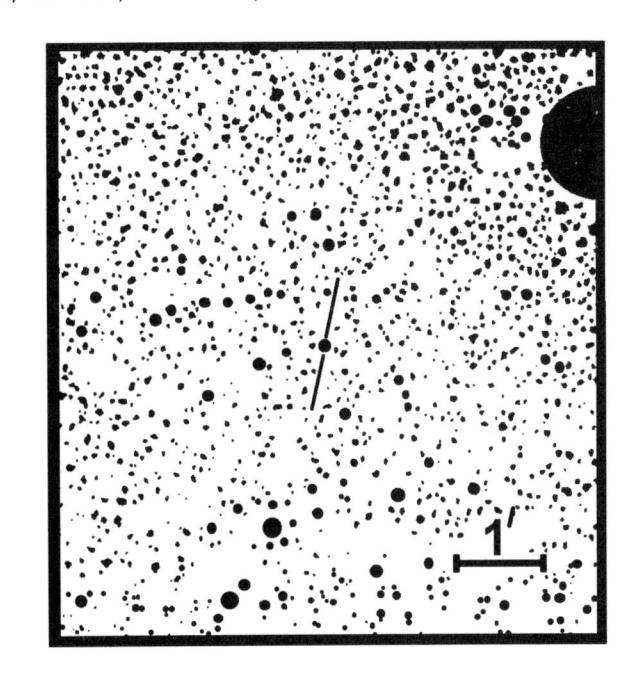

FG Ser/AS 296/SS 148/IRAS 18125−0019

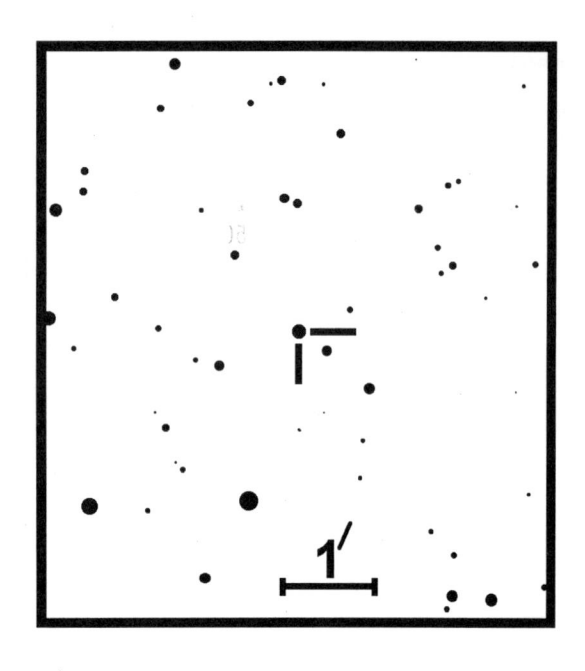

α (2000.0)=$18^h15^m07^s.17$
δ (2000.0)=$-00°18'52''.52$
  Type: S
  B=10.4–13.9
  Spectrum: M5.3 III+wd? A0?
  $P=650^d$? or 1200^d?

FG Ser is known as an emission line object [220] and as a symbiotic nova [140]. Two eclipses separated by 650 days were observed. The main outburst was in 1988. Inverse P Cyg profiles in Fe II, Cr II, Ti II, Sc II are present.

References: [4, 140, 220, 239]

HM Sge/IRAS 19396+1637

α (2000.0)=$19^h41^m57^s.08$
δ (2000.0)=$16°44'39''.73$
  Type: D
    N, Mira
  B=11.1–18
  Spectrum: M5–9 III
  $P=527 \pm 2^d$
  Max.phot.=JD 2440310±30

HM Sge underwent an outburst to V=11 mag in 1975 [98]. It has a bipolar nebula, similar to the R Aqr nebula. The finding chart corresponds to an epoch before the outburst.

References: [4, 98]

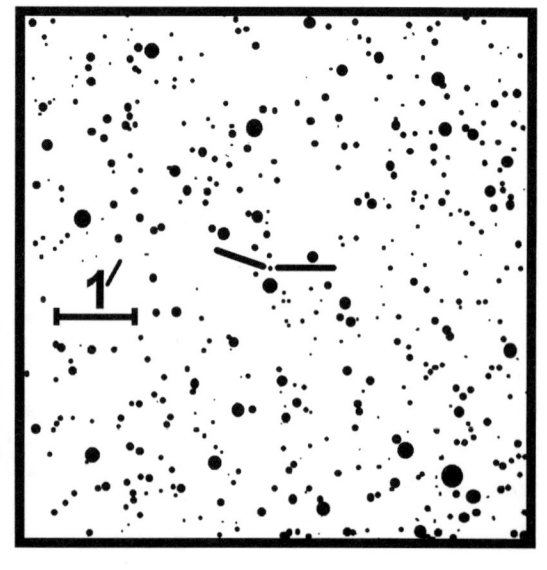

V 2601 Sgr/AS 313

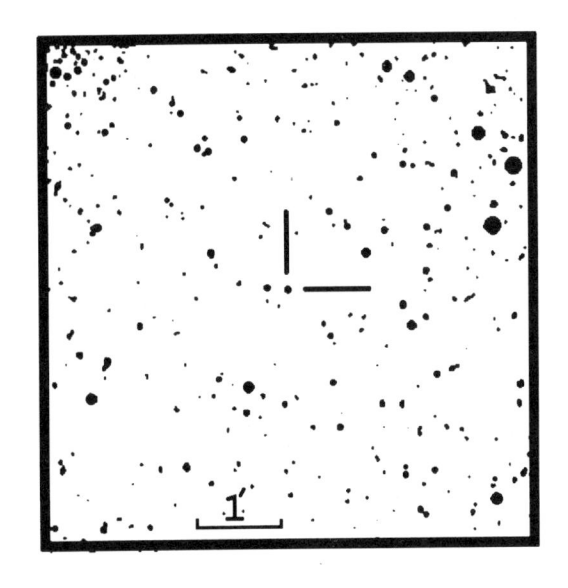

α (2000.0)=$18^h38^m02^s.11$
δ (2000.0)=$-22°41'50''.97$
Type: S
Pg=14.0–15.3
Spectrum: M1–5 III
P=850^d
Min.phot.=JD 2429850

References: [4]

V 2756 Sgr/AS 293

α (2000.0)=$18^h14^m34^s.54$
δ (2000.0)=$-29°49'23''.30$
Type: S
Pg=13.2–15.2
Spectrum: M2e
P=243^d
Min.phot.=JD 2437485

References: [4, 98]

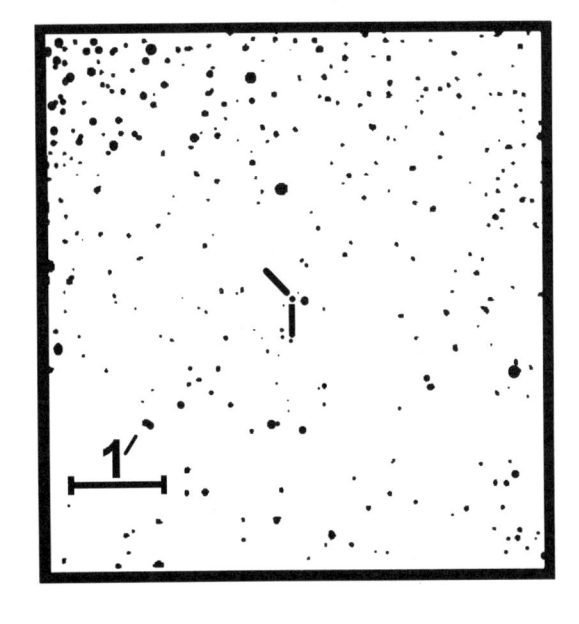

RR Tel/IRAS 20003–5552/Hen 1811

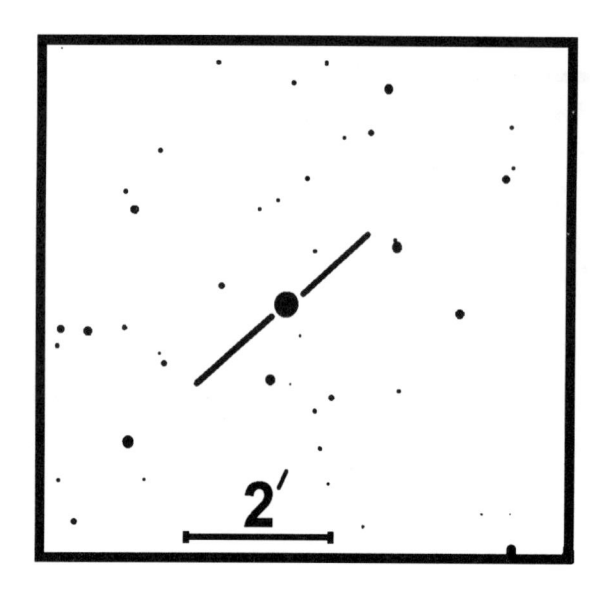

α (2000.0)=$20^h04^m18^s.52$
δ (2000.0)=$-55°43'34''.10$
   Type: S
      N, Mira
      Pg=6.8–16
   Spectrum: later than M5
      P=374.2 ± $3^d.8$
   Min.phot.=JD 2442550.7±18.7

RR Tel is a prototype of the symbiotic novae, known before the outburst

References: [4]

PU Vul/Kuwano object/IRAS 20189+2124

α (2000.0)=$20^h21^m13^s.35$
δ (2000.0)=$21°34'18''.16$
   Type: N, S
      B=9–16.5
Spectrum: M4–5+F–A I?wd?

PU Vul is a symbiotic nova. It increased its brightness in 1977. There was a very deep minimum in 1980 with amplitude $\sim 5^m$, underwent an eclipse. It is possible that the outburst of PU Vul is due to the process of nuclear shell burning. The F-type spectrum is suggested to be that of the expanding envelope around the white dwarf [319].

References: [4, 319]

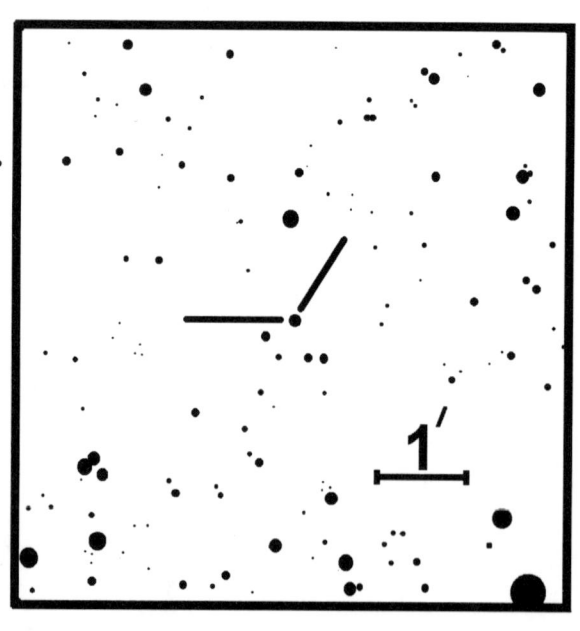

CATACLYSMIC VARIABLES.

NOVAE, DWARF NOVAE AND RELATED OBJECTS

This part of the Atlas contains three groups of cataclysmic variables — bursting stars of U Gem, Z Cam, and UX UMa types; SU UMa subtype stars; and Novae, nova-like stars, VY Scl subtype stars.

For the types of the systems, the Ritter's Catalog designations were used:

DN—dwarf nova;

N—classical nova of undetermined speed class;

NA—fast nova;

NB—slow nova;

NC—extremely slow nova;

NL—novalike variables;

NR—recurrent nova;

SH—non-SU UMa star showing either permanent or transient superhumps;

SU—SU UMa star (with normal outbursts, superoutbursts, and superhumps);

SW—SW Sex star (with intermittent absorption events near phase 0.5, large spectroscopic phase offset and high-excitation lines);

UG—dwarf nova, either U Gem or SS Cyg subtype;

VY—VY Scl star (antidwarf nova; NL with unpredictable low states);

XS—X-ray source;

ZC—Z Cam star, subtype of DN.

Here we give:

S.C.—the epoch of the optical star's superior configuration;

I.C.—the epoch of the optical star's inferior conjunction;

Max—the epoch of the system's maximum brightness;

Min—the epoch of the system's minimum brightness.

RX And

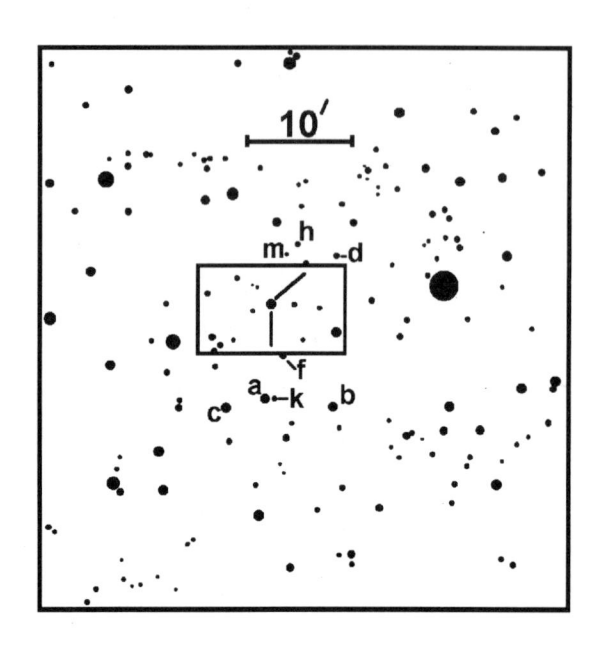

α (2000.0)=$01^h04^m35^s.49$
δ (2000.0)=$41°17'58''.17$
Type: DN ZC XS
V=10.3–14.0
Spectrum: M5 V
$P=0^d.2098930$
S.C.=JD 2447041.932
$P_{cycle}\sim 14^d$

The comparison stars are:

Star	V	B
a	10.59(2)	10.81(2)
b	10.67(2)	11.67(2)
c	11.03(2)	11.21(2)
d	11.29(2)	12.37(2)
e	11.85(2)	12.88(2)
f	12.04(2)	12.38(2)
g	12.14(2)	13.37(2)
h	12.44(2)	13.33(2)
k	12.47(2)	13.08(2)
l	13.05(3)	13.51(3)
m	13.67(4)	14.32(4)
n	14.44(7)	14.93(7)

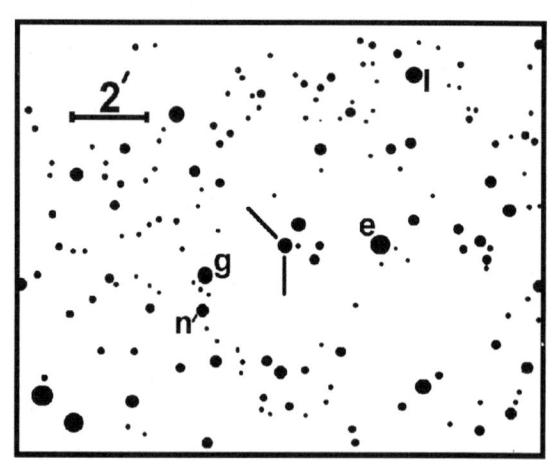

Cyclic variations with $P \sim 38 - 54$ s were observed after a normal maximum; after a short maximum no periodicity was found.

References: [3, 48, 52, 176, 190, 332]

AR And

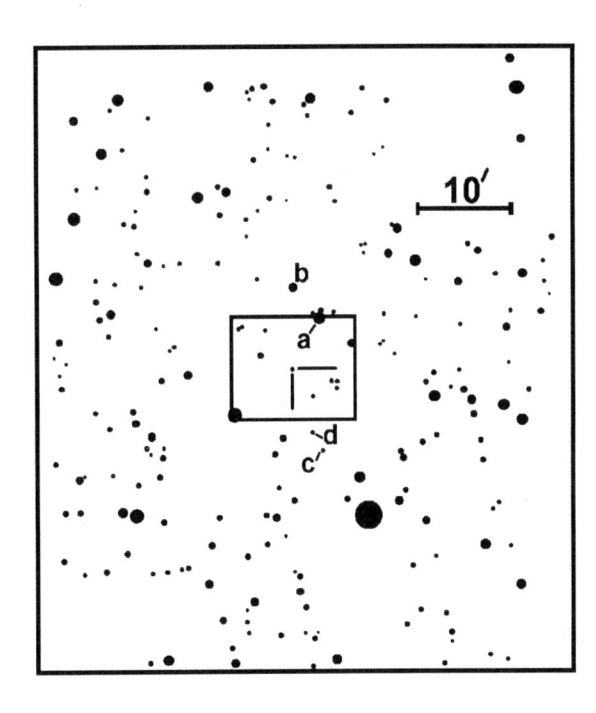

α (2000.0)=$01^h 45^m 03^s.322$
δ (2000.0)=$37°56'34''.35$
Type: DN UG
Pg=11.0–17.6
$P=0^d.1639$
$P_{cycle}\sim 25^d$

The comparison stars are:
(Photographic magnitudes)

Star	Pg
a	11.5
b	12.2
c	12.9
d	13.6

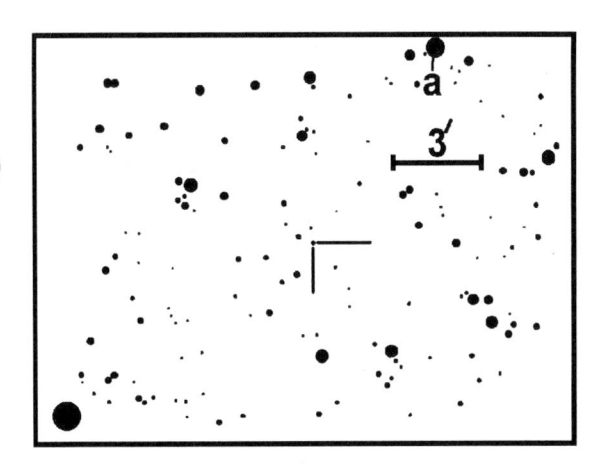

Photographic photometry is published in [225].

References: [52, 176, 205, 225, 402]

DX And

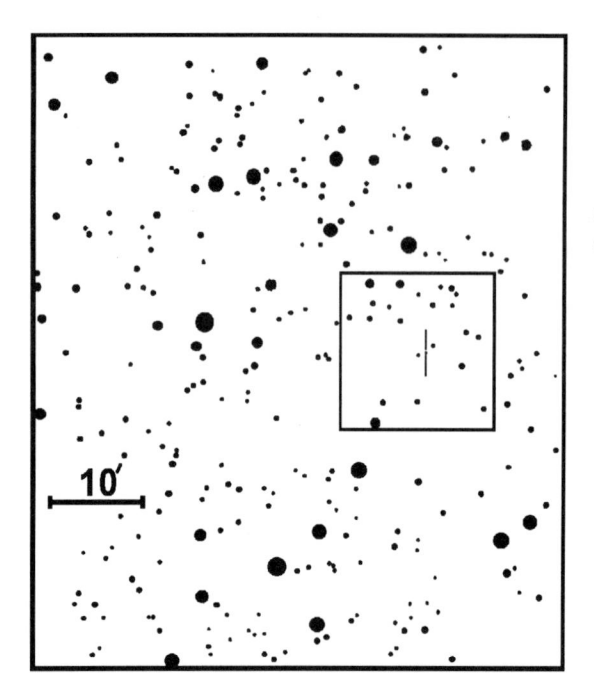

α (2000.0)=$23^h 29^m 46^s.82$
δ (2000.0)=$43°45'03''.4$
Type: DN UG
Pg=10.9–16.4
Spectrum: K1 V
$P=0^d.44167$
$P_{cycle}\sim 214^d$

The comparison stars are:
(Photographic magnitudes)

Star	Pg	Star	Pg
1	10.7	6	13.1
2	11.4	7	13.8
3	11.6	8	14.4
4	11.9	9	14.6
5	12.6		

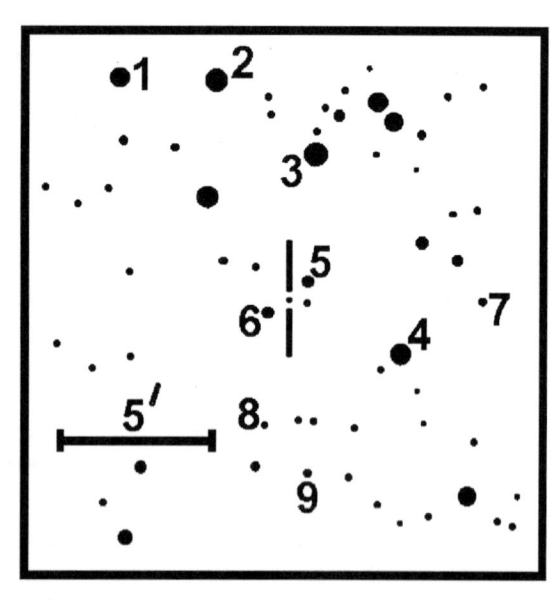

A mean duration of the flare is 11^d.

References: [52, 176, 233, 286, 287]

FO And

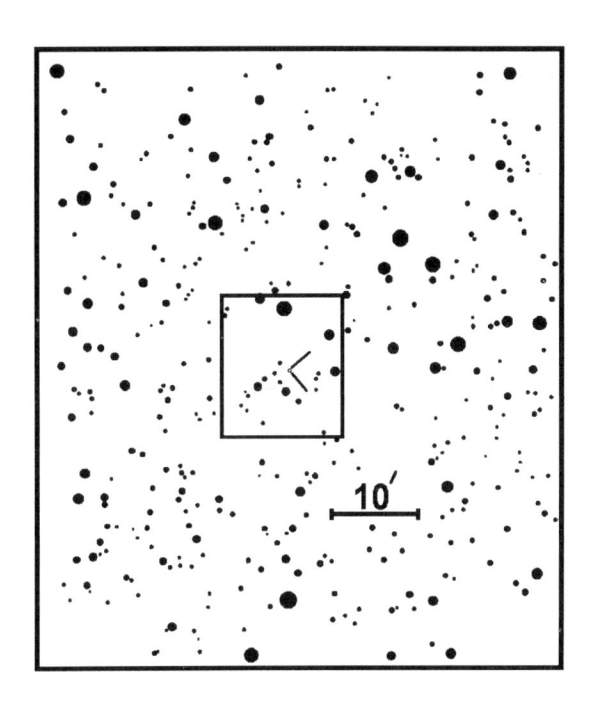

α (2000.0)=$01^h15^m32^s.137$
δ (2000.0)=$37°37'35''.94$
Type: DN SU
Pg=13.5–17.5
P=$0^d.073$

The comparison stars are:
(Photographic magnitudes)

Star	Pg(B)
1	14.5
2	15.6
3	15.8
4	16.3
5	16.8

The superhump period is 104 min.

References: [52, 176, 217, 371]

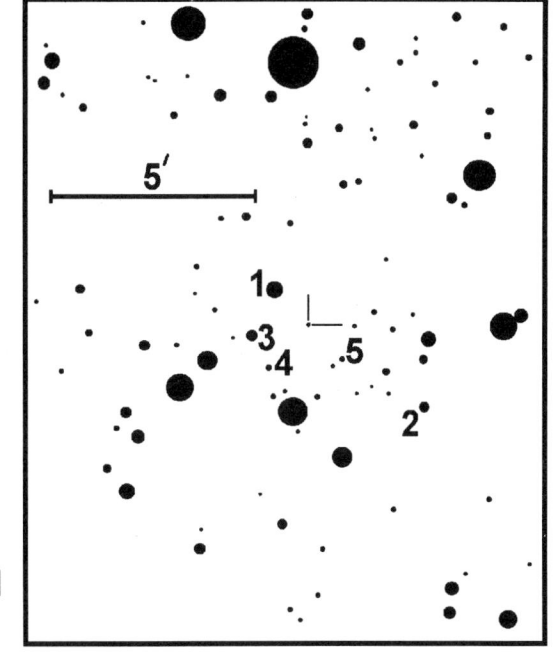

KV And

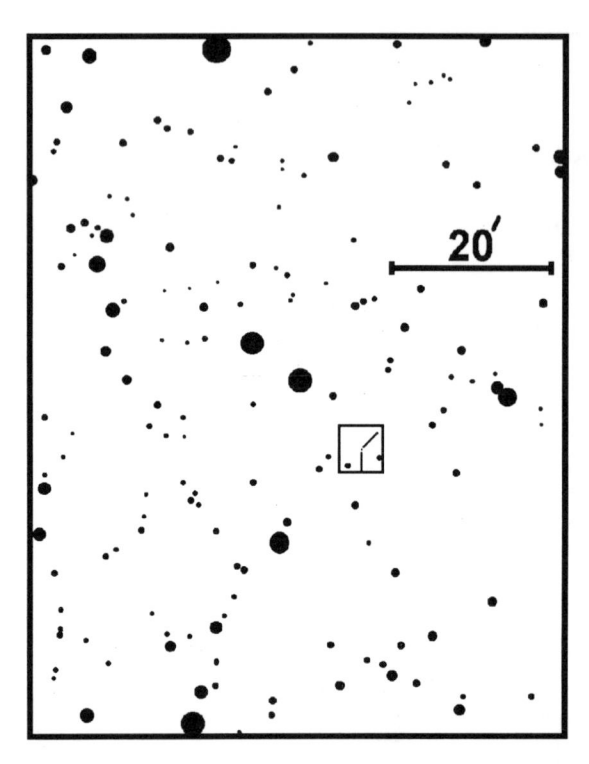

α (2000.0)=$02^h17^m14^s$
δ (2000.0)=$40°41'30''$
Type: DN SU
Pg=14.6–22
$P=0^d.069$:
$P_{cycle}\sim 112^d$

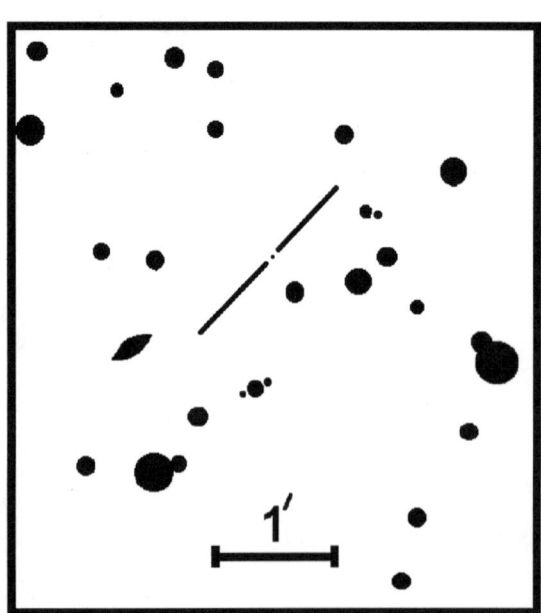

References: [176, 187]

LL And

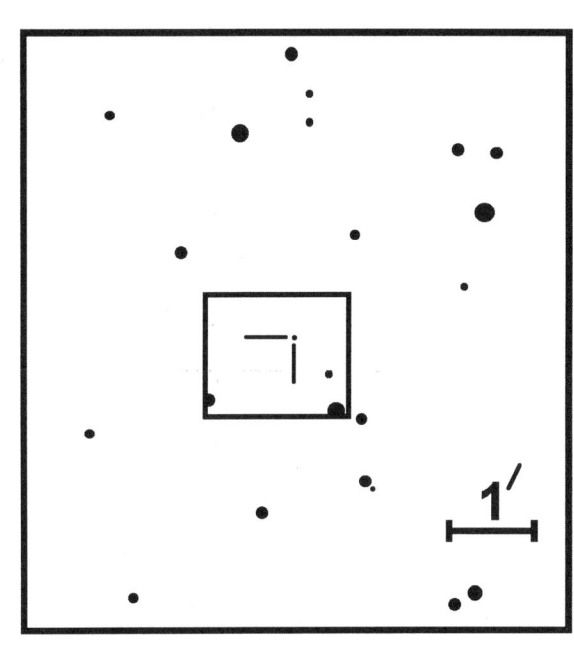

α (2000.0)=$00^h41^m51^s$
δ (2000.0)=$26°37'25''$
Type: DN, SU
Pg=13–17
$P=0^d.055$:
$P_{cycle}\sim 5000^d$:

References: [103, 158, 161]

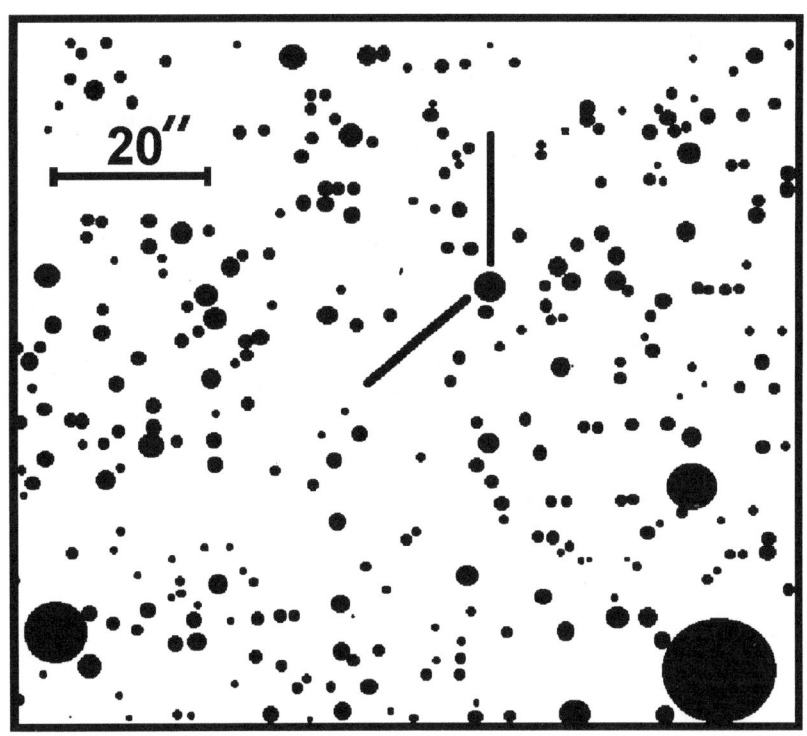

LT And

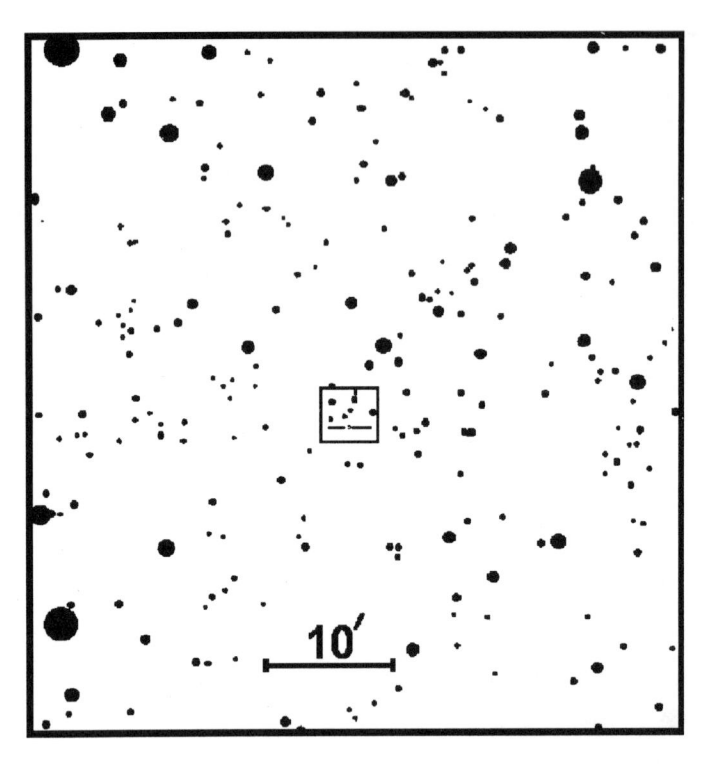

α (2000.0)=$00^h35^m24^s.49$
δ (2000.0)=$39°46'14''.6$
Type: UG
Pg=18.1–19.4

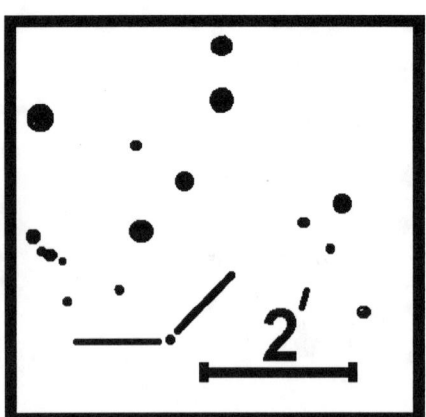

In minimum V=19.03, $B - V$=+0.44, $U - B$=−0.95.

References: [45, 176, 266]

FO Aql

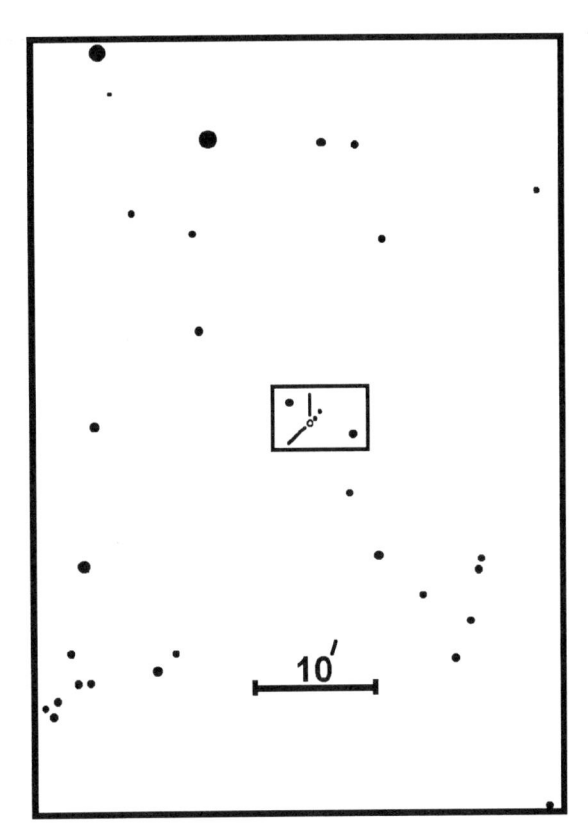

α (2000.0)=$19^h16^m32^s.20$
δ (2000.0)=$00°07'36''.7$
Type: DN SU?
B=13.8–18
P=$0^d.2$
$P_{cycle} \sim 14 - 23^d$

The comparison stars are:

Star	V	B	U
a	10.65	11.33	11.39
b	10.87	11.44	11.58
c	16.2	17.1	17.8

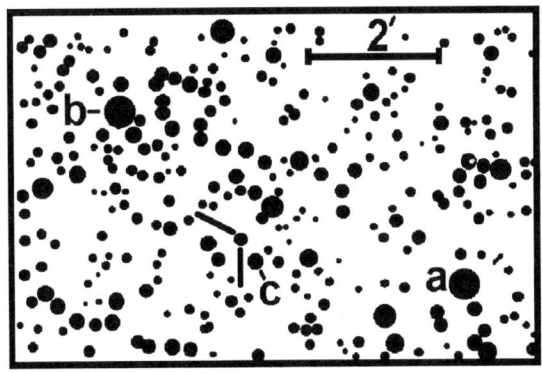

Photometric observations are published in [50, 362]. Large interstellar reddening. Superoutbursts have $B = 13^m.8$ and duration $\sim 20^d$. Smaller outbursts have $B = 14.^m3$ and duration $\sim 9^d$.

References: [50, 176, 360, 362]

UU Aql

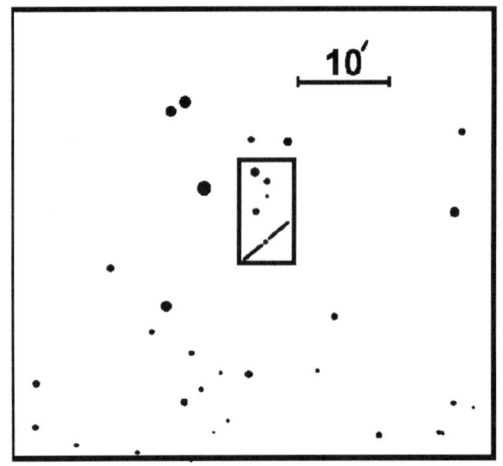

α (2000.0)=$19^h57^m18^s.64$
δ (2000.0)=$-9°19'20''.5$
Type: DN UG
Pg=11.0–16.8
$P=0^d.14049$:
$P_{cycle}\sim 49^d.5$:

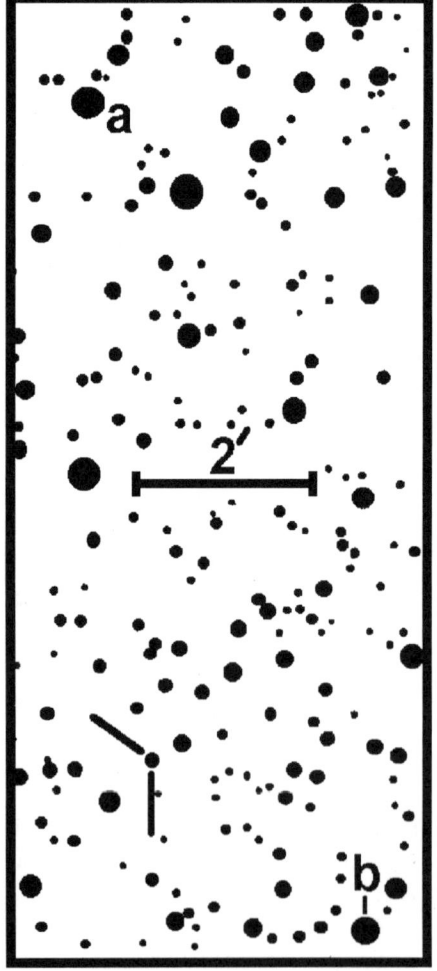

The comparison stars are:

Star	V	B	U
a	11.00	11.49	11.48
b	12.39	13.15	13.46

Photoelectric observations are published in [50, 362]. Mean cycle according to [270] is $71^d.3$, according to [212], 56^d.

References: [50, 176, 212, 270, 360, 362, 364, 402]

AT Ara

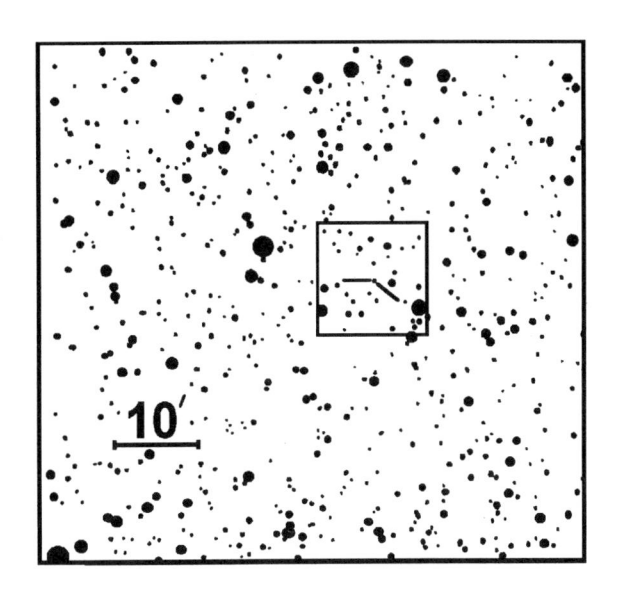

α (2000.0)=$17^h 30^m 34^s.06$
δ (2000.0) = $-46°05'58''.5$
Type: NL?
Vis=11.5–14.9
$P=0^d.333$
$P_{cycle}=61^d.8$

The comparison stars are:
(Visual magnitudes)

Star	Vis
1	10.5
2	11.7
3	12.0
4	13.0
5	13.4
6	13.5
7	13.6
8	14.2
9	14.6

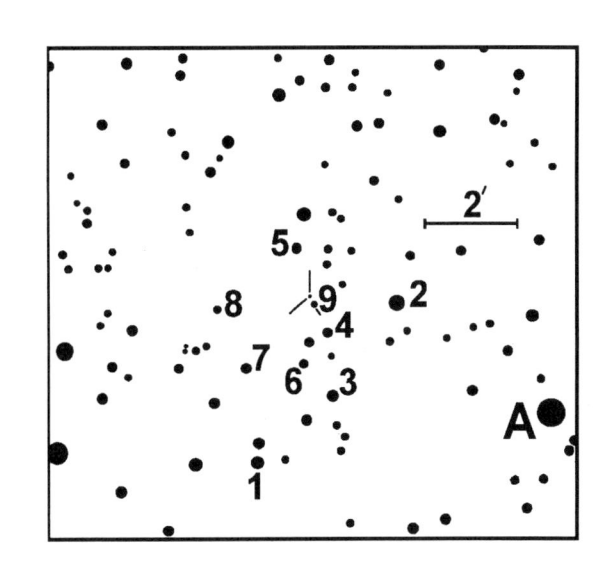

A = CPD−46°8682. The spectrum of AT Ara reveals absorption lines of a late type secondary. Photometric observations are published in [50, 362].

References: [6, 50, 176, 358, 362, 364, 380]

TT Ari

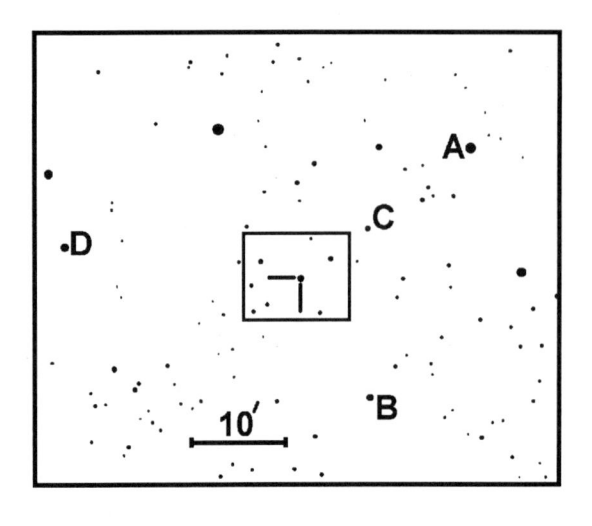

α (2000.0)=$02^h06^m53^s.15$
δ (2000.0)=$15°17'42''.0$
Type: NL VY XS SH?
B=9.5–16.3
P=$0^d.1327710$
Max.phot.=JD 2437646.665

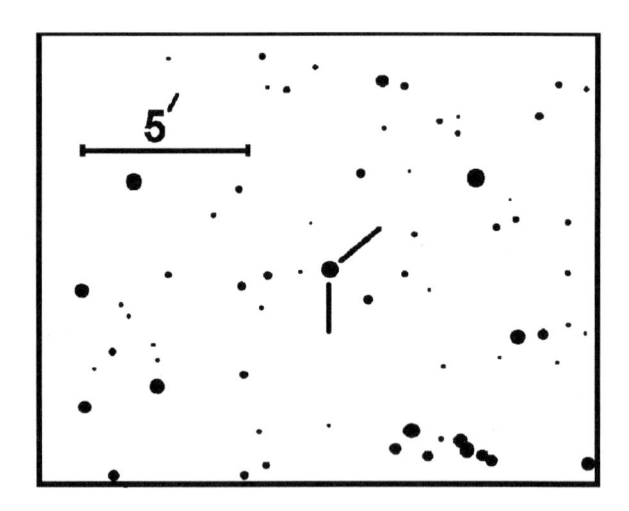

A = BD+14°337,
B = BD+14°339,
D = BD+14°342,
C is a constant star.

QPOs in X-ray range with $P = 9, 12, 32$ s. Irregular flickering in optics up to $0^m.1$ in 1 min. A 15.3 min period with $A \sim 0^m.04$ was found.

References: [176, 201, 217, 322, 402]

SS Aur

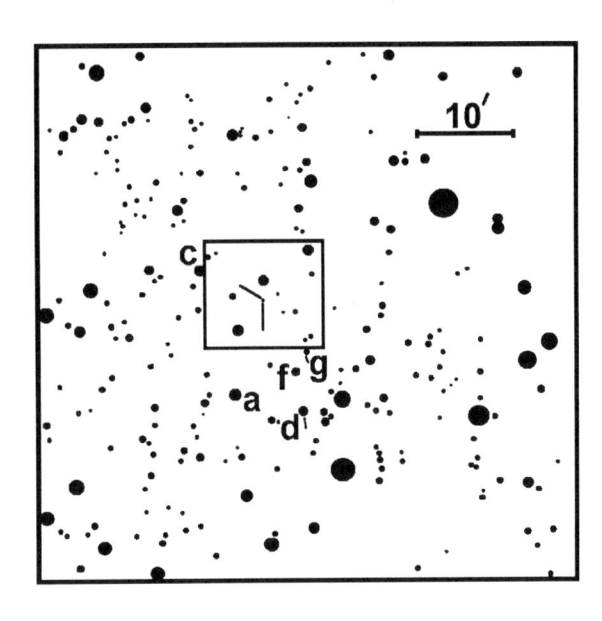

α (2000.0)=$06^h13^m22^s.0$
δ (2000.0)=$47°44'24''$
Type: DN UG
V=10.3–15.8
Spectrum: M1–3 V
$P=0^d.1828$
$I.C.$=JD 2446864.825
$P_{cycle} \sim 55^d.5$

The comparison stars are:

Star	V	B
a	10.61(2)	11.08(2)
b	10.85(2)	11.30(2)
c	10.85(2)	11.81(2)
d	10.95(2)	11.23(2)
e	10.98(2)	11.07(2)
f	11.19(2)	11.94(2)
g	11.30(2)	11.58(2)
h	12.10(2)	13.10(2)
k	12.53(3)	12.95(3)
l	13.18(3)	14.21(3)
m	13.21(3)	13.70(3)
n	13.93(5)	14.78(5)
p	16.0(2)	16.9(2)

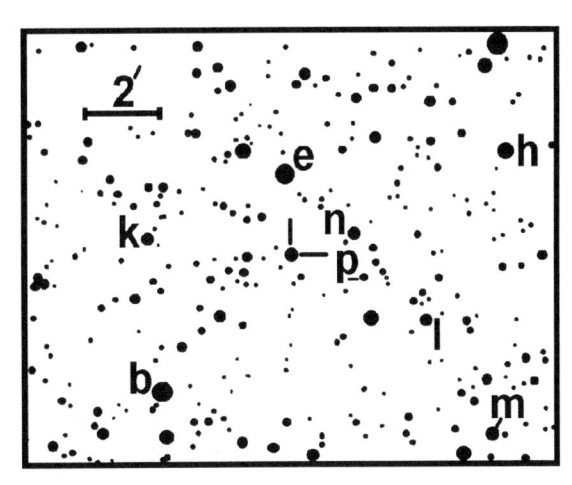

The cycles of 2.32 and 1.2 years were found. Sometimes abnormal behavior of the star is being observed when comparatively numerous and small light maxima appear.

References: [52, 156, 176, 190, 402]

FS Aur

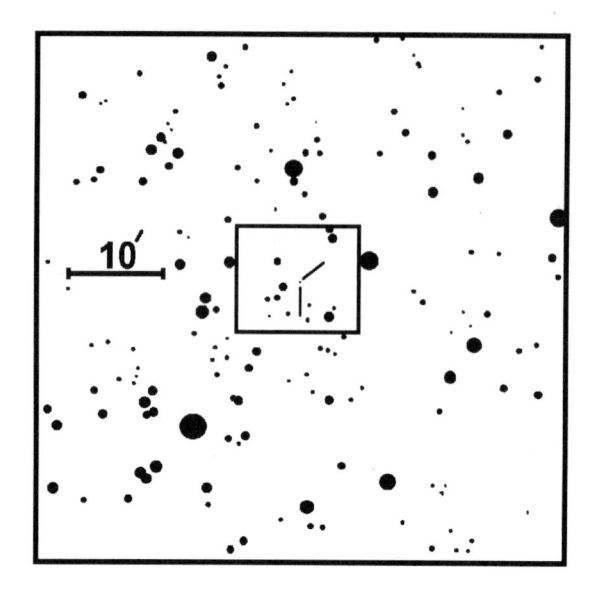

α (2000.0)=05^{h}47^{m}48^s.37
δ (2000.0)=28°35′11″.2
Type: DN UG
Pg=14.0–17.2
P=0^d.059 :
P$_{cycle}$~12^d

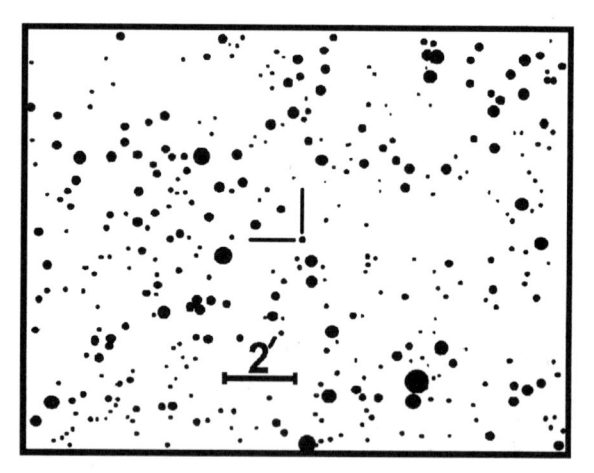

Alternative value of the orbital period is 0^d.067.

References: [176, 201, 217, 322, 338, 402]

KR Aur

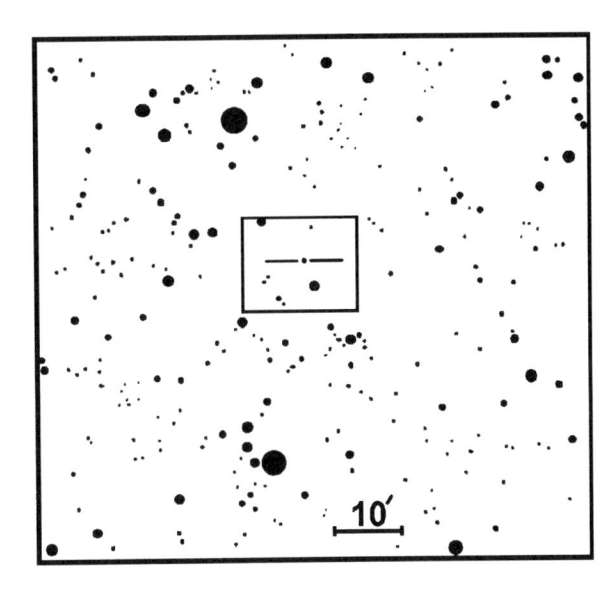

α (2000.0)=$06^h15^m43^s.7$
δ (2000.0)=$28°35'08''.1$
Type: NL VY XS
V=11.3->18
Spectrum: later than M3 V
$P=0^d.16279$
I.C.=JD 2445000.7144

A=BD+28°1054.

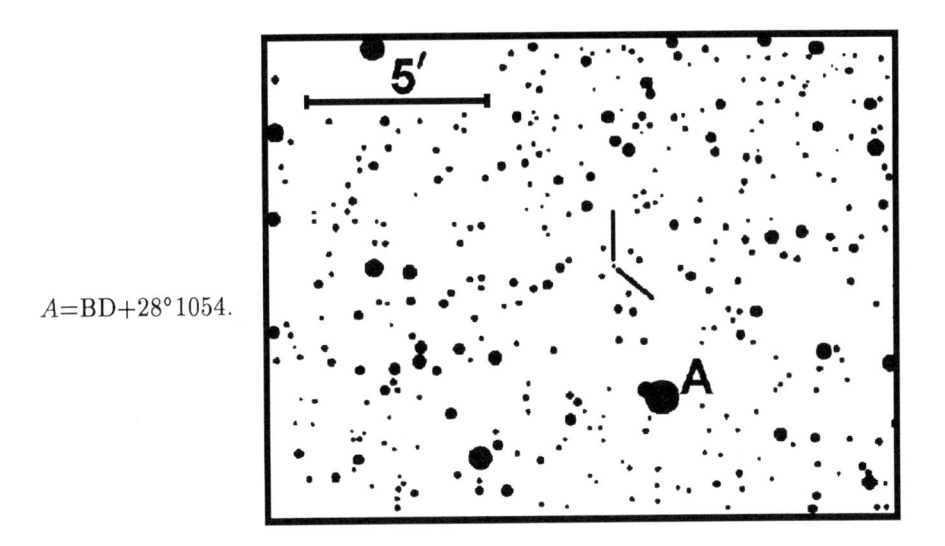

Cyclic variability with P~490-780 and 1000-1200 s. Long maxima and minima lasting for years are superimposed by rapid light variations of 1-2 mag in tens of seconds. Soft X-ray source.

References: [100, 176, 234, 272, 274, 402]

TT Boo

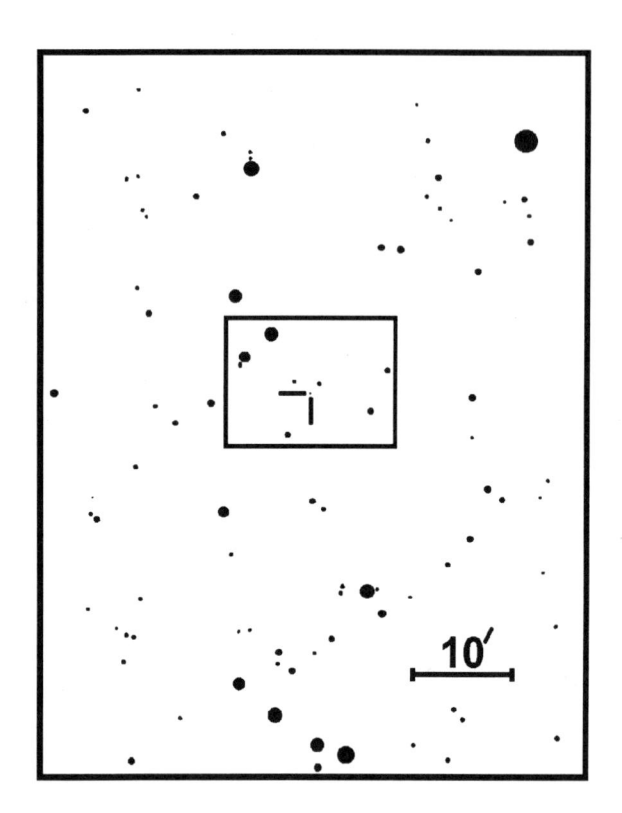

α (2000.0)=$14^h57^m44^s.839$
δ (2000.0)=$40°43'41''.68$
Type: DN SU
Pg=12.7->15.6
P=$0^d.07523$
$Max.phot.$=JD 2449088.110
$P_{cycle}\sim 45^d$

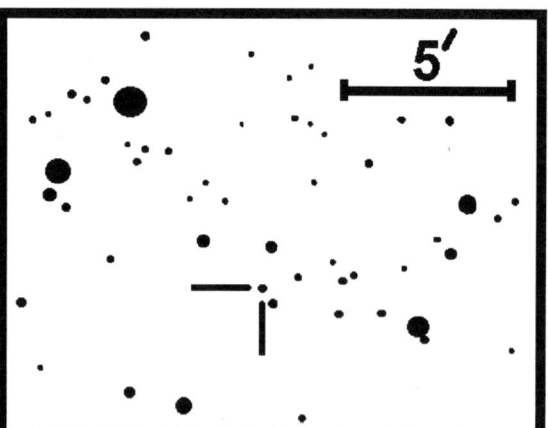

Other possible periods of the orbital variability are $0^d.067$ or $0^d.077$ with the amplitude of $0^m.2$.

References: [52, 176, 380]

UZ Boo

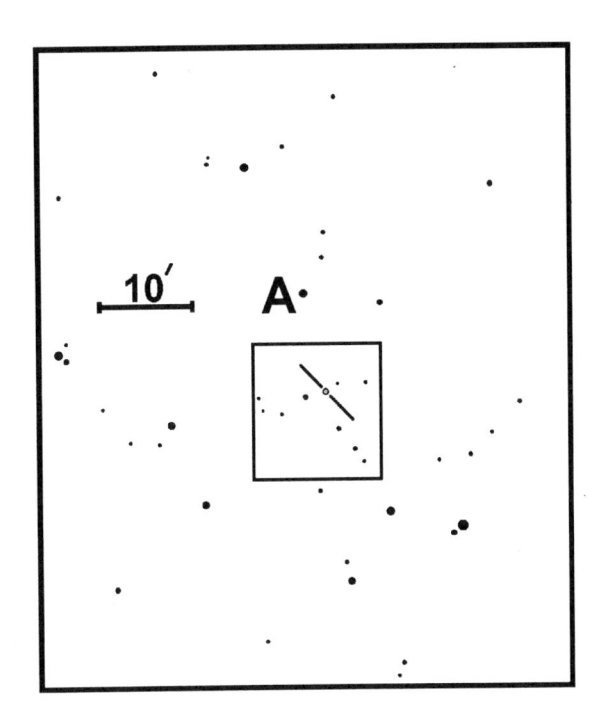

α (2000.0)=$14^h44^m01^s.35$
δ (2000.0)=$22°00'56''.2$
Type: DN
V=11.5–19
P=$0^d.125$:
P_{cycle}~2191?^{d}

The comparison stars are:

Star	V	B	U
a	13.18	14.13	14.58
b	14.42	15.06	15.22

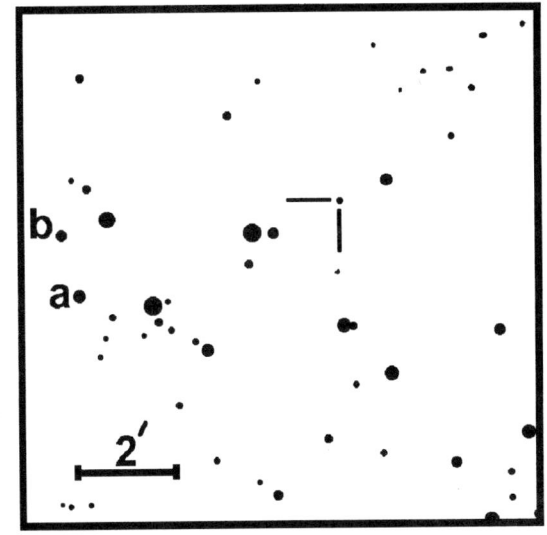

A=BD+22°2742. Possibly, a recurrent nova with the cycle 2191^d. The cycle of QPOs is 1490 s.

References: [176, 188, 334, 364, 380]

Z Cam

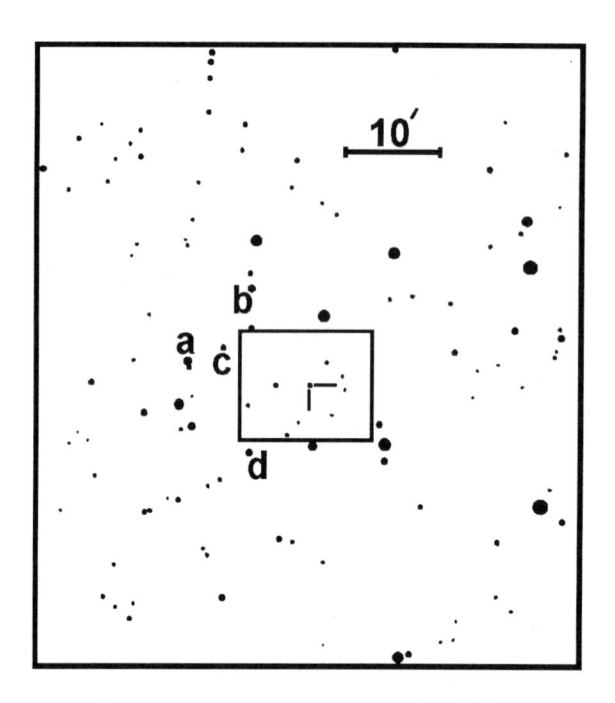

α (2000.0)=$08^h25^m15^s.105$
δ (2000.0) =$73°26'39''.52$
Type: DN ZC XS
V=10.0–14.8
Spectrum: K7 V
P=$0^d.289840$
$I.C.$=JD 2438470.841
P_{cycle}~22^d

The comparison stars are:

Star	V	B
a	10.16(2)	10.68(2)
b	10.70(2)	11.08(2)
c	10.76(2)	11.16(2)
d	11.00(2)	11.47(2)
e	11.90(2)	12.37(2)
f	12.56(2)	13.52(2)
g	13.16(2)	13.95(3)
h	13.67(4)	14.61(3)

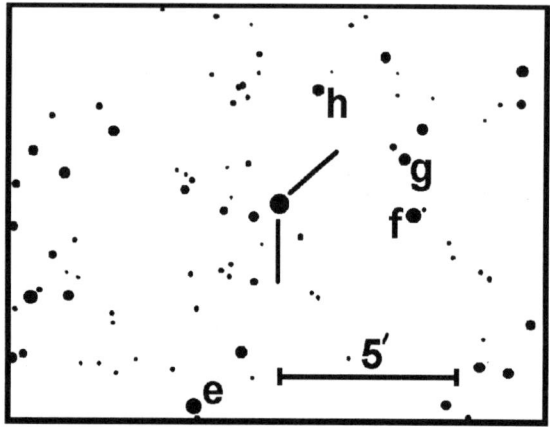

The 8.84 year cycle was found. Outside maxima of the light curve, there is flickering up to $0^m.5$. Brightness fluctuations during maxima are small. The epoch of the lower conjunction of the K component is given. Noneclipsing binary.

References: [6, 176, 402]

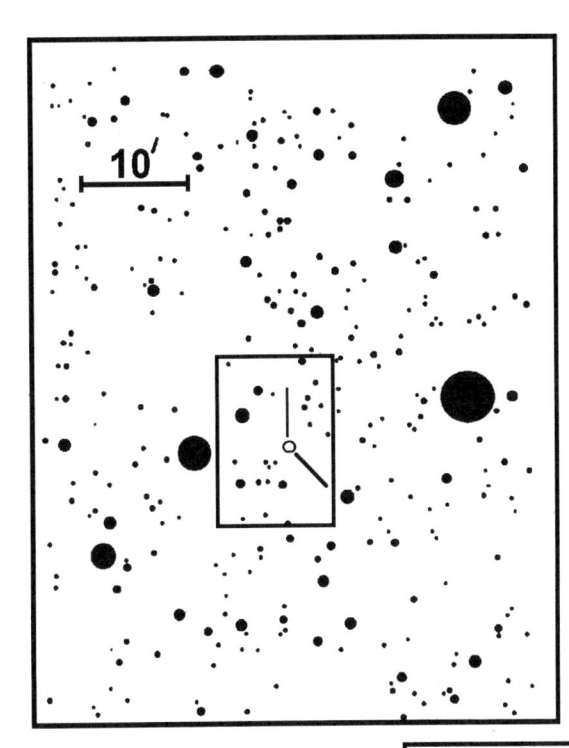

AF Cam

α (2000.0)=$03^h32^m15^s.46$
δ (2000.0)=$58°47'21''.7$
Type: DN
Pg=13.4–17.6
$P=0^d.2306$
$P_{cycle}\sim 74^d.9$

The comparison stars are:

Star	Pg
1	12.81
2	14.00
3	14.51
4	15.09
5	15.81

References: [52, 176, 184, 334]

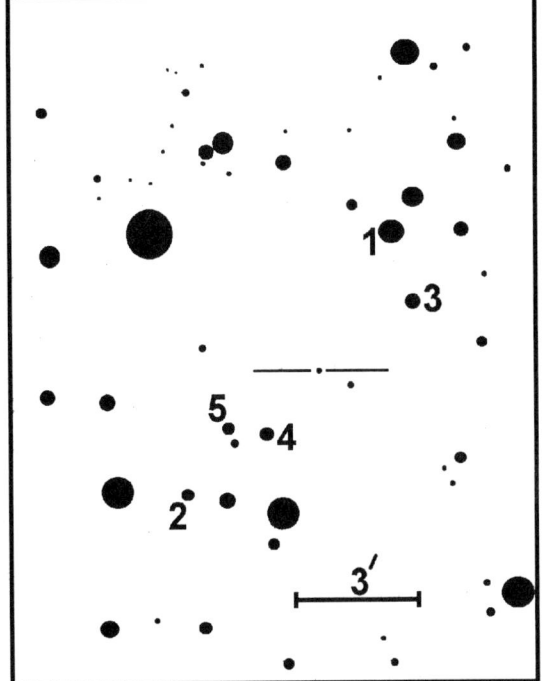

BZ Cam

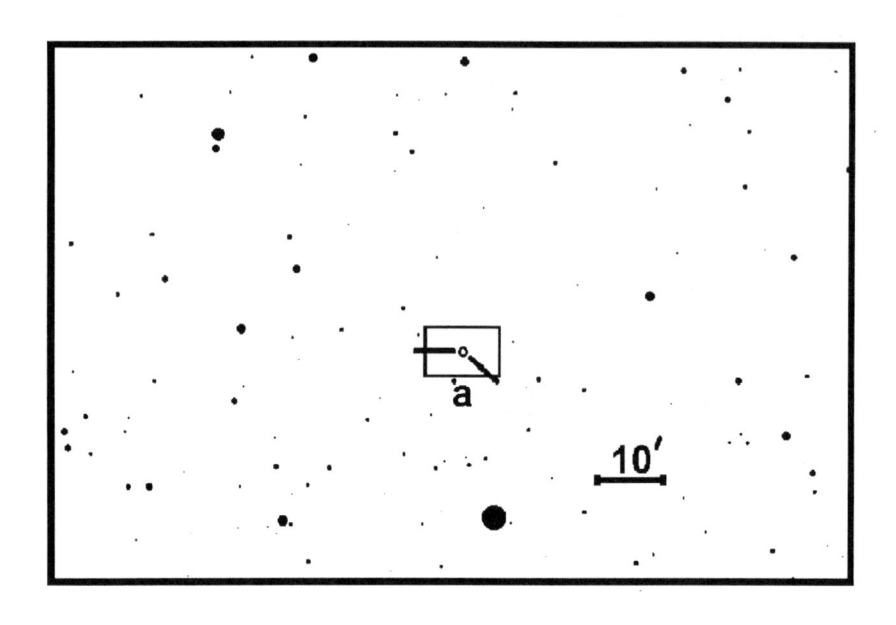

α (2000.0)=$06^h29^m34^s.12$
δ (2000.0)=$71°04'36''.2$
  Type: NL VY
    B=12.5–14.1
    P=$0^d.1390$:
    I.C.=JD 2446073.836

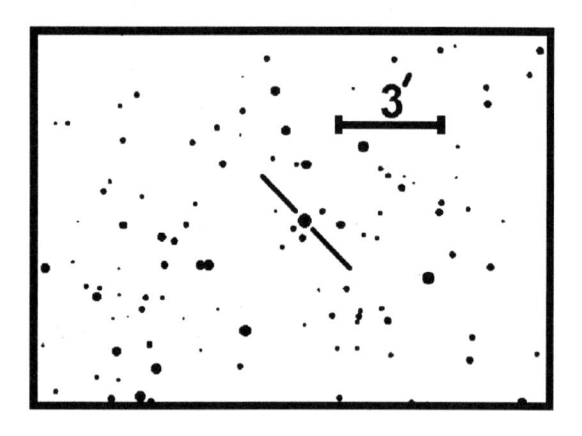

The system is surrounded with an emission nebula. The constant star *a* may be used as a comparison star.

References: [281, 402, 404]

SY Cnc

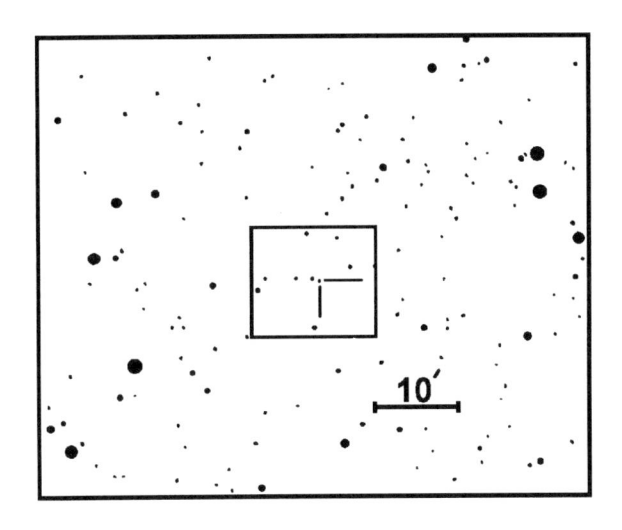

α (2000.0)=$09^h01^m03^s.48$
δ (2000.0) =$17°53'56''.7$
Type: DN ZC
Spectrum: G8-9 V
Pg=10.8–14.5
$P=0^d.380$
$P_{cycle}=27^d$

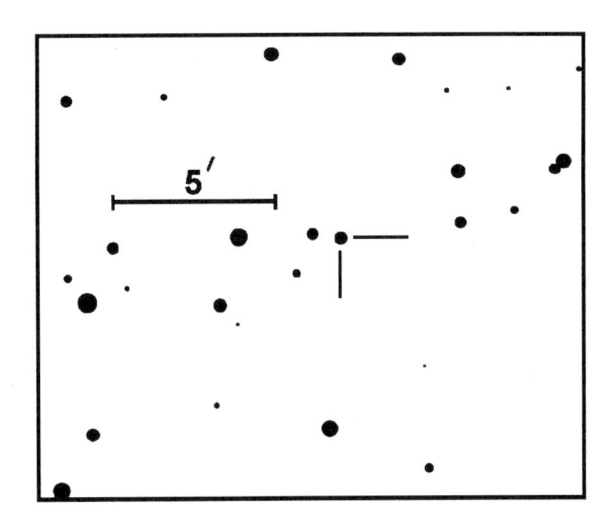

Flickering with periods 17, 22, 25 s. Radial velocity semi-amplitude was $K = 90$: km s^{-1}. Photometric observations are published in [222, 244, 257, 282].

References: [48, 59, 176, 222, 244, 257, 282, 307]

YZ Cnc

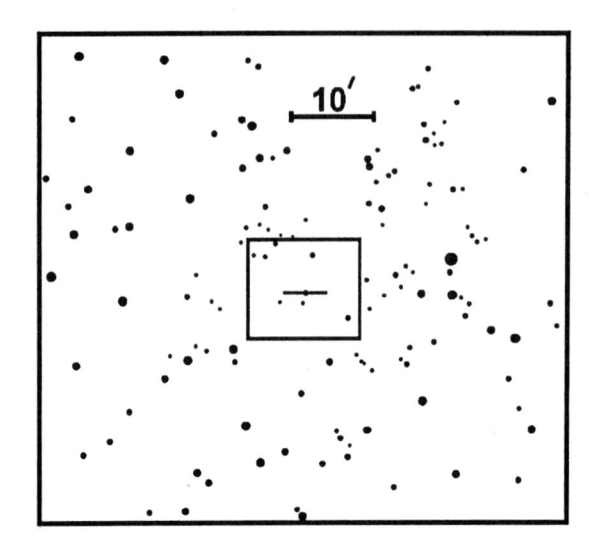

α (2000.0)=$08^h10^m56^s.66$
δ (2000.0)=$28°08'35''.5$
Type: DN SU XS
Pg=10.2–14.6
$P=0^d.0868$
$P_{cycle}=11^d.3$

QPOs with period 227 s in X-ray range. Flickering with 26 s and 75–90 s periods. Superhump period is $P_{hump} = 0^d.09204$.

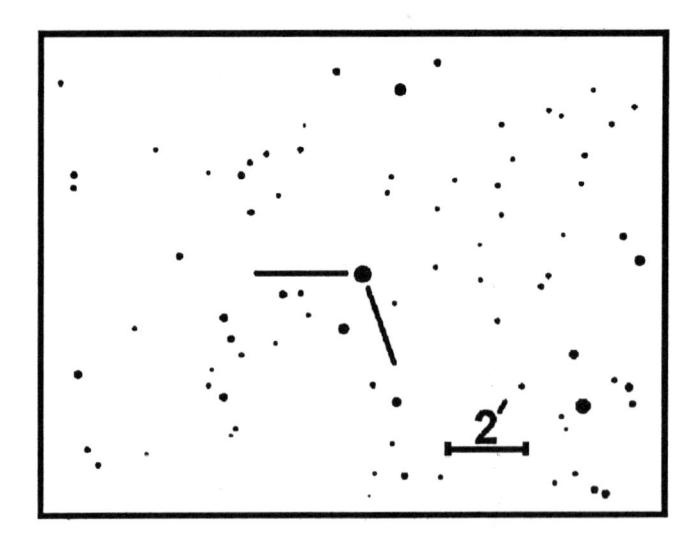

References: [6, 176, 255, 402]

AR Cnc

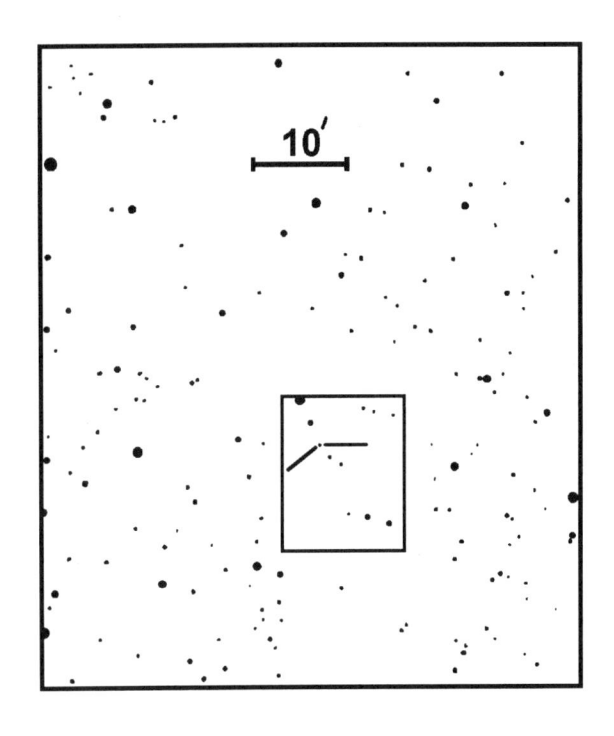

α (2000.0)=09^{h}22^{m}07^s.62
δ (2000.0)=31°03'13".1
Type: DN UG?
Pg=15.3–>21
Spectrum: M4–5 V
P=0^d.2146

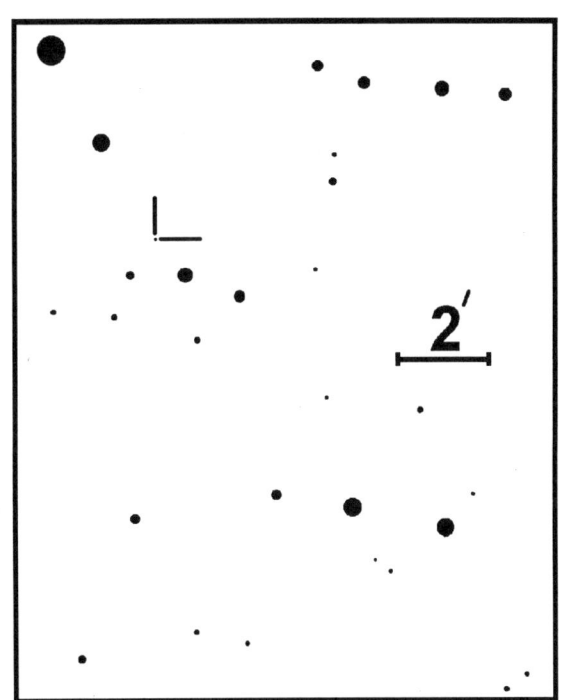

The depth of eclipses
is $> 3^m$ B, V.

References: [52, 165, 176]

AT Cnc

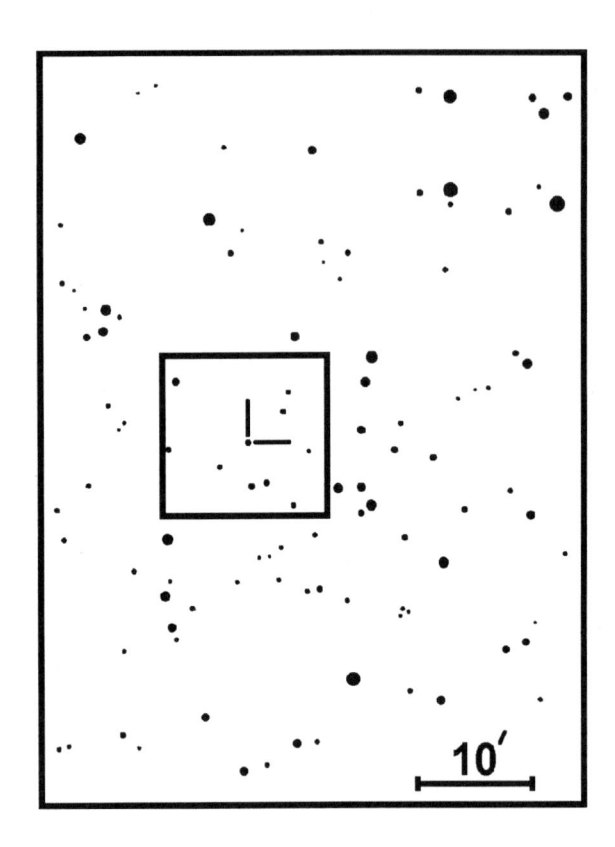

α (2000.0)=$08^h28^m36^s.99$
δ (2000.0)=$25°20'01''.8$
Type: DN ZC
B=12.29–16.38
P=$0^d.2386913$
Min.phot.=JD 2446110.504
$P_{cycle}\sim 14^d$

The comparison stars are:
(Photographic magnitudes)

Star	Pg
a	12.25
b	12.93
c	13.62
d	14.30
e	14.97
f	15.67
g	16.45

References: [129, 176, 207, 288]

CC Cnc

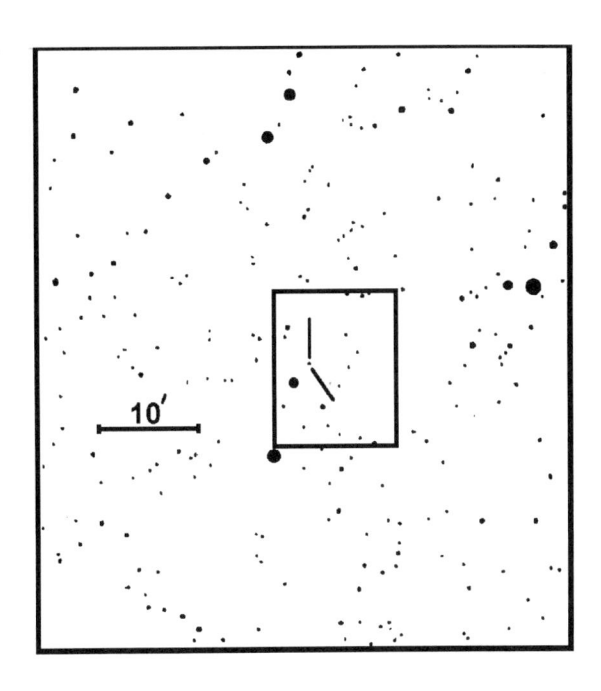

α (2000.0)=$08^h36^m19^s.4$
δ (2000.0) =$21°21'06''$
Type: DN UG
B=13.1–17.6
P=$0^d.094$
$P_{cycle}<90^d$

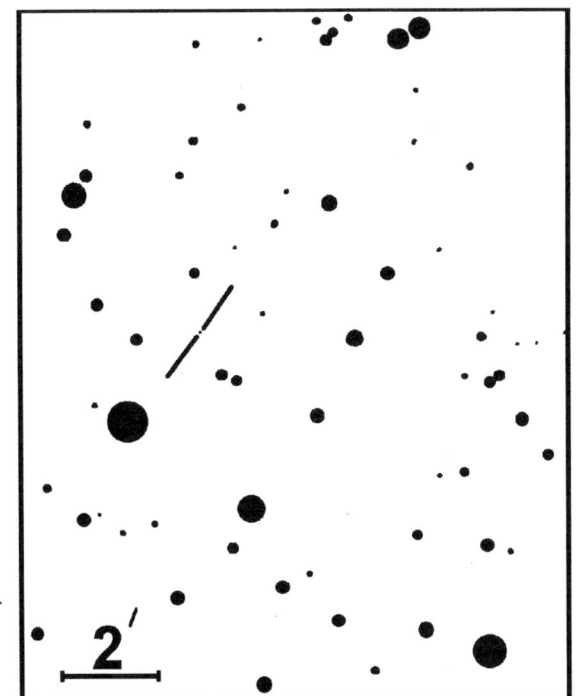

In the region of the open
cluster NGC 2632 ($d = 90$ pc).

References: [167, 176]

SV CMi

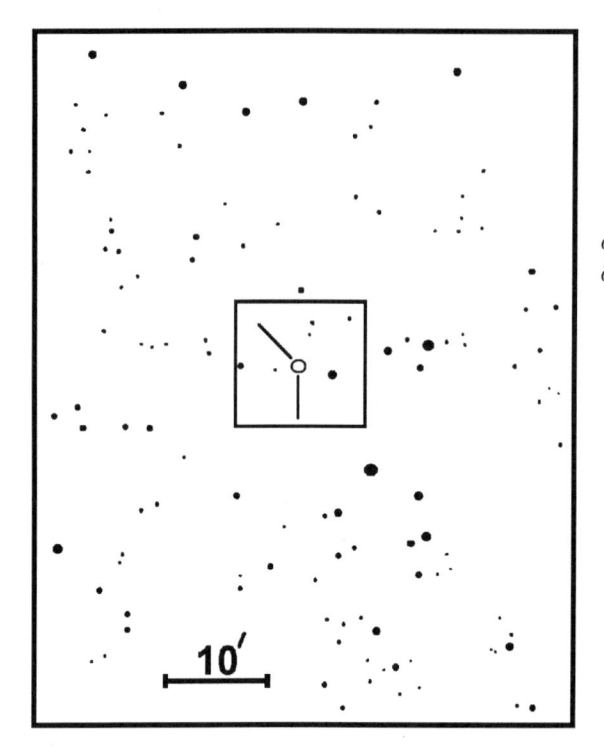

α (2000.0)=$07^h31^m08^s.5$
δ (2000.0)=$-05°58'47''$
Type: DV ZC XS?
B=13.0–16.3
P=$1^d.14433$:
$P_{cycle}\sim 16^d$

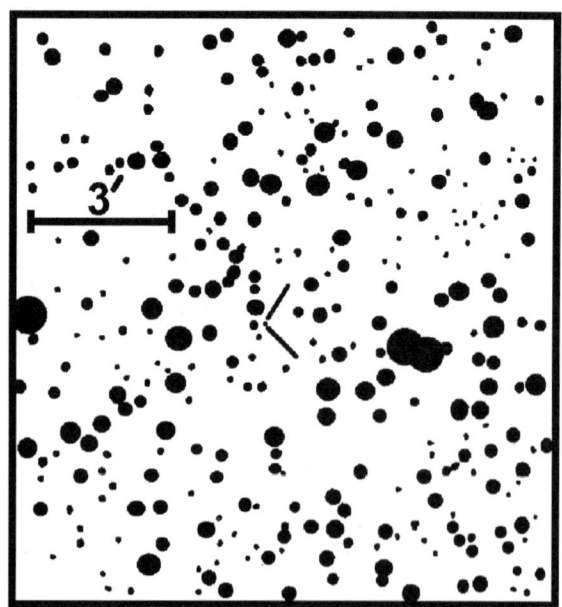

A possible source of hard X-rays (> 2 keV). Soft X-ray flux during outbursts was observed (0.2–0.5 keV).

References: [81, 176, 362, 364, 369, 390]

OY Car

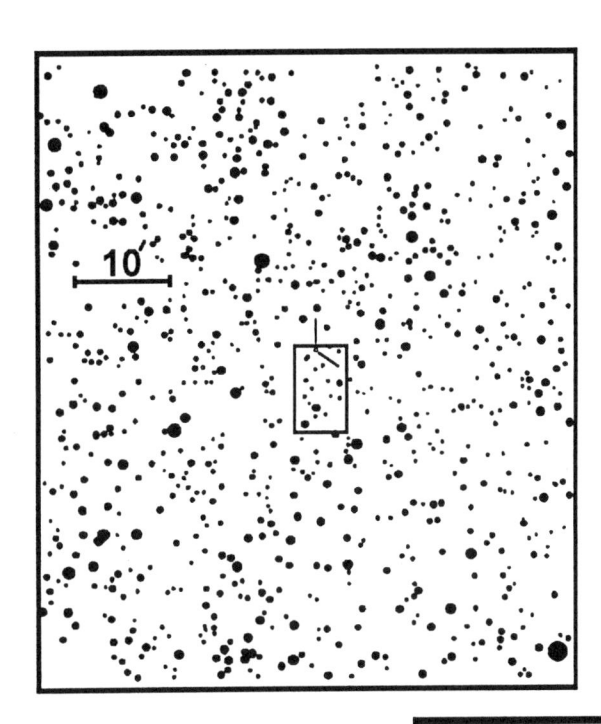

α (2000.0)=$10^h06^m22^s.36$
δ (2000.0)=$-70°14'04''.9$
Type: DN SU
B=12.2–17
Spectrum: M5 V
P=$0^d.0631209239$
$Min.phot.$=JD 2443993.55384
P_{cycle}~260^d

The comparison stars are:

Star	V	B	U
a	11.72	12.44	12.72
b	12.01	12.29	12.53
c	12.47	14.75	16.05
d	13.36	14.04	15.84

Eclipsing binary. The depth of minimum I is $1.5 - 2.5^m$, its duration is equal to $0^p.05$. Superhump period is equal to $P_{hump} = 0^d.064631$.

References: [18, 176, 360, 363, 364]

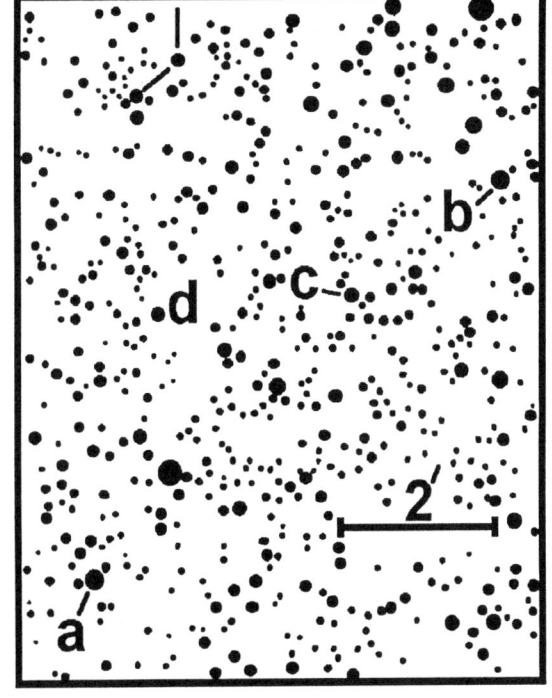

HT Cas

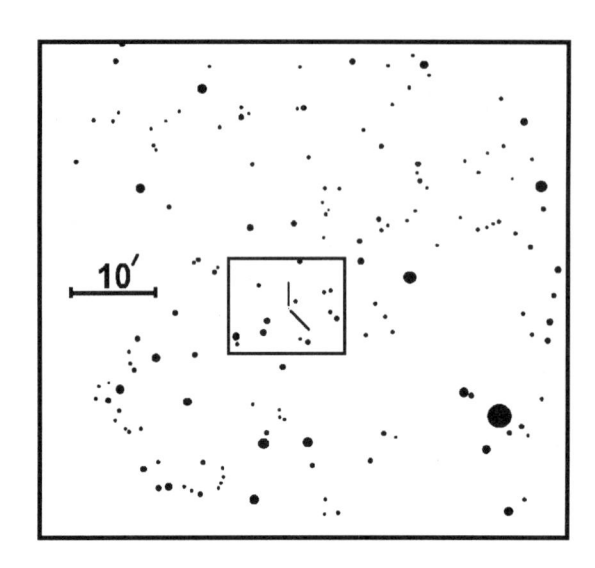

α (2000.0)=$01^h 10^m 13^s.1$
δ (2000.0)=$60°04'36''.6$
Type: DN SU XS
V=12.6–19.3
Spectrum: M5.4 V
P=$0^d.0736472039$
Min.phot.=JD 2443727.93721
P_{cycle}~70^d :

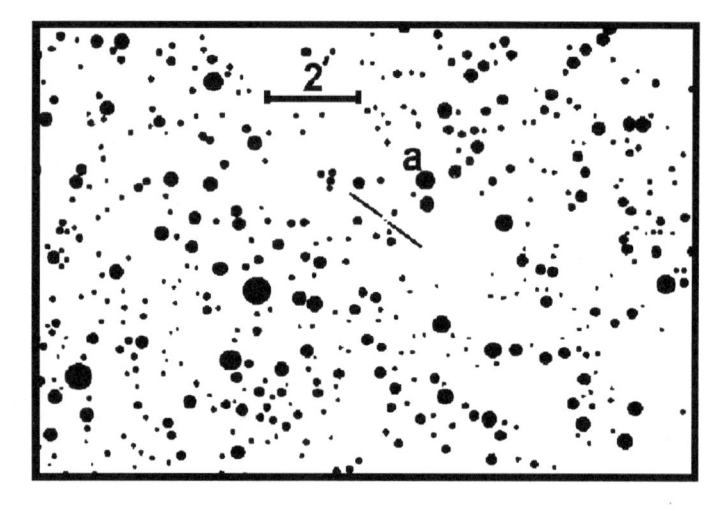

A comparison star is a:
V=13.61(4), B=14.45(5), R=12.85(5).

Eclipsing binary. The depth of eclipses is $2^m.4V$. The radius of the disc in outburst is $0.27R_\odot$. Variations of X-ray flux by 67±24% in 21.9 s were observed.

References: [52, 176, 232, 364, 407, 409]

BV Cen

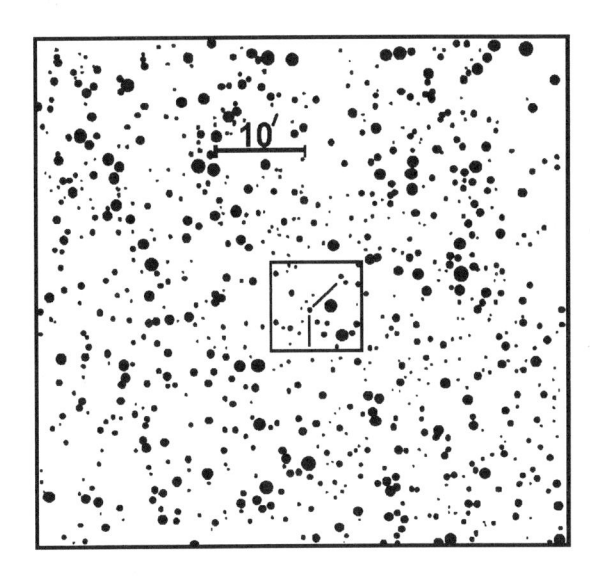

α (2000.0)=$13^h 31^m 19^s.5$
δ (2000.0)=$-54°58'32''.3$
Type: DN UG XS
V=10.5–14.2
Spectrum: G5–8 V
$P=0^d.610108(4)$
Max.phot.=JD 2420264.780
$P_{cycle}<149^d$

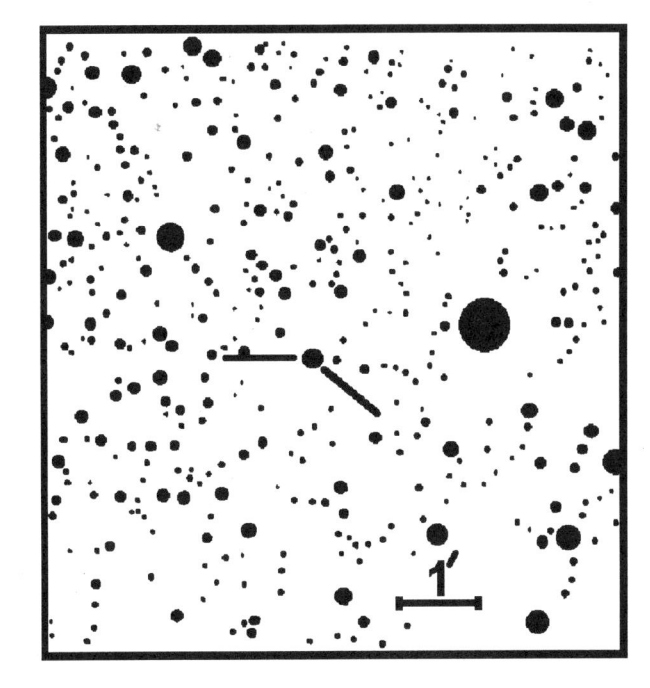

Eclipsing binary ($12^m.8 - 13^m.3$). Soft X-ray source.

References: [48, 82, 176, 351, 364]

MU Cen

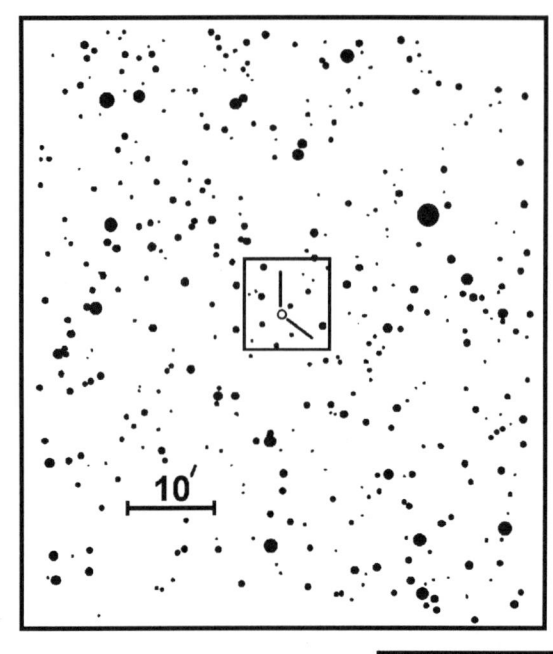

α (2000.0)=$12^h12^m53^s.82$
δ (2000.0)=$-44°28'15''.2$
Type: DN UG
V=11.8–15.0
Spectrum: G3–7 V
P=$0^d.342$
I.C.=JD 2446953.632
P_{cycle}=$44^d.9$

The comparison stars are:

Star	V	B	U
a	12.24	13.39	14.43
b	13.25	13.91	14.01
c	13.31	14.02	14.33
d	13.99	14.81	15.08
e	14.15	14.94	15.30

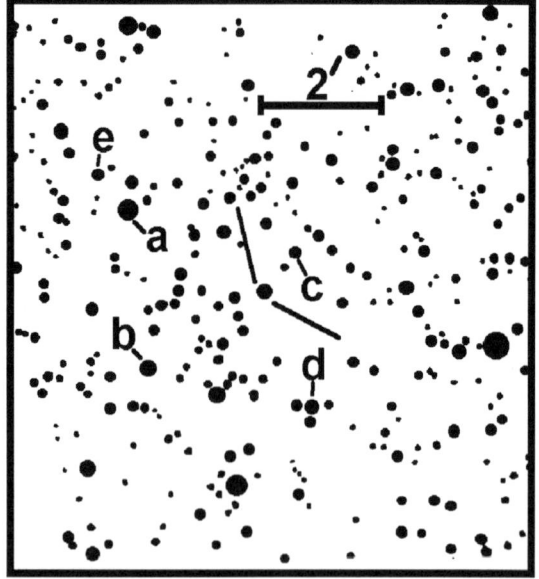

Flickering by $0^m.01$ in 6 min. Photoelectric observations are published in [237, 362], reproduction of the spectrum is in [358]. Radial velocity curve semi-amplitudes are $K_1 = 60 - 120$, $K_2 = 60 - 120$ km s^{-1}.

References: [48, 176, 237, 358, 360, 362, 364, 383]

V 436 Cen

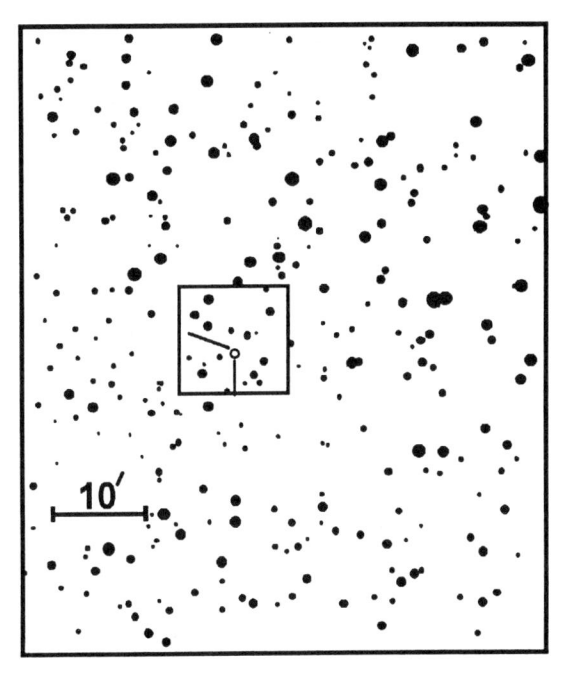

α (2000.0)=$11^h14^m00^s.10$
δ (2000.0)=$-37°40'47''.3$
Type: DN SU XS
B=11.5–15.2
$P=0^d.062501$
Min.phot.=JD 2444321.459
$P_{cycle}\sim 32^d$

The comparison stars are:

Star	V	B	U
a	11.05	12.73	14.77
b	11.75	12.36	12.39
c	12.01	12.64	12.71
d	12.26	12.86	12.86
e	12.53	13.42	13.91
f	12.93	13.74	14.10
g	13.19	13.89	14.13
h	13.56	14.63	15.31
k	14.09	15.09	15.66
l	14.81	15.53	15.70

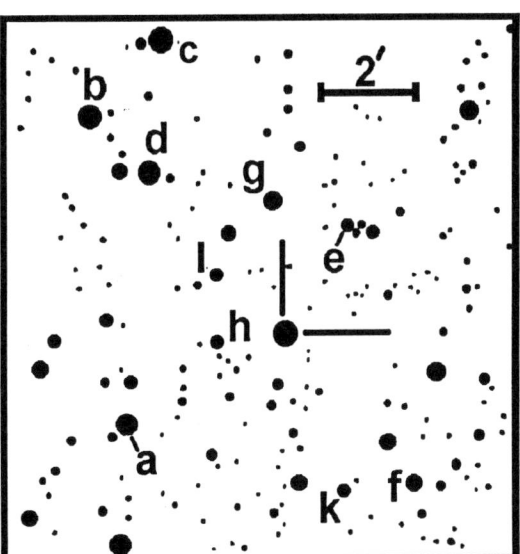

A=CoD$-36°7038$. The northern component of a close pair. The photometric period during supermaxima is $P = 0^d.06383$. Rapid oscillations in supermaxima with $\Delta m \sim 0.002$, $P \sim 19.6$ s. X-ray source, superhump period is $P_{hump} = 0^d.063785$.

References: [82, 123, 176, 304, 364, 380, 388]

V 442 Cen

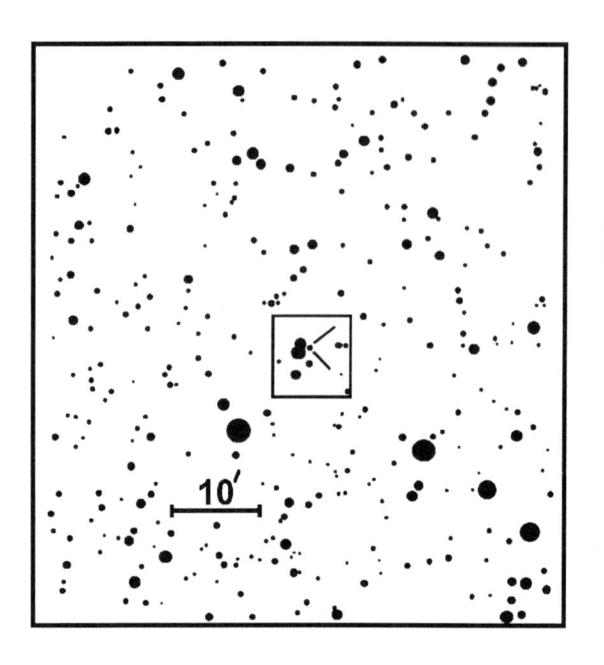

α (2000.0)=$11^h24^m51^s.96$
δ (2000.0)=$-35°54'37''.8$
Type: DN UG
V=11.9->16.5
P=$0^d.46$:
$P_{cycle}\sim25^d$

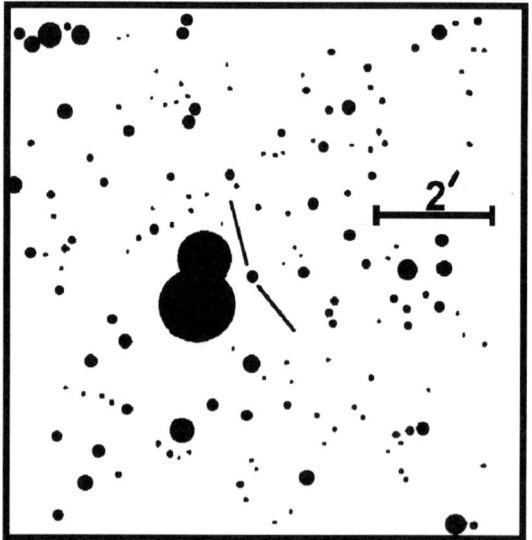

A quite active dwarf nova. Photoelectric observations in outburst are published in [28, 362, 384]. $A_v = 0^m.48$. In maximum $V = 11.95$, $B - V = -0.02- +0.01$, $U - B = -(0.69 - 0.65)$. In minimum $V = 16.24$, $B - V = 0.05-0.28$, $U - B = -(1.05-0.86)$.

References: [28, 51, 176, 351, 362, 364, 384]

V 485 Cen

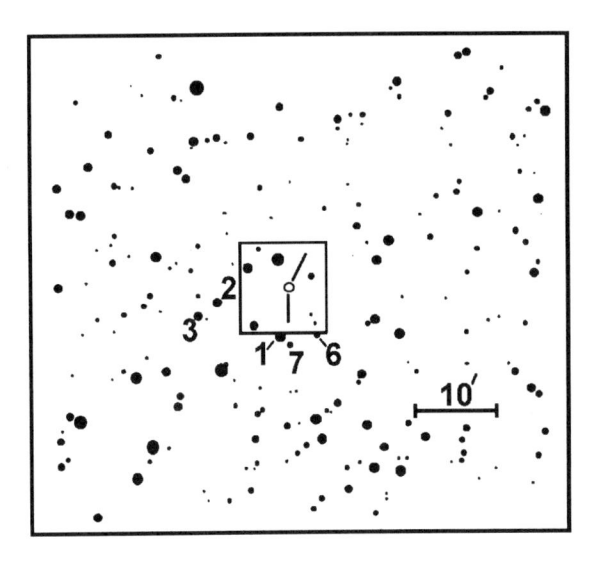

α (2000.0)=$12^h 57^m 23^s.45$
δ (2000.0)=$-33°12'06''.7$
Type: DN UG
Pg=12.9->16.5
$P=0^d.041$:
P_{cycle}~$12^d : -14^d$

The comparison stars are:
(Photographic V magnitudes)

Star	Pg(V)
1	10.0
2	10.8
3	11.1
4	12.1
5	12.4
6	12.8
7	12.9
8	13.9
9	14.0
10	14.8

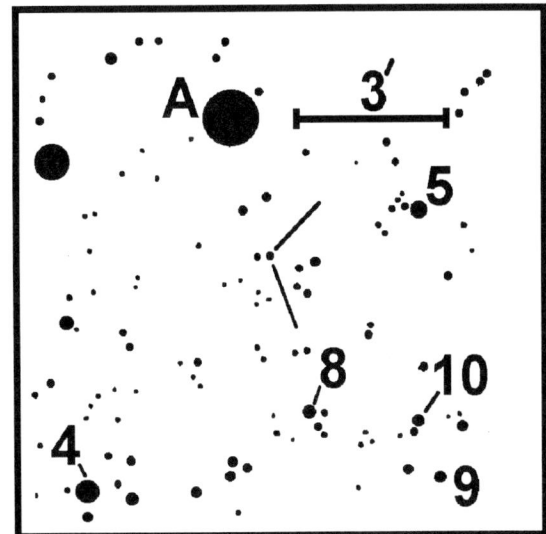

A=CoD$-32°9042$. Another possible value of the cycle length is $P_{cycle} = 148^d$.

References: [29, 176, 359, 362, 364, 380]

WW Cet

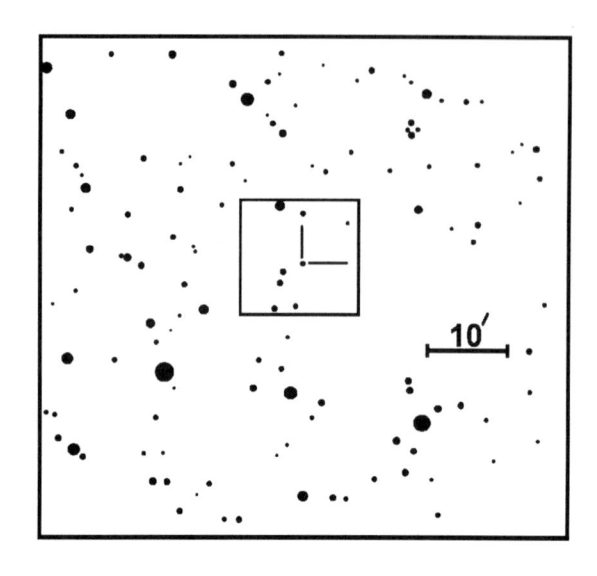

α (2000.0)=$00^h11^m24^s.85$
δ (2000.0)=$-11°28'44''.0$
Type: DN ZC XS
Pg=9.3–16.8
Spectrum: later than M3 V
$P=0^d.1681$ or $0^d.169$
$I.C.$=JD 2445938.9083
$P_{cycle}\sim31^d.2$

The comparison stars are:

Star	V	B	U
a	10.95	11.59	11.71
b	12.51	13.71	14.87
c	13.12	13.90	14.24
d	13.39	13.95	13.87
e	13.45	14.03	13.97
f	13.68	14.21	14.16
g	14.38	15.27	15.8:

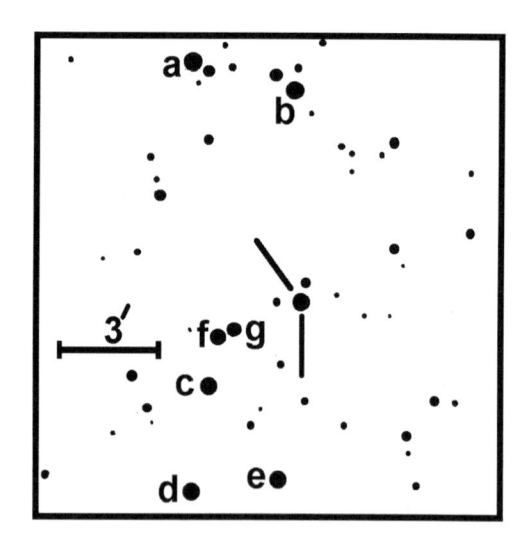

Radial velocities are variable with the period $P = 0^d.160$, $K = 108\pm7$ km s^{-1}. Possibly the period should be doubled. Rapid oscillations of brightness with a cycle of 10 min are observed.

References: [147, 176, 343, 360, 364, 402]

WX Cet

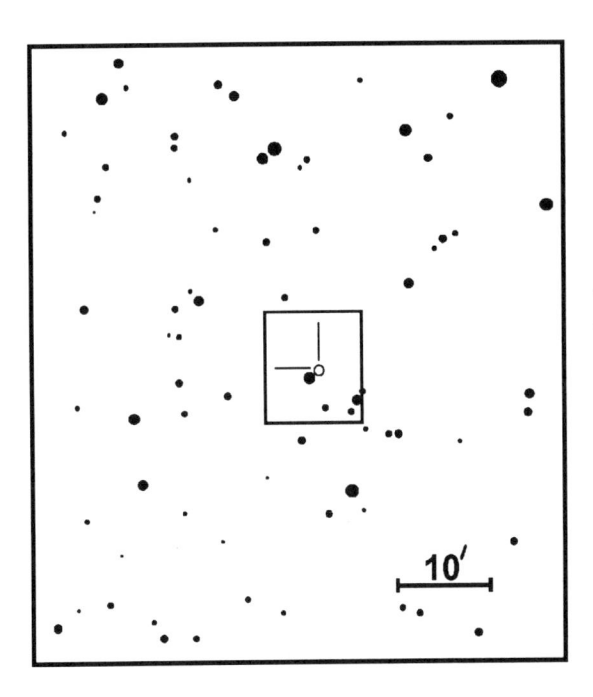

α (2000.0)=$01^h17^m04^s.23$
δ (2000.0)=$-17°56'22''.7$
Type: DN SU
Pg=9.5–18.5
$P=0^d.079$:
$P_{cycle}\sim 450^d$:

The comparison stars are:

Star	V	B	U
1	10.59	11.20	11.25
2	11.88	12.81	13.46
3	13.16	13.77	13.76
4	11.86	12.54	12.76
5	13.91	14.70	15.23

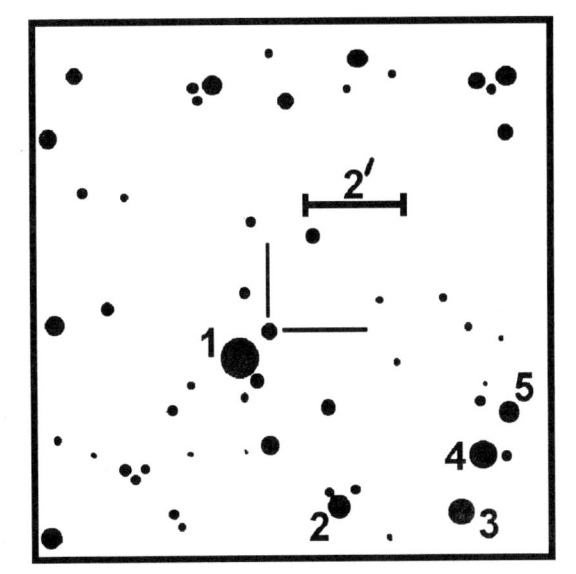

Other possible values of the period are $0^d.0302$ and $0^d.063$.

References: [107, 176, 360, 364, 406]

Z Cha

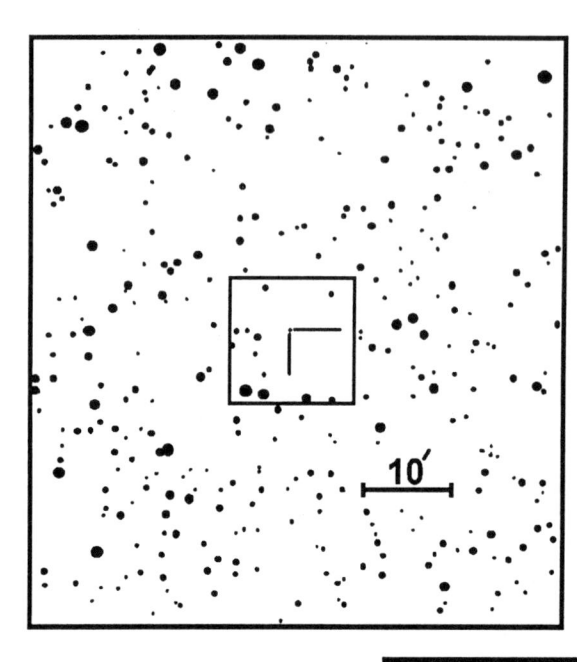

α (2000.0)=$08^h 07^m 28^s.2$
δ (2000.0)=$-76°32'01''$
Type: DN SU XS
V=11.5–16.2
Spectrum: M5.5 V
P=$0^d.074499167$
Min.phot.=JD 2440264.68336
P_{cycle}~$104^d.1$

The comparison stars are:

Star	V	B	U
a	8.87	9.81	10.51
b	11.40	11.72	11.90

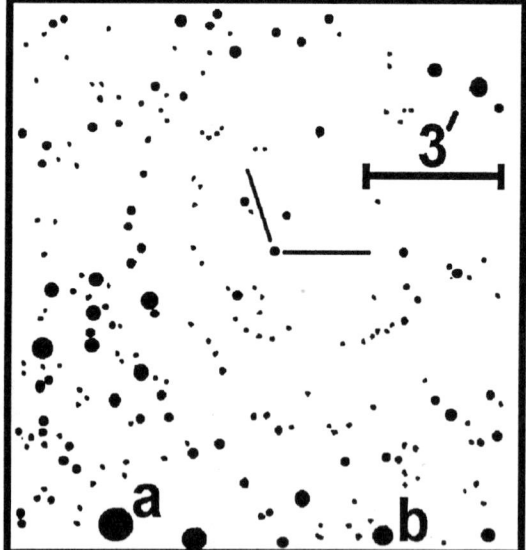

This is the western component of a close pair (the faint component lies $5''.6$E). *Eclipsing* binary, Min = HJD $2440264.68232 + 0^d.0744992195 \cdot E + 7^d.88 \times 10^{-13} \cdot E^2$ [78]. Depth of Min is $1^m.4 - 2^m.1$ B. Variations with $P = 27.7$ s during a supermaximum are observed. Superhump period is $P_{hump} = 0^d.07749$.

References: [48, 78, 113, 176, 237, 364]

GO Com

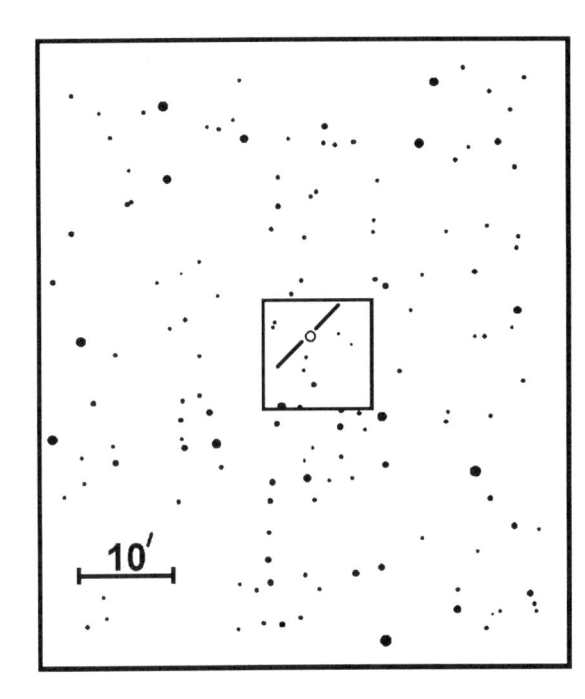

α (2000.0)=$12^h56^m37^s.18$
δ (2000.0)=$26°36'42''.6$
Type: DN WZ:
Pg=12.7–20
P=$0^d.0658$

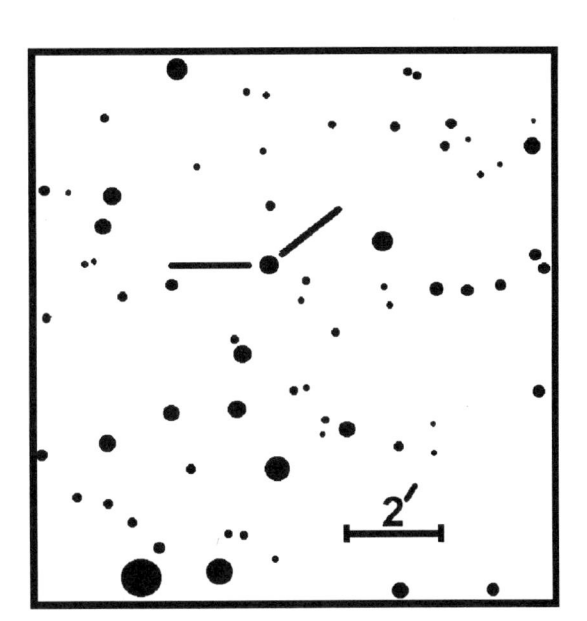

In the region of the Comae
Berenices cluster.

References: [48, 176, 364]

TT Crt

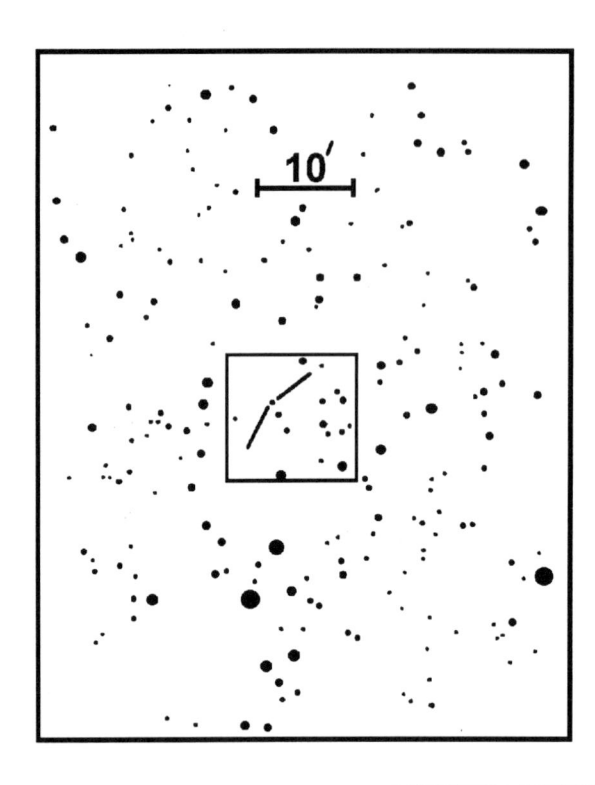

α (2000.0)=$11^h34^m47^s.17$
δ (2000.0)=$-11°45'28''.5$
Type: DN
Spectrum: K5-M0 V
V=13.0–16.2
P=$0^d.30428$
Max.phot.=JD 2448306.8965
P_{cycle}=100^d

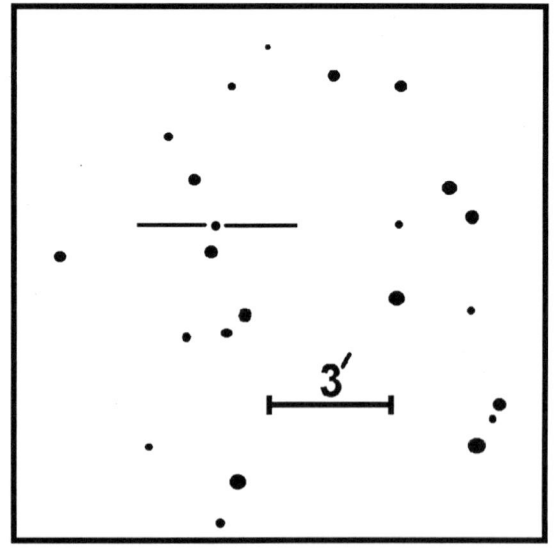

Duration of outbursts is $\sim 10^d$, a characteristic time between outbursts is $\sim 100^d$.

References: [176, 318]

EM Cyg

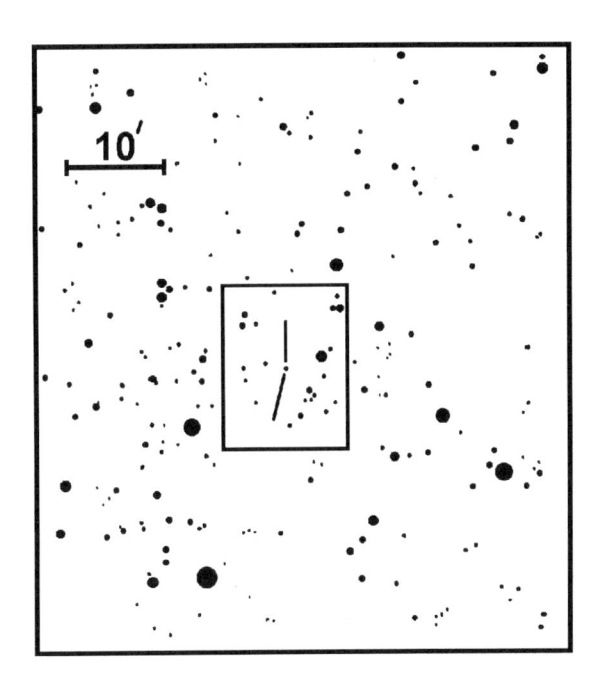

α (2000.0)=$19^h 38^m 40^s.16$
δ (2000.0)=$30°30'28''.0$
Type: DN ZC XS
Pg=11.9–14.4
P=$0^d.2909095$
Min.phot.=JD 2437882.8595

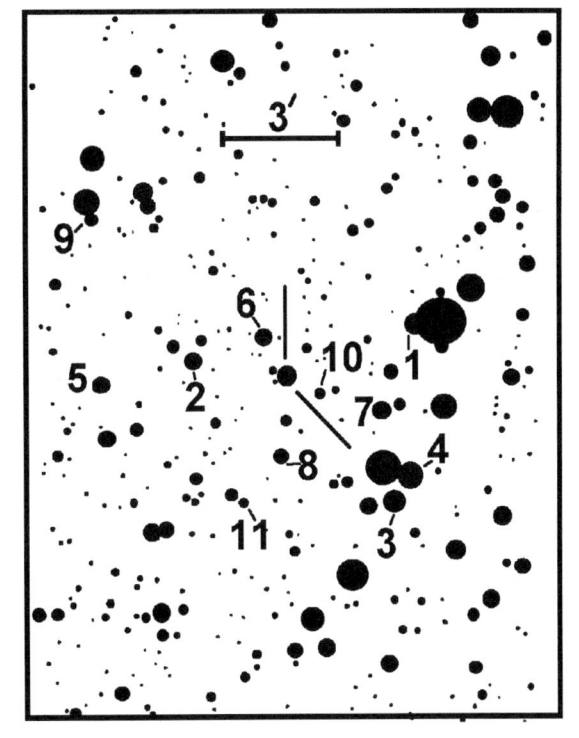

The comparison stars are:
(Photographic magnitudes)

Star	Pg	Star	Pg
1	11.83(3)	7	12.98(6)
2	12.18(7)	8	13.52(7)
3	12.49(7)	9	13.84(3)
4	12.59(6)	10	14.00(7)
5	12.75(6)	11	14.41(6)
6	12.95(6)		

QPOs with cycles 14–21 s and $A \sim 0^m.04$. An *eclipsing* binary with depth of minima $0^m.25V$.

References: [52, 176, 351, 375]

EY Cyg

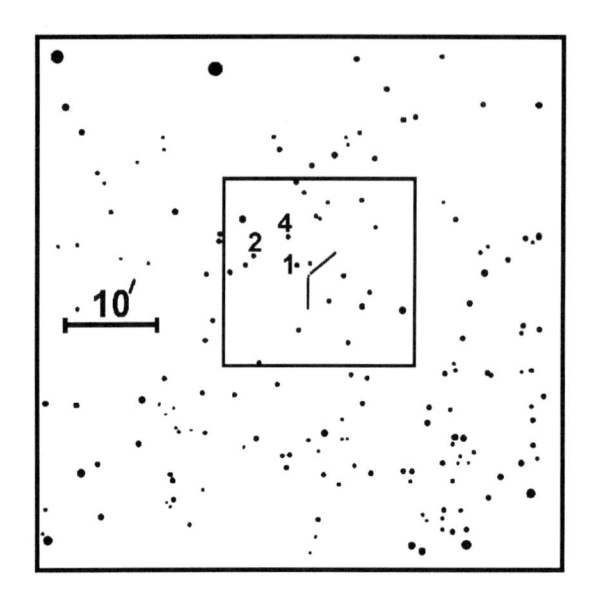

α (2000.0)$=19^h54^m40^s.76$
δ (2000.0)$=32°21'55''.43$
Type: DN UG XS
Spectrum: K0 V
$Pg=11.4-15.7$
$P=0^d.181228$
$Min.phot.=$JD 2446595.323
$P_{cycle}=240^d$

The comparison stars are:
(Photographic magnitudes)

Star	Pg	Star	Pg
1	11.35(4)	6	13.39(2)
2	12.08(2)	7	13.83(2)
3	12.32(3)	8	14.24(4)
4	12.55(3)	9	14.51(4)
5	13.00(3)	10	14.89(6)

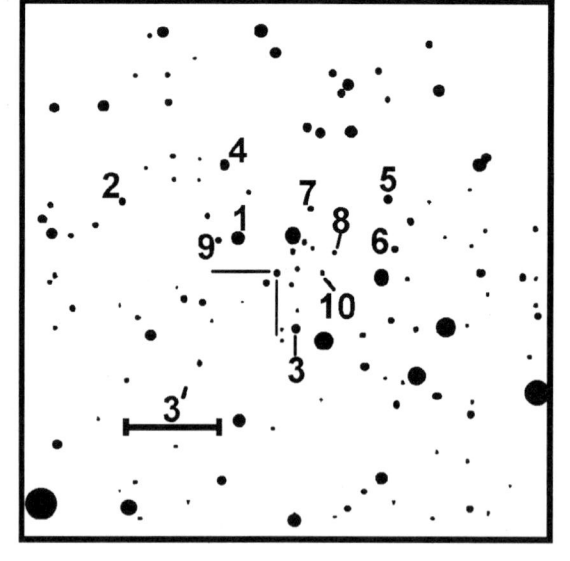

Light maxima were observed on JD 2425144, 26323, 27947, 27950, 32554, 34455, 34457.

References: [48, 52, 176, 376]

V 503 Cyg

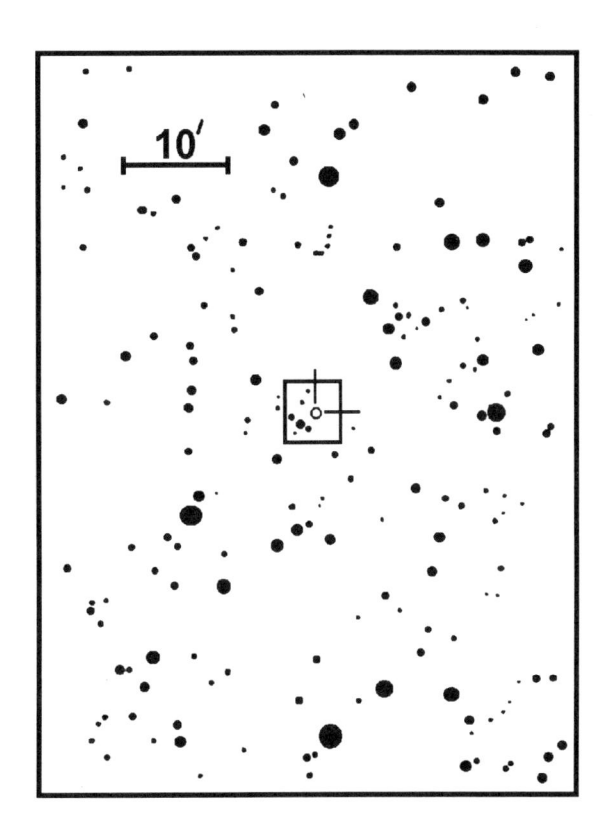

α (2000.0)=$20^h27^m17^s.49$
δ (2000.0)=$43°41'23''.2$
Type: DN SU
Pg=13.4–17.0
P=$0^d.07598$
$P_{cycle}=28^d$

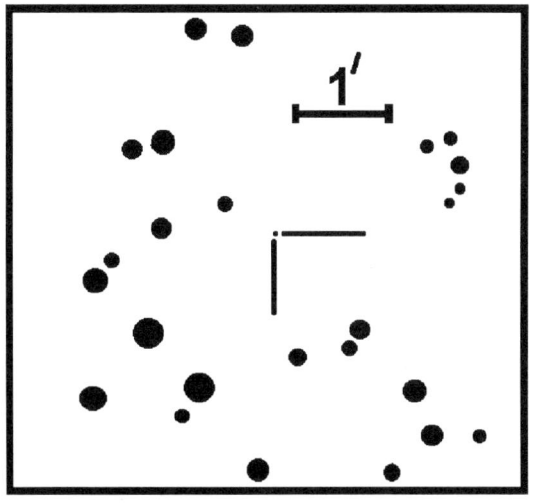

The contribution of the secondary is noticeable in the IR light. In light minimum $V = 17.40$, $U - B = -0.94$, $B - V = +0.50$, $V - J = 3.0$. Variability with a time-scale of 90–120 min.

References: [48, 52, 176, 334, 373]

V 792 Cyg

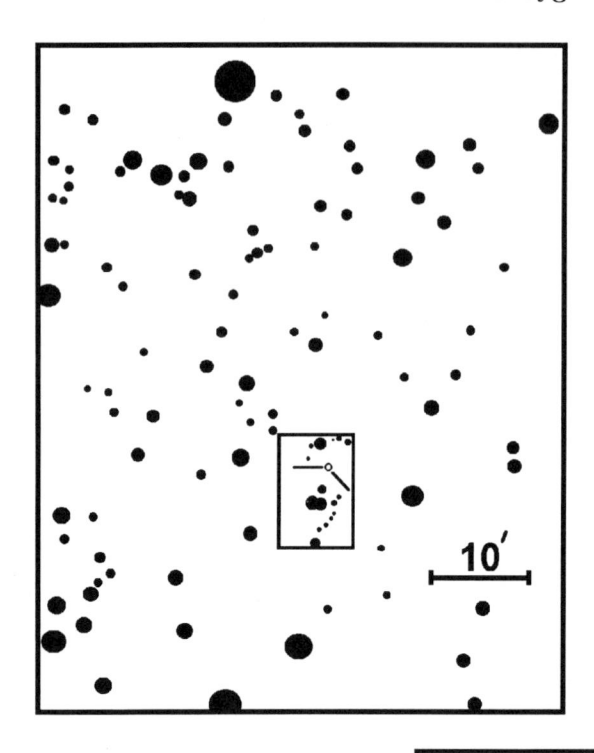

α (2000.0)=$19^h11^m01^s.11$
δ (2000.0)=$33°47'04''.0$
Type: DN UG
B=14.0–17.0
$P_{cycle}=24^d - 33^d$

The comparison stars are:
(Photographic magnitudes)

Star	Pg	Star	Pg
1	14.07(4)	6	15.65(2)
2	14.34(2)	7	15.88(2)
3	14.59(2)	8	16.19(2)
4	14.80(2)	9	16.75(4)
5	15.08(7)	10	16.98(2)

A=BD+33°3483 ($Pg = 9^m.5$),
B=BD+33°3482 ($Pg - 9^m.2$).
Superoutbursts have $B = 14^m$ and duration $\sim 15^d$. Smaller outbursts have B=14.5 and duration $\sim 11^d$.

References: [52, 176, 382]

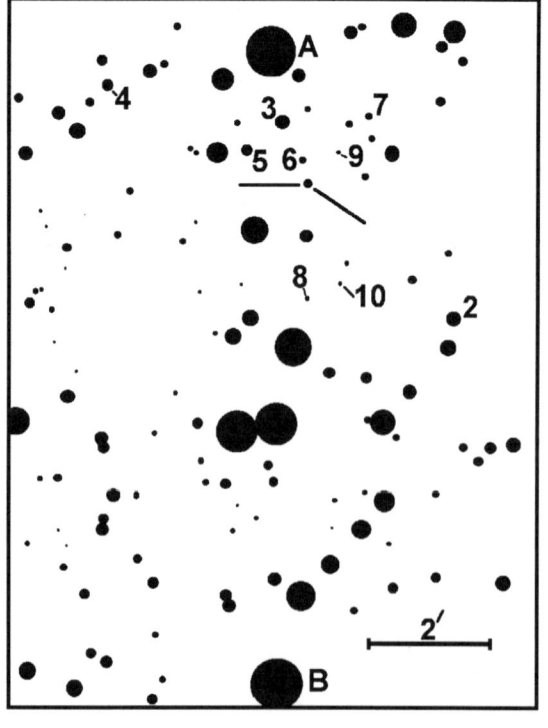

V 1251 Cyg

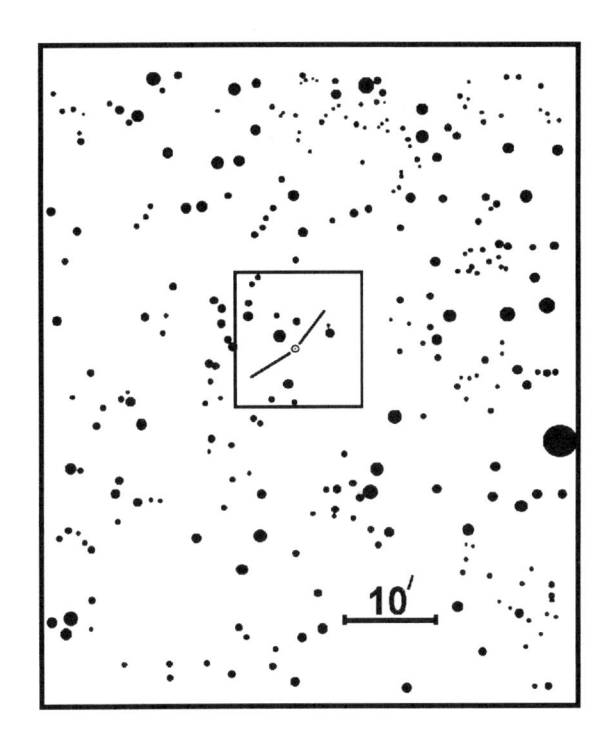

α (2000.0)=$21^h 40^m 52^s.65$
δ (2000.0)=$48°39'52''.6$
Type: DN SU
Pg=12.5–18.5
$P=0^d.074$

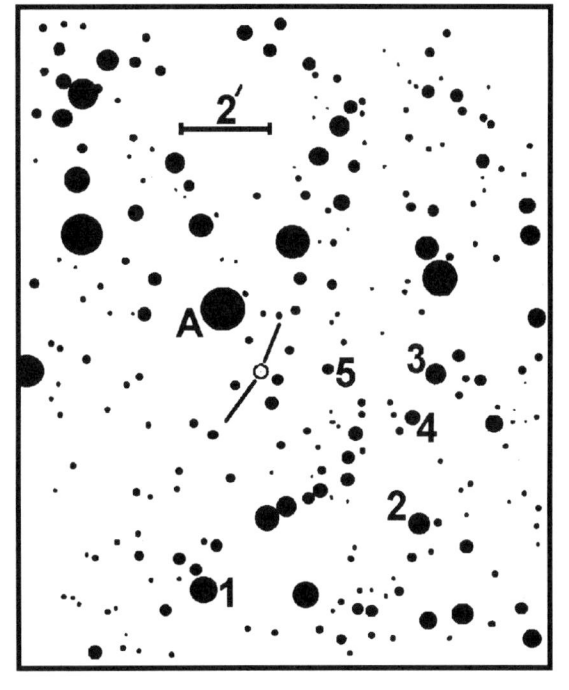

The comparison stars are:
(Photographic magnitudes)

Star	Pg
1	12.9
2	13.3
3	14.0
4	14.7
5	15.2

A=BD+48°3463.

References: [176, 391]

V 1504 Cyg

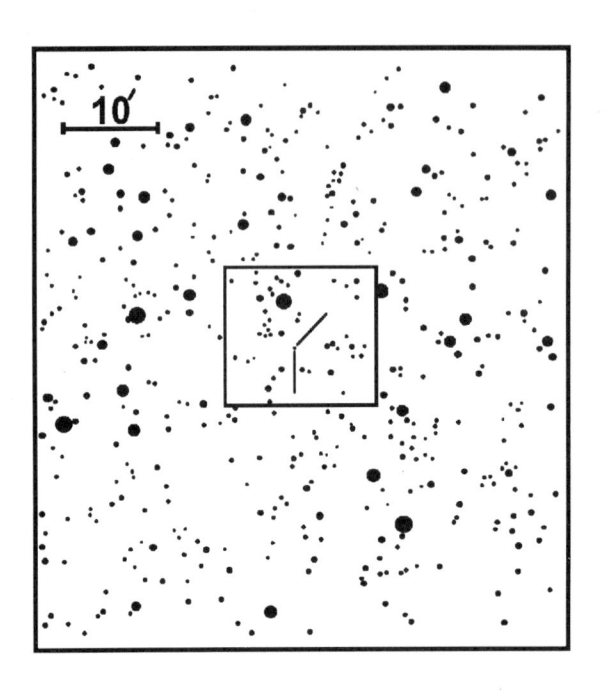

α (2000.0)=$19^h28^m57^s$
δ (2000.0)=$43°05'36''$
Type: DN SU
Pg=13.7–18.0
$P_{cycle}=5^d.8-19^d$

The comparison stars are:
(Photographic magnitudes)

Star	Pg
1	13.63
2	14.46
3	14.92
4	15.47
5	15.71
6	16.18
7	16.60

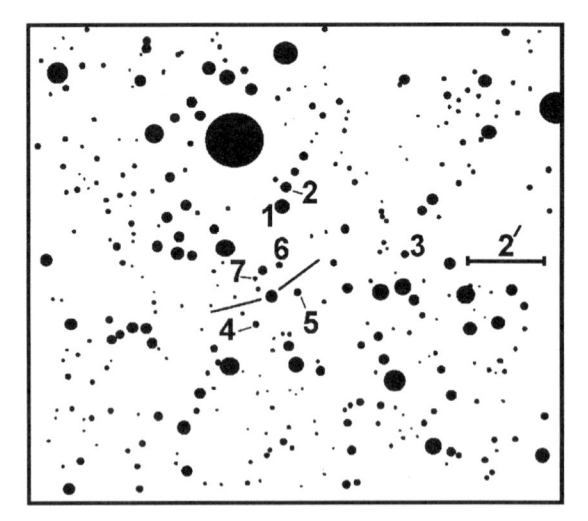

Photographic observations are published in [314]. Superoutbursts have $B=13^m.7$ and duration $\sim 14^d$. Smaller outbursts have $B=14^m.4$ and duration $\sim 3^d$.

References: [89, 176, 314]

CM Del

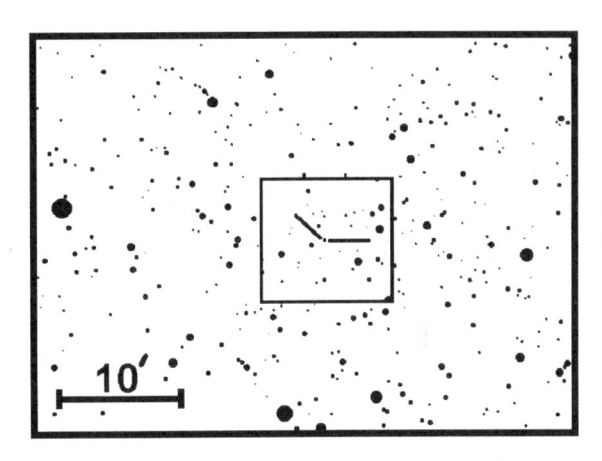

α (2000.0)=$20^h24^m56^s.93$
δ (2000.0)=$17°17'55''.7$
Type: NL UX
Pg=13.4–15.3
$P=0^d.162$:

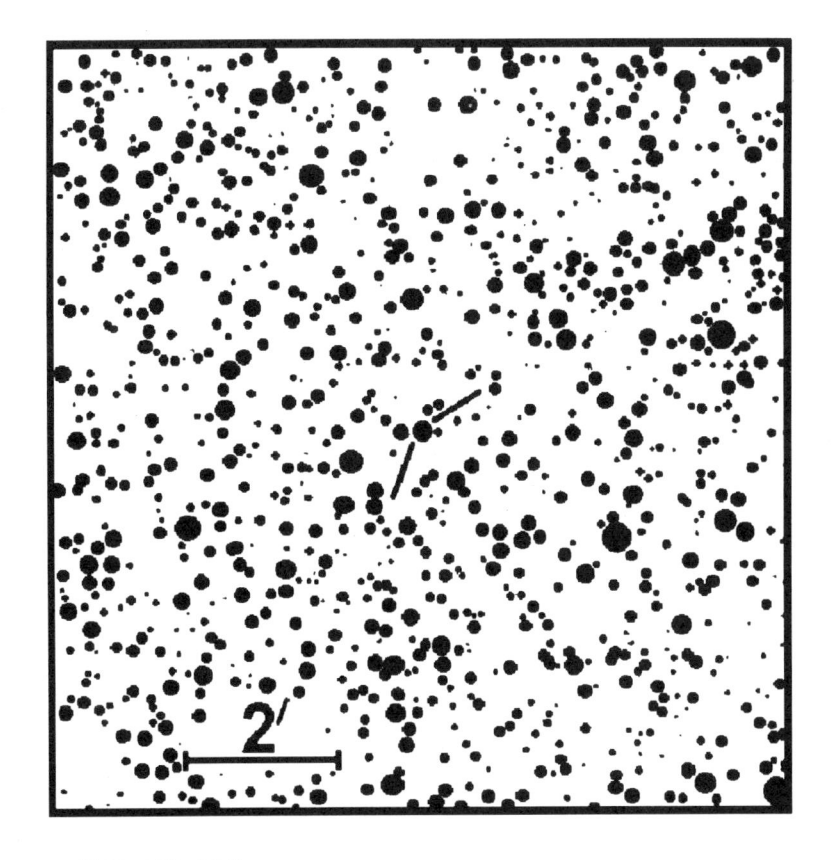

References: [102, 176, 364]

AB Dra

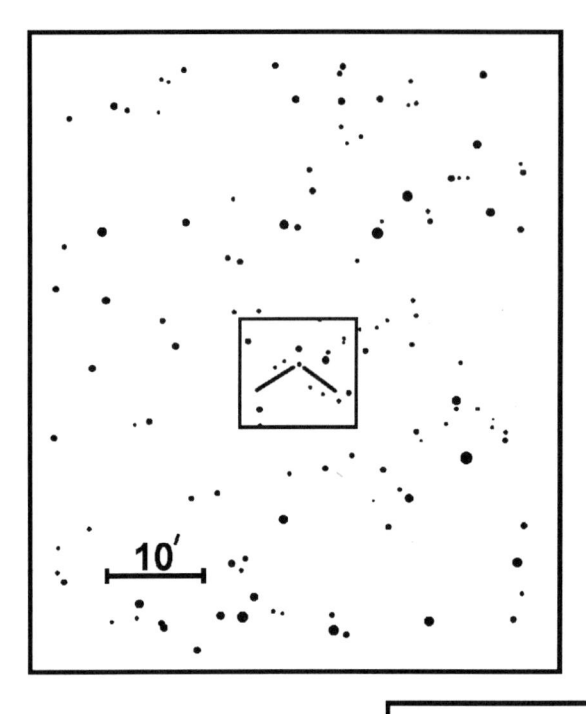

α (2000.0)=$19^h 49^m 06^s.2$
δ (2000.0)=$77°44'22''.9$
Type: DN ZC XS
V=11.0–15.3
P=$0^d.151622$
Max.phot.=JD 2445938.7427
$P_{cycle}=13^d.4$

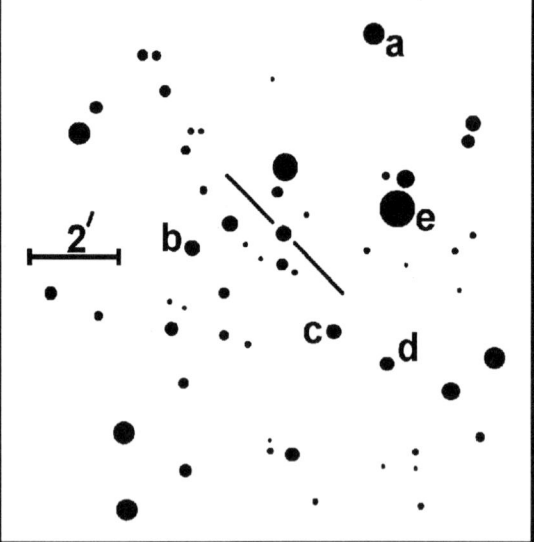

The comparison stars are:

Star	V	B
a	12.60(2)	12.96(2)
b	12.96(2)	14.13(2)
c	13.79(4)	14.49(4)
d	14.14(7)	14.95(7)
e	10.00(2)	10.91(2)

Max.phot. corresponds to transition of radial valocity from negative to positive values. In X-range (0.15–4.5 keV), $P = 290$ s and $A = 48\%$.

References: [52, 176, 190, 367]

DM Dra

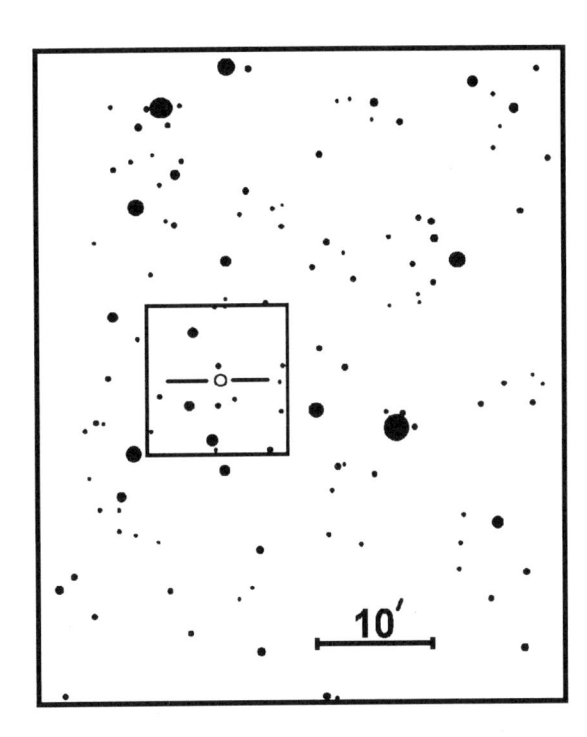

α (2000.0)=$15^h 34^m 15^s.9$
δ (2000.0)=$59°48'13''$
Type: DN UG
V=15.5–20.8
P=$0^d.087$:

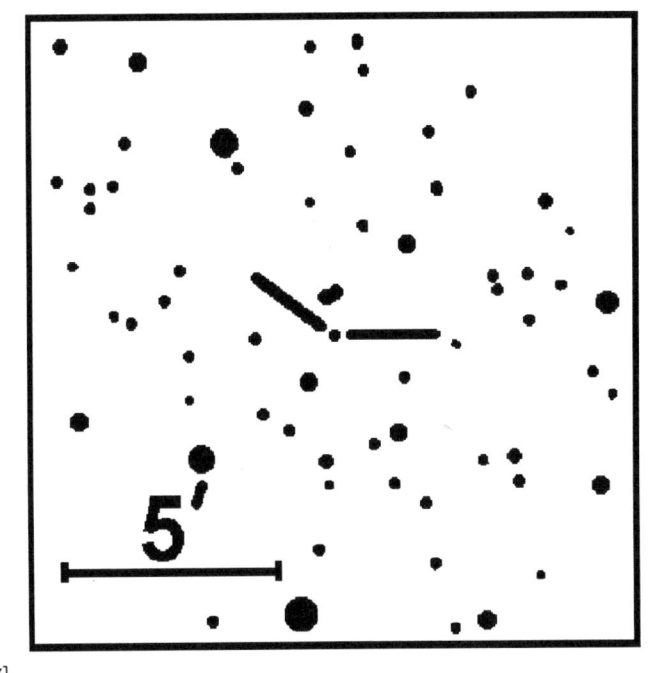

References: [176, 327]

AH Eri

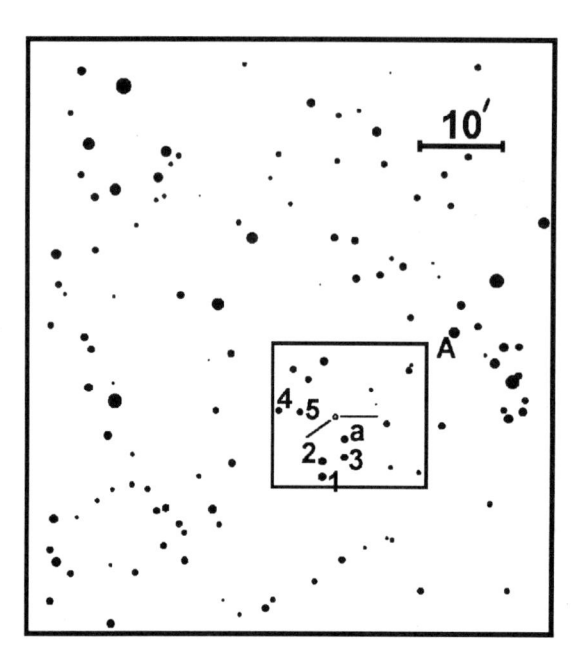

α (2000.0)=04^{h}22^{m}38^s.15
δ (2000.0)=−13°21′29″
Type: DN UG XR
Pg=13.5-17.5
P=0^d.029?
P_{cycle}∼24^d :

Comparison stars are:
(Visual magnitudes)

Star	Vis
1	10.9
2	11.6
3	12.2
4	13.1
5	13.5
6	16.5

Star a has

V= 12.57
B= 12.68
U= 13.30

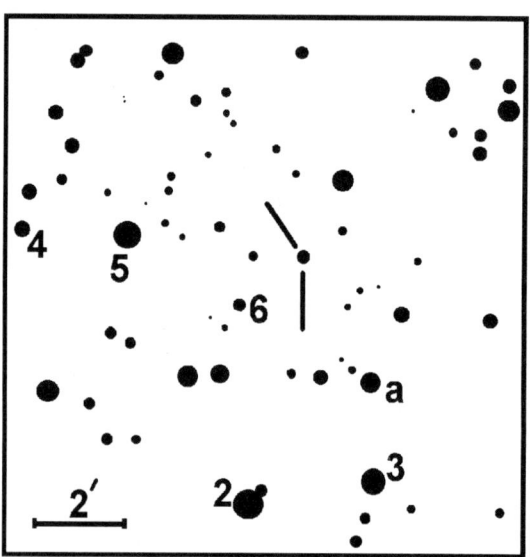

A = BD−13°869. The system is a soft X-ray source. In the light minimum V = 17.71, $B-V$ = +0.45, $U-B$ = −1.06.

References: [27, 176, 351, 360, 362, 364, 380]

AQ Eri

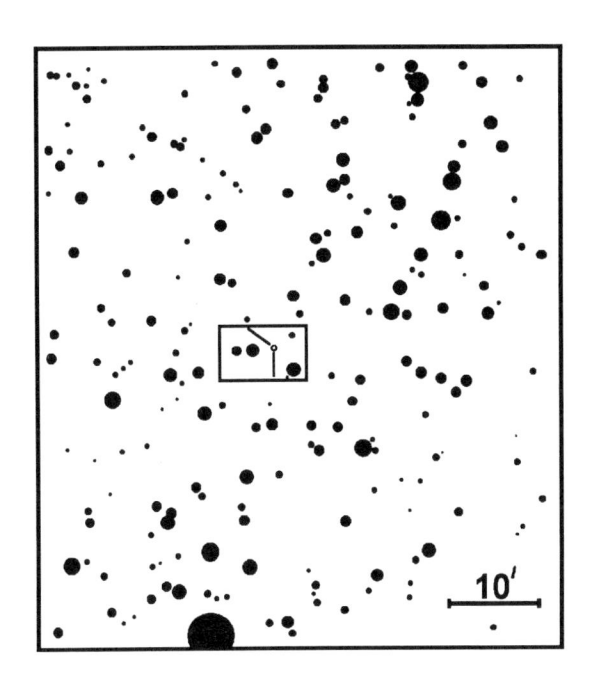

α (2000.0)=$05^h06^m12^s.99$
δ (2000.0)=$-04°08'05''.2$
Type: DN SU
Pg=12.5–16.5
$P=0^d.06225$
$P_{cycle}=76^d$

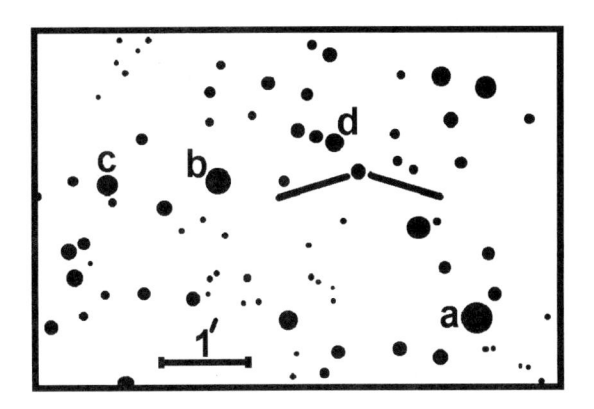

The comparison stars are:

Star	V	B	U
a	10.67	11.34	11.60
b	10.98	12.24	13.62
c	12.96	13.46	13.55
d	15.28	15.96	16.16

Another possible period is $0^d.05854$. In the light minimum $V = 17.52$, $B-V=+0.06$, $U-B=-1.03$.

References: [176, 360, 364]

U Gem

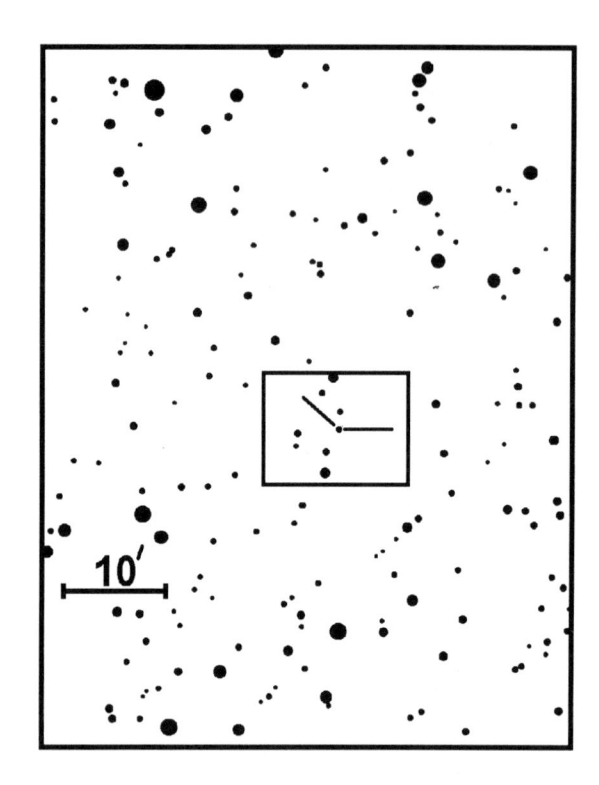

α (2000.0)=$07^h55^m05^s.38$
δ (2000.0)=$22°00'06''.4$
Type: DN UG XS
V=8.2–14.9
Spectrum: M4.5 V
P=$0^d.1769062$
Min.phot.=JD 2437638.82652
$P_{cycle}=105^d.2$

The comparison stars are:

Star	V	B
a	14.70(2)	(Sp G8 V)
b	11.45(2)	11.93(2)
c	11.64(2)	12.07(2)
d	11.70(2)	13.31(2)
e	12.02(2)	12.47(2)
f	13.13(5)	13.89(5)
g	13.76(9)	14.48(7)

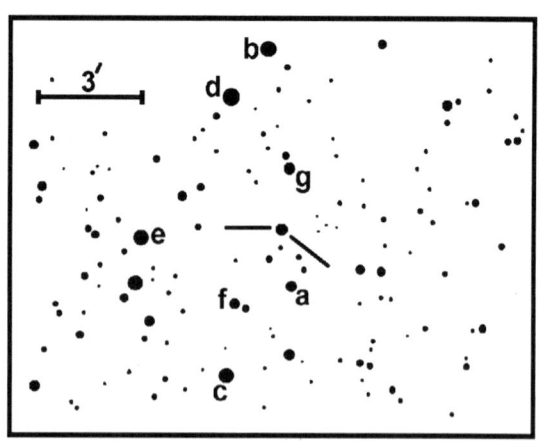

Eclipsing binary. The depth of minima is $0^m.5 - 1^m.0$. X-ray oscilations with the periods 121 s ($A = 35\%$), 135 s ($A = 35\%$), and 585 s ($A = 12\%$) were found.

References: [15, 52, 176, 190, 324, 402, 405]

AW Gem

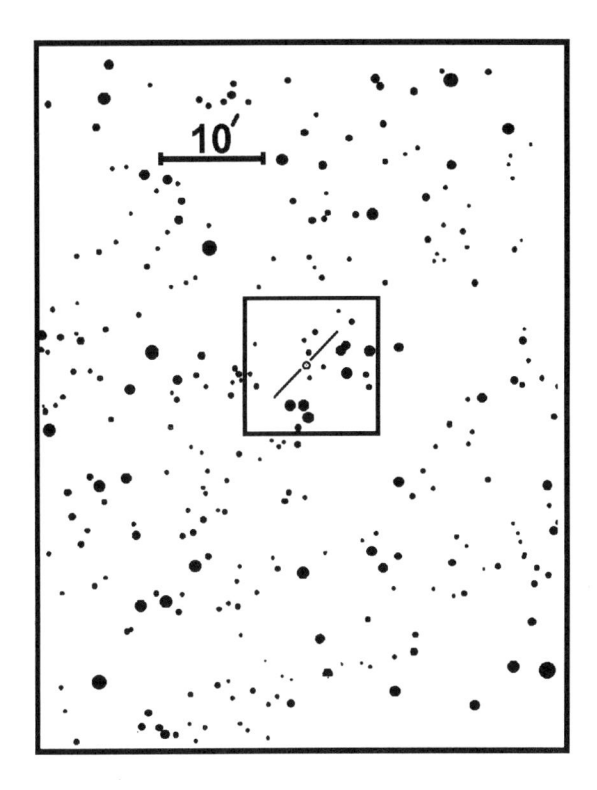

α (2000.0)=$07^h22^m40^s.1$
δ (2000.0)=$28°30'15''$
Type: DN SU
Pg=13.1-19.4
P=$0^d.073$:
P_{cycle}=330^d :

The comparison stars are:
(Photographic magnitudes)

Star	Pg
1	12.62
2	13.13
3	13.54
4	14.16
5	14.43

(Visual magnitudes)

Star	Vis	Star	Vis
6	11.3	10	13.9
7	12.2	11	14.0
8	12.8	12	14.5
9	12.8	13	16.0

References:[6, 48, 176, 185, 364]

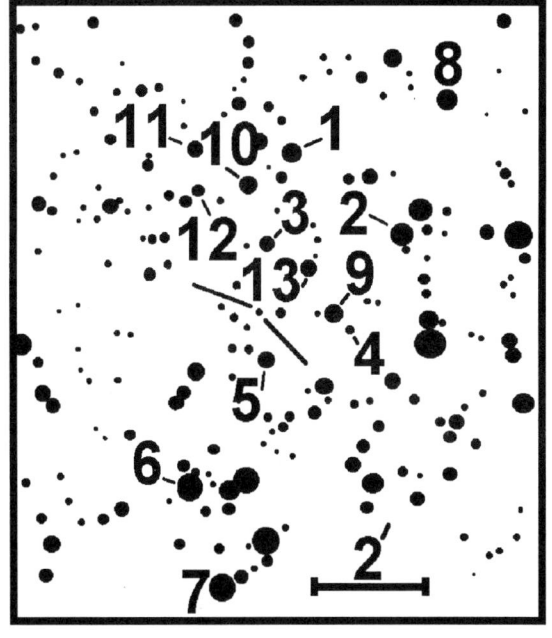

IR Gem

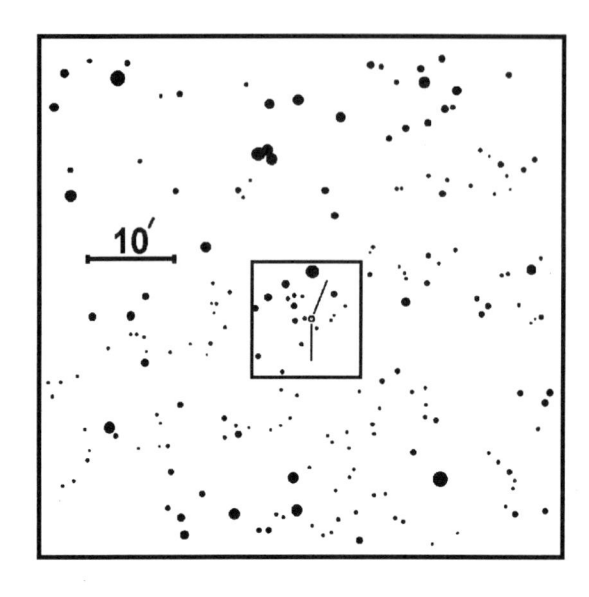

α (2000.0)=$06^h47^m34^s.56$
δ (2000.0) =$28°06'22''.8$
Type: DN SU
V=10.7–17.5
Spectrum: later than M6 V
P_{orb}=$0^d.06837$
$I.C.$=JD 2446808.803
P_{cycle}=75^d

The comparison stars are:
(Photographic magnitudes)

Star	Pg
1	10.4
2	11.2
3	11.8
4	12.2

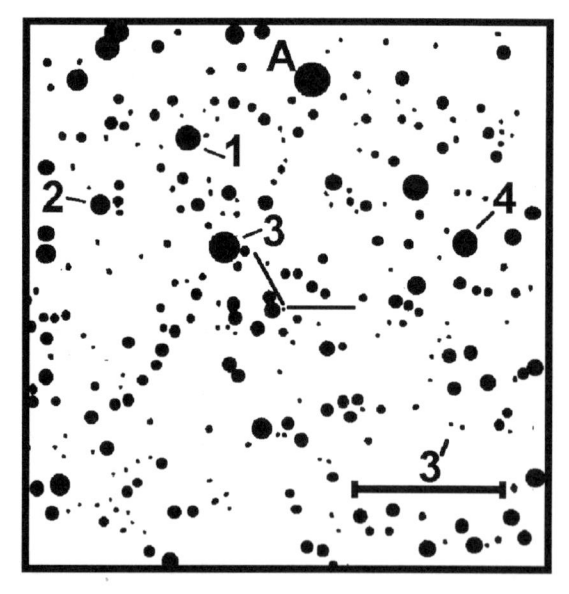

A = BD+28°1234. P_{orb} is the spectroscopic orbital period, $K = 30\pm4$ km s^{-1}. The photometric period value is $0^d.07076$ [335]. In supermaxima there are superhumps with $P = 102$ min. The system consists of a massive white dwarf and a degenerate low-mass secondary.

References: [176, 273, 333, 335, 364, 402]

AH Her

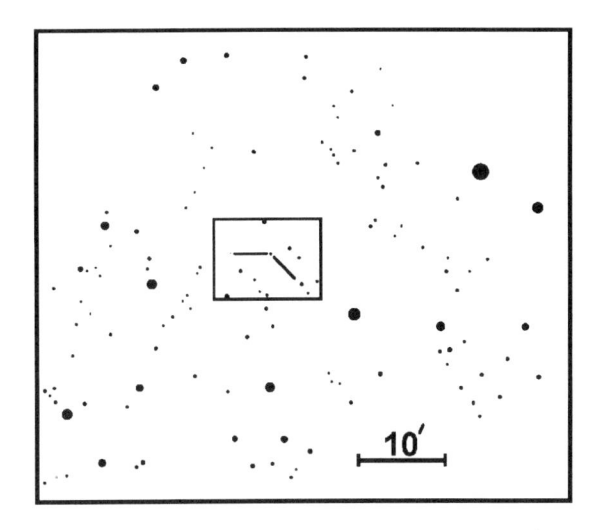

α (2000.0)=$16^h 44^m 10^s.7$
δ (2000.0)=$25°15'01''.1$
Type: DN ZC XS
Pg=10.9–14.7
Spectrum: K2–M0 V
P_{orb}=$0^d.258116(4)$
Max. VR=JD 2445491.328
P_{cycle}=$19^d.8$

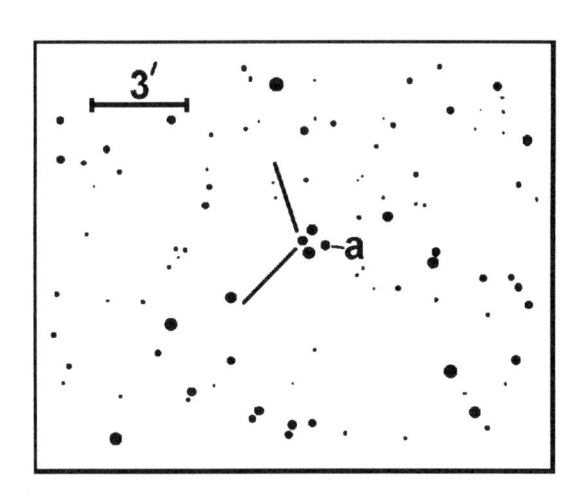

The comparison star a has

V=12.64
B=13.14
U=13.08

Rapid variability with $P \sim 60$ s. P_{orb} is the spectroscopic orbital period, K=126±4 km s^{-1} from emission lines and 158±8 km s^{-1} from absorption lines. The photometric orbital elements are $Min.phot. = $ JD $2445501.441(5) + 0^d.2440(2) \cdot E$, the amplitude of the variability is $B \sim 0.4$. Flickering in maxima and minima up to $0^m.4$ in 2–8 min. X-ray source.

References: [48, 82, 176, 364, 366, 378, 402]

V 544 Her

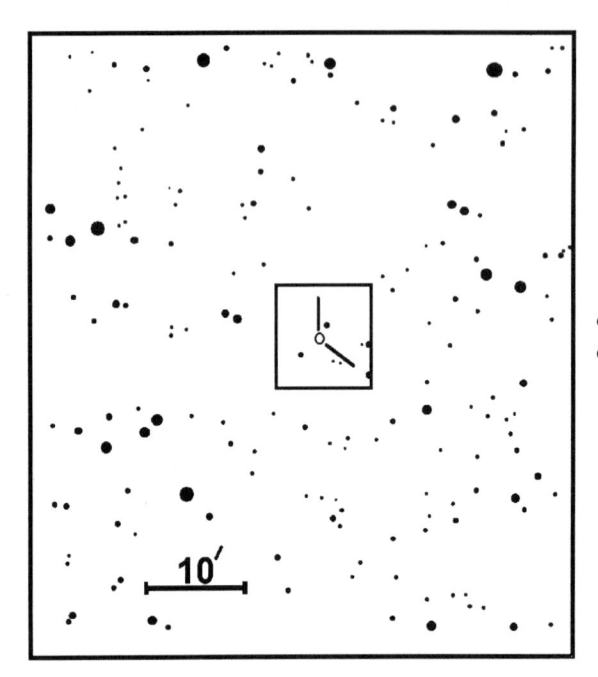

α (2000.0)=$16^h38^m05^s.48$
δ (2000.0)=$08°37'57''.4$
Type: DN UG?
Pg=14.5–20
$P=0^d.069$

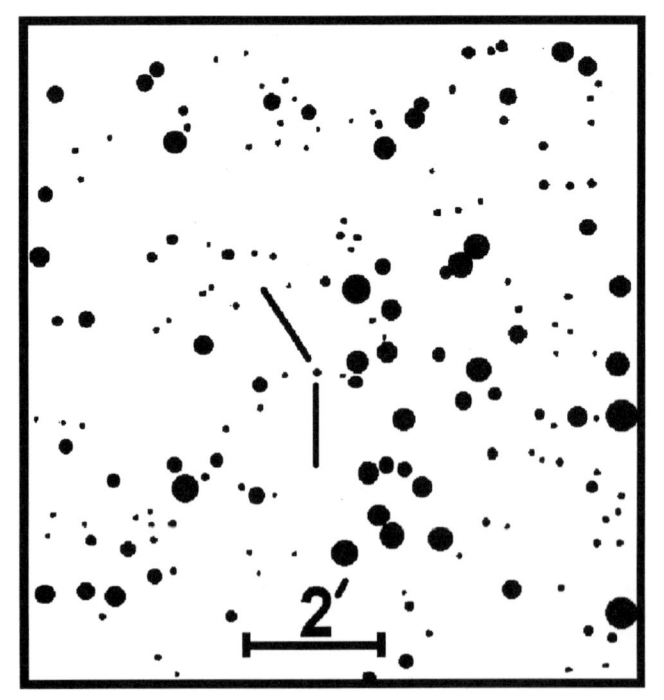

References: [176, 364]

VW Hyi

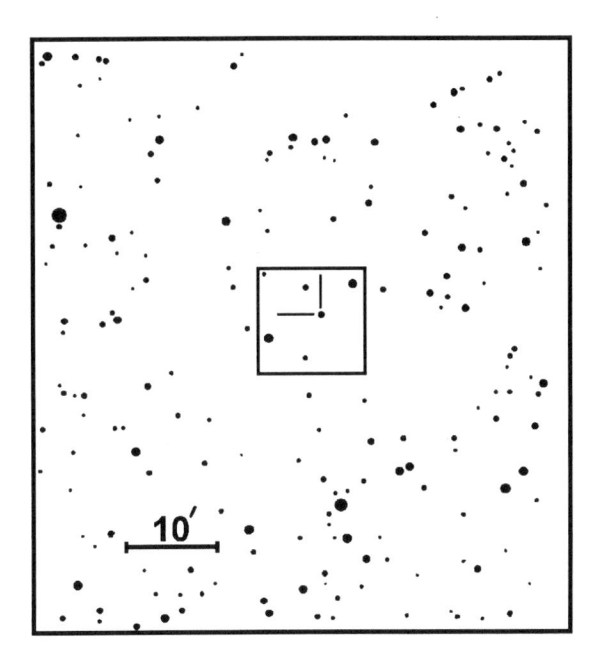

α (2000.0)=$04^h09^m11^s.1$
δ (2000.0)=$-71°17'42''.2$
Type: DN SU XS
V=8.4–14.4
P=$0^d.0748134$
Max.phot.=JD 2440128.0241
$P_{cycle}=27^d.3$

The comparison stars are:
(Photographic magnitudes)

Star	Pg
1	9.4
2	10.4
3	12.4
4	12.9
5	13.5
6	14.1

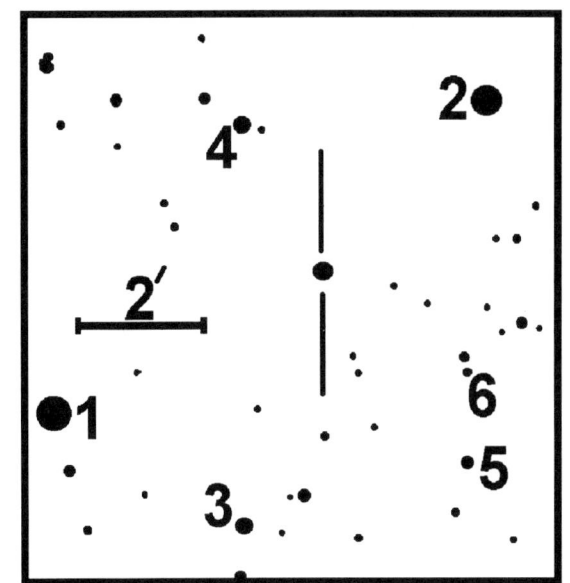

Eclipsing binary. X-ray source.
During the superoutburst of 1983, QPOs with $P = 14.06$ s were observed in X-ray range. Oscillations with the amplitude $A = 0^m.02$ and cycles 28–34 s are observed.

References: [6, 48, 176, 351, 355, 361, 364, 387]

WX Hyi

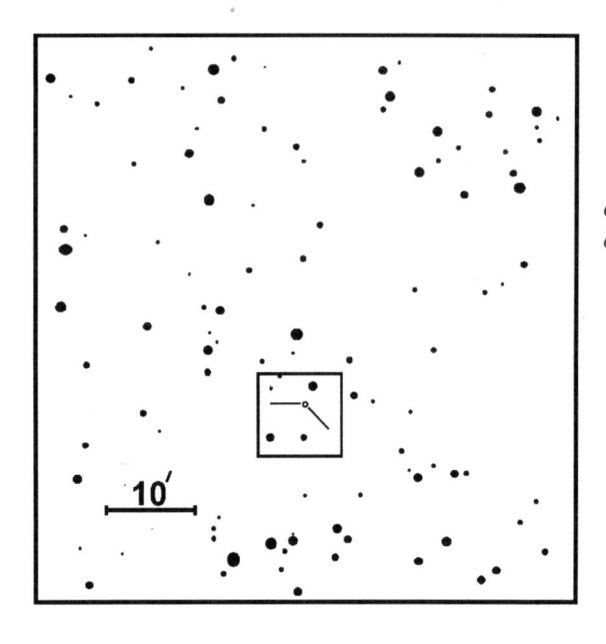

α (2000.0)=$02^h09^m51^s.8$
δ (2000.0)=$-63°18'40''$
Type: DN SU XS
V=9.6–15.0
P=$0^d.0748134$
I.C.=JD 2443819.56457
$P_{cycle}=13^d.7$

QPOs with $P = 19$ min, 26 min are observed. The mean interval between supermaxima is $141^d.5$. The period between humps during supermaxima is $0^d.07712$. X-ray source.

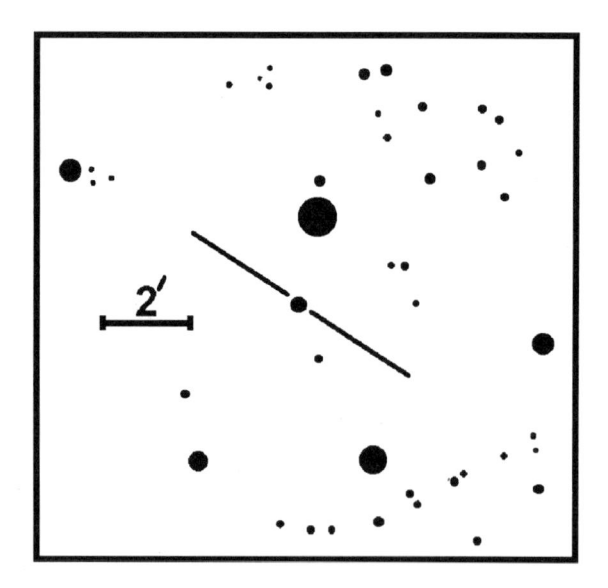

References: [17, 82, 176, 300, 364]

T Leo

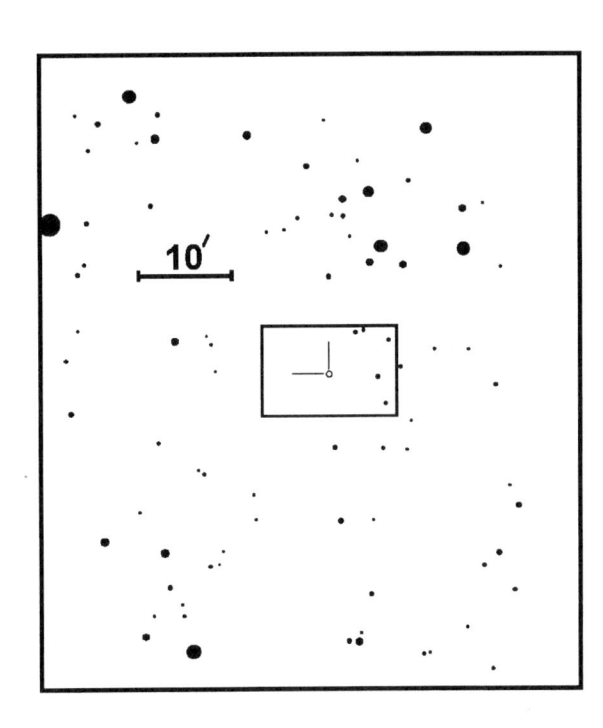

α (2000.0)=$11^h38^m27^s.9$
δ (2000.0)=$03°22'08''$
Type: DN SU
B=10–15.71
Spectrum: K0–M5 V
$P_{orb}=0^d.05881873$
I.C.=JD 2444974.0099
$P_{cycle}\sim 420^d$

The comparison star *1* has

Vis=13.7

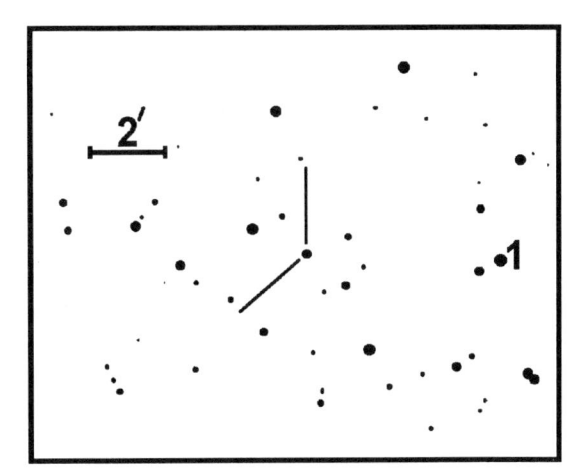

P_{orb} is the spectroscopic period, $K = 135\pm 8$ km s^{-1}. This is a noneclipsing binary.

References: [48, 176, 309, 310, 364, 397, 398]

X Leo

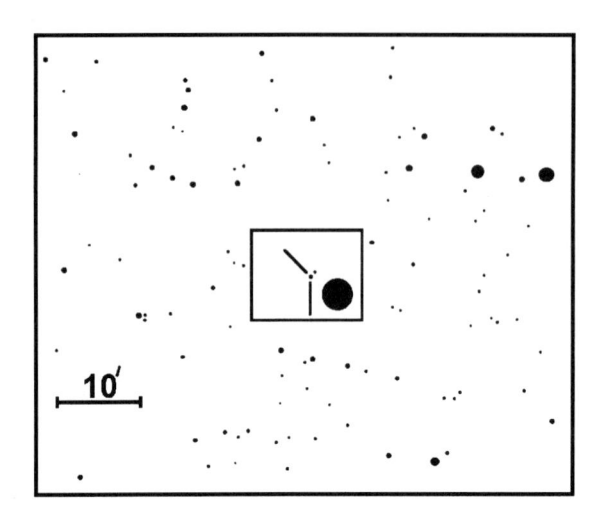

α (2000.0)=$09^h51^m01^s.4$
δ (2000.0)=$11°52'31''$
Type: DN UG
V=11.1–15.7
$P_{orb}=0^d.1644$
I.C.=JD 2446434.843
$P_{cycle}=20^d.9$

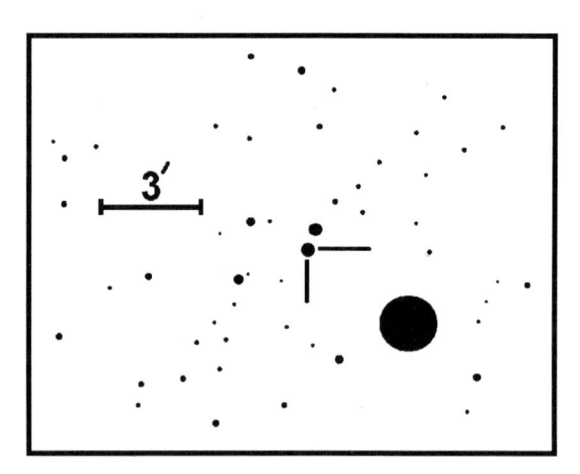

P_{orb} is the spectroscopic period, $K = 105\pm10$ km s^{-1} from the Hα emission line. Secondary component's lines have not been observed. $F_x < 1.35 \times 10^{-13}$ erg cm^{-2}s^{-1} in the X-ray range (0.2–4.5 keV).

References: [48, 176, 311, 364, 402]

RU LMi

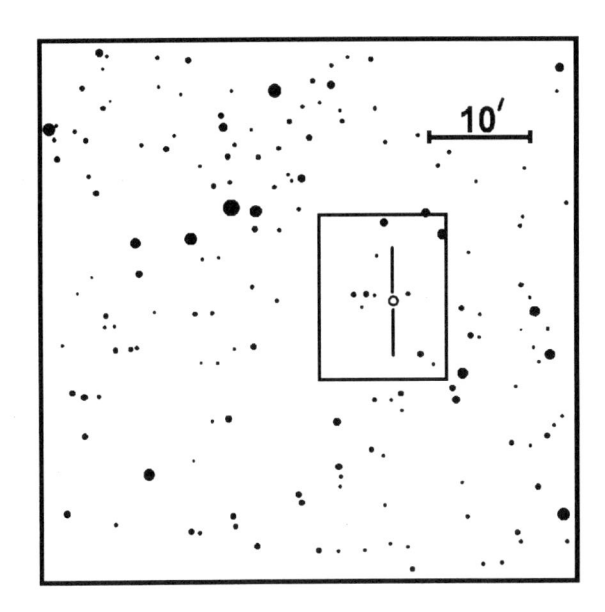

α (2000.0)=$10^h02^m07^s.3$
δ (2000.0)=$33°51'00''$
Type: DN
Pg=13.8–19.5
P=$0^d.245$
P_{cycle}=25^d :

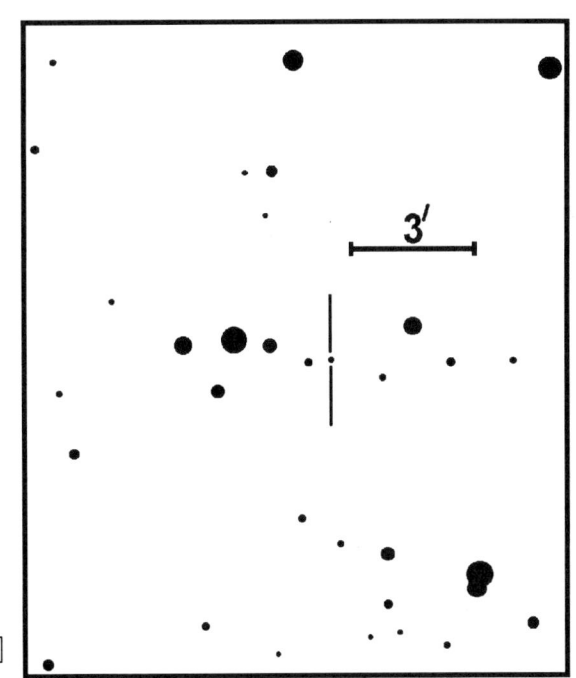

References: [52, 65, 131, 176]

BR Lup

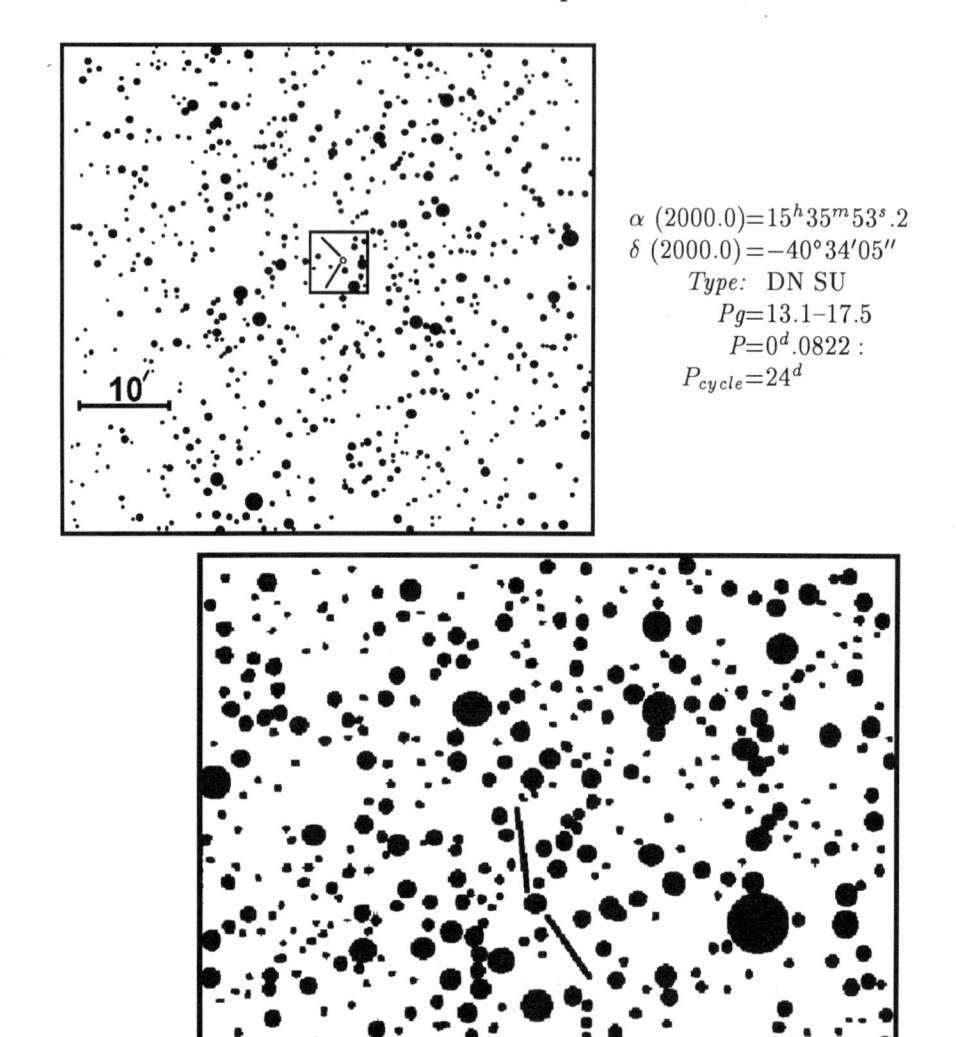

α (2000.0)=$15^h 35^m 53^s.2$
δ (2000.0)=$-40°34'05''$
Type: DN SU
Pg=13.1–17.5
$P=0^d.0822$:
$P_{cycle}=24^d$

A = CoD−40°9768. The system is a south-eastern component of a close pair. Superhump period is $P_{hump} = 0^d.08216$.

References: [48, 52, 99, 176, 351, 364, 380]

AY Lyr

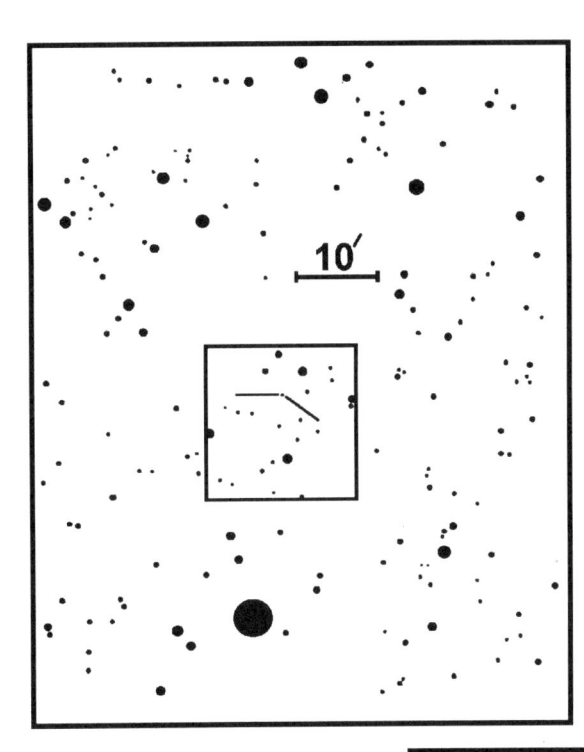

α (2000.0)=$18^h44^m26^s.79$
δ (2000.0)=$37°59'51''.5$
Type: DN SU XS
B=12.5–18.4
P=$0^d.0734$
Min.phot.=JD 2447060.3527
P_{cycle}=24^d

The comparison stars are:
(Visual magnitudes)

Star	Vis
1	12.21
2	12.60
3	13.34

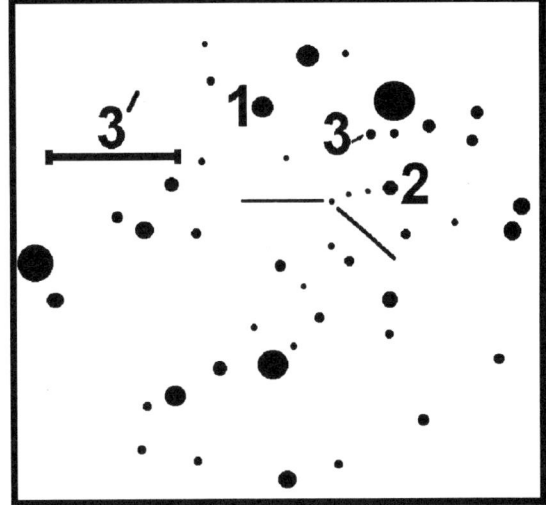

The soft X-ray source H 1839+37. Cycles of supermaxima are $188^d.9 - 212^d.8$. Superhump period is $P_{hump} = 0^d.07552$.

References: [6, 52, 80, 176, 255, 352, 410]

TU Men

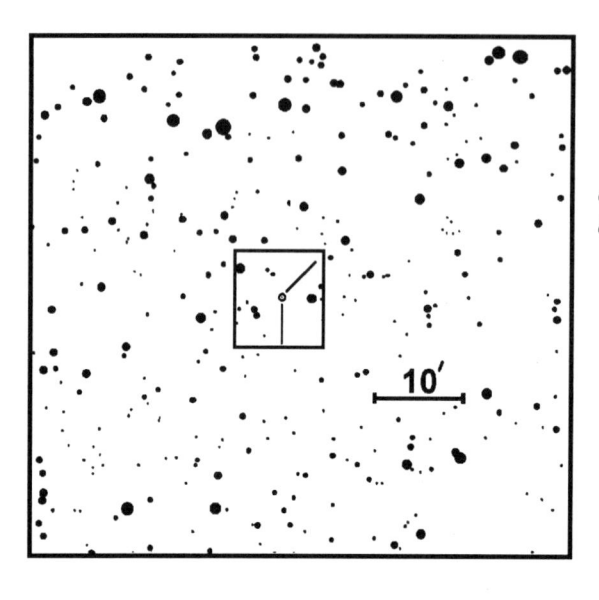

α (2000.0)=$04^h41^m40^s.1$
δ (2000.0)=$-76°36'46''.3$
Type: DN SU
Pg=11.4–17
Spectrum: M4.5 V
P=$0^d.1176$
I.C.=JD 2444574.561
$P_{cycle}=68^d$

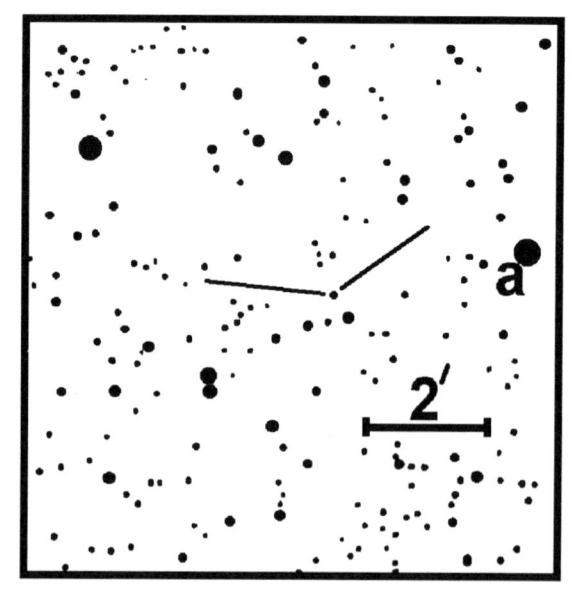

The comparison star a has

V=10.95
B=11.31
U=11.42

Superhump period is $P_s = 0^d.12625$.

References: [176, 330, 360, 364]

BT Mon

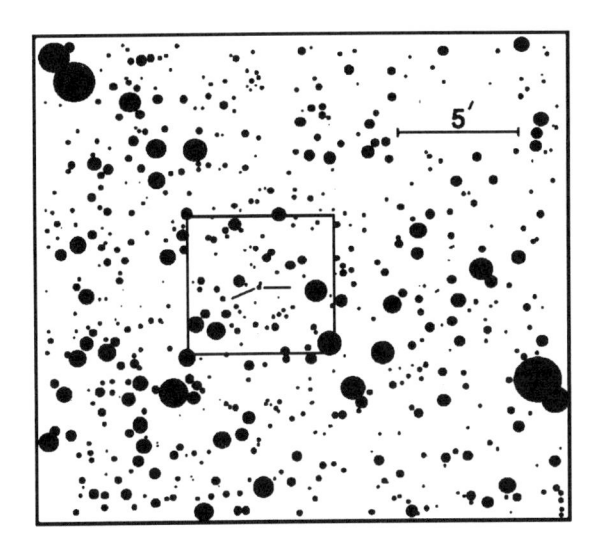

α (2000.0)=$06^h43^m47^s.9$
δ (2000.0)=$-02°01'13''.6$
Type: NA
Pg=4.0–18.7
Spectrum: K5–7 V
P=$0^d.333814$
Min.phot.=JD 2443491.7155

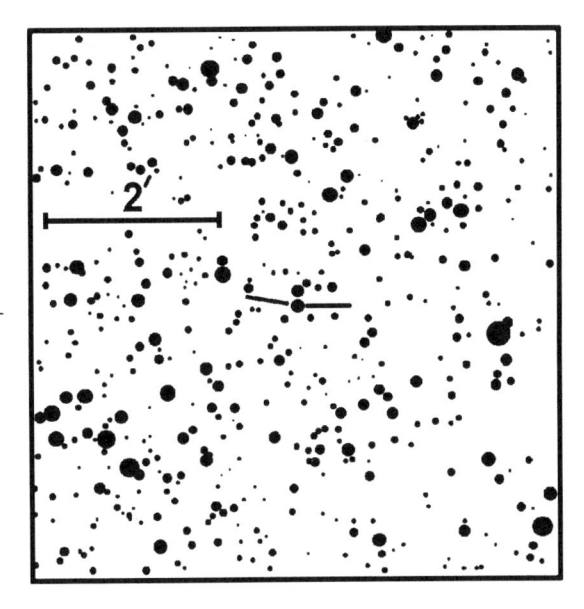

Before the outburst in 1939, the following elements were valid: Min =JD 2429500.207 + $0^d.3338010 \cdot E$.

References: [318]

CW Mon

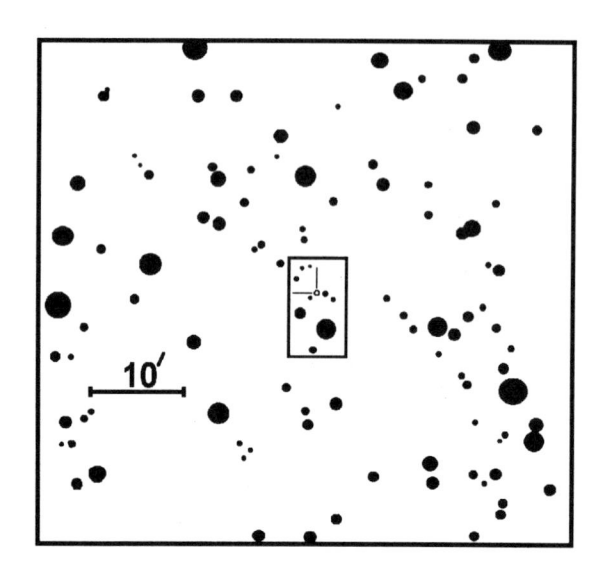

α (2000.0)=$06^h36^m54^s.4$
δ (2000.0)=$00°02'16''$
Type: DN UG
Pg=11.9–16.3
Spectrum: M3 V
$P=0^d.1762(3)$
$P_{cycle}=121^d.7$

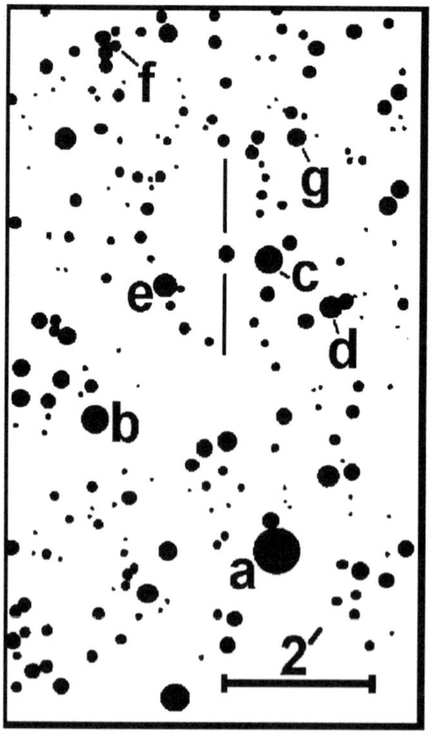

The comparison stars are:

Star	V	B	U
a	9.22	9.44	9.63
b	11.39	11.90	11.82
c	11.78	12.43	12.59
d	12.56	13.86	15.00
e	12.78	13.39	13.44
f	13.58	14.42	14.62
g	13.89	15.50	16.4:

The ellipsoidal effect was observed with the amplitude of $0^m.2$.

References: [48, 176, 336, 360, 364]

HP Nor

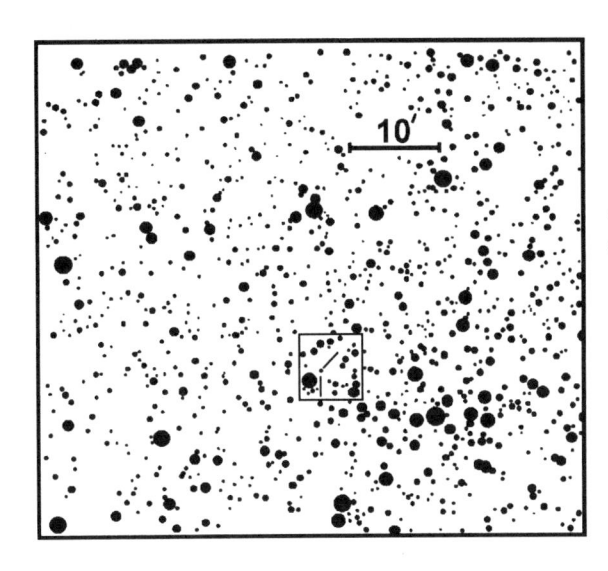

α (2000.0)=$16^h20^m49^s.64$
δ (2000.0)=$-54°53'23''.3$
Type: UG
V=12.8–16.41
P=$0^d.183$
P_{cycle}=$17^d.3$

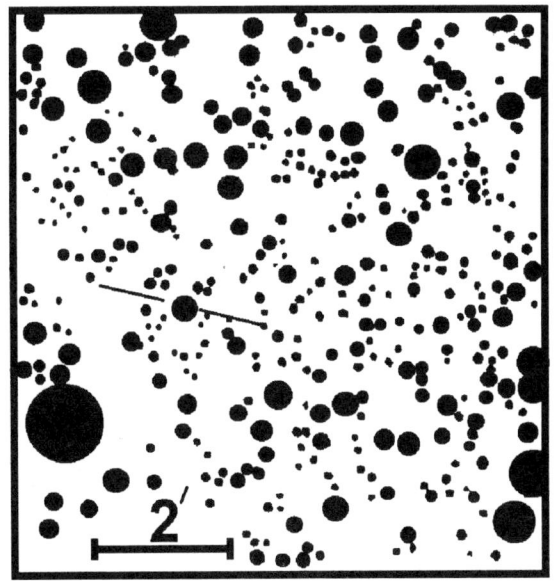

A strong flickering activity with the amplitude of $A \sim 0^m.4 - 0^m.8$ was observed.

References: [48, 176, 362, 364]

CN Ori

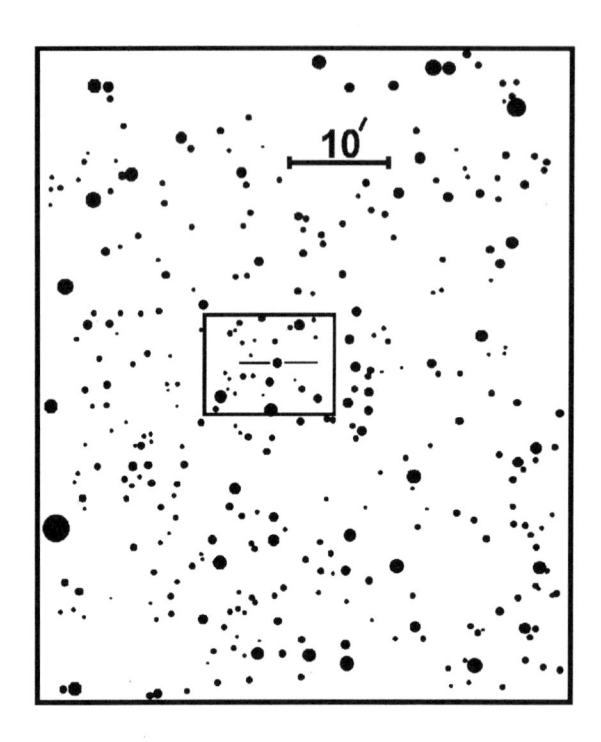

α (2000.0) = $05^h52^m07^s.80$
δ (2000.0) = $-05°25'00''.8$
Type: DN UG
V = 11.0–16.2
Spectrum: M4 V
$P = 0^d.163199$
$I.C. =$ JD 2446456.929
$P_{cycle} = 15^d.85$

The comparison stars are:
(Photovisual magnitudes)

Star	Vis
1	9.9
2	11.2
3	13.3
4	13.6
5	13.8
6	14.0
7	14.3
8	14.6

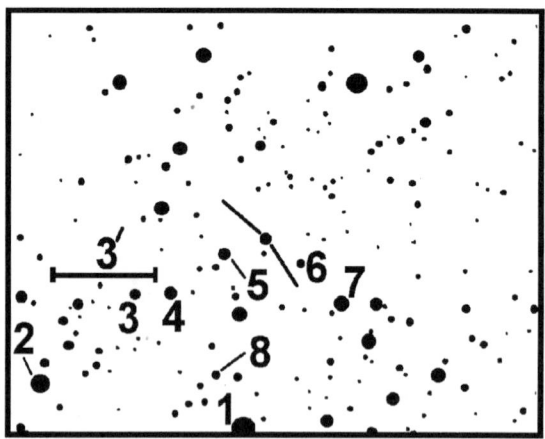

Sometimes rapid light oscillation are also observed with a period of 24–30 s ($A < 0^m.01$). Another possible period is $0^d.15946$. Photometric observations are published in [236, 244, 299, 385, 388], spectrophotometry in [247], the proper motion is given in [179].

References: [48, 124, 176, 179, 236, 244, 247, 299, 362, 364, 385, 388, 402]

CZ Ori

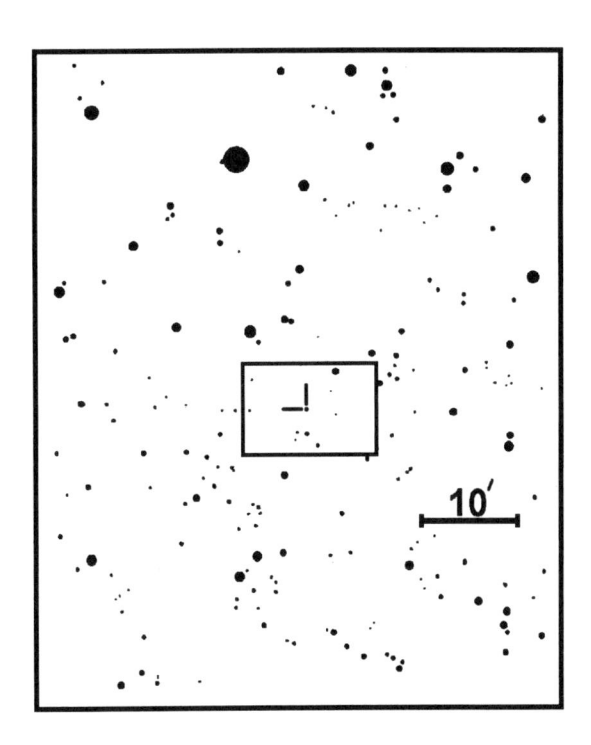

α (2000.0)=$06^h16^m43^s.20$
δ (2000.0)=$15°24'11''.1$
Type: DN UG?
V=11.2–15.6
$P=0^d.18$:
$P_{cycle}=25^d.73$

The comparison stars are:
(Photovisual magnitudes)

Star	Vis
1	10.6
2	11.8
3	12.6
4	12.9
5	13.3
6	13.3
7	13.9
8	14.1
9	14.4
10	14.7
11	15.0

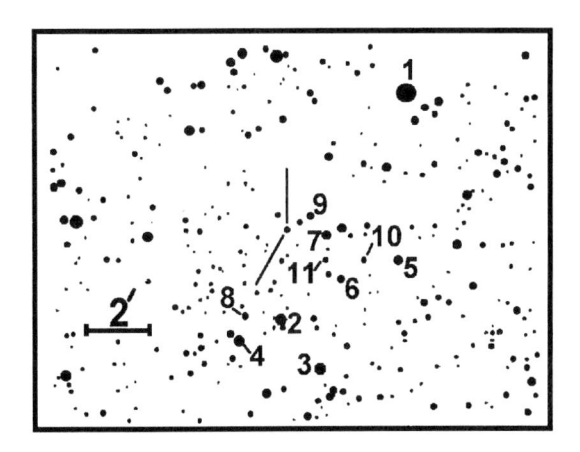

References: [48, 124, 176, 364, 402]

BD Pav

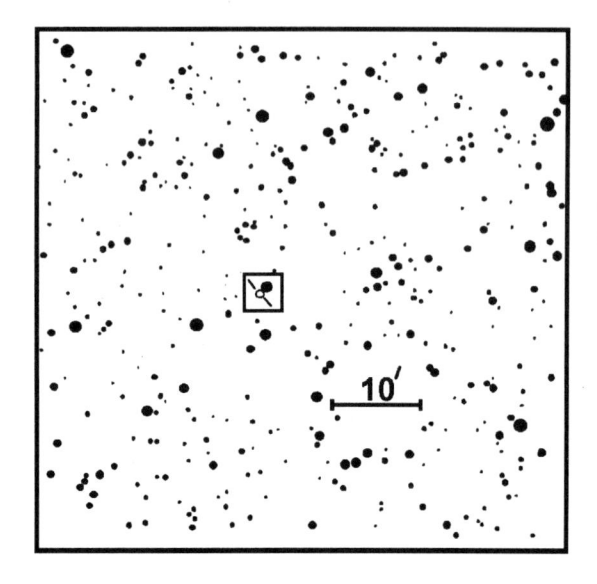

α (2000.0)=$18^h43^m11^s.1$
δ (2000.0)=$-57°30'44''$
Type: DN UG
B=12.4–16.2
Spectrum: M1–5 V
$P=0^d.17930$
Min.phot.=JD 2444787.5965

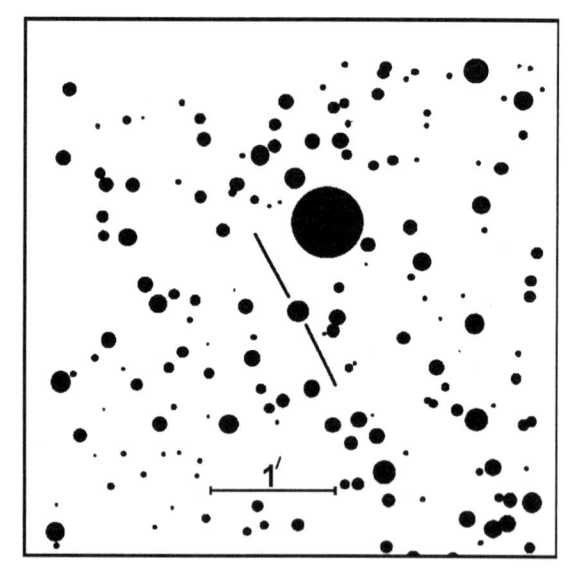

Eclipsing binary, $A \sim 0^m.5$. Flickering is observed with timescale of minutes. The spectrum is characterized by broad Balmer emission lines with complex structures.

References: [23, 24, 107, 176, 351, 406]

EF Peg

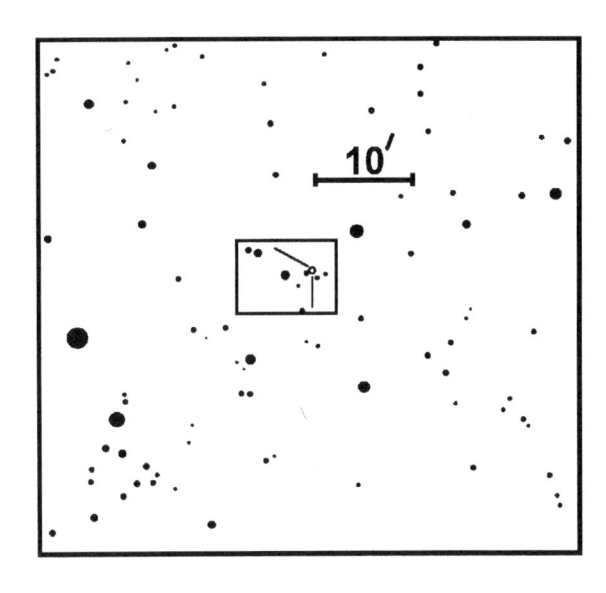

α (2000.0)=$21^h17^m26^s$
δ (2000.0)=$14°16'24''$
Type: UG SU
Pg=10.7–18.5
$P=0^d.08696$
Max.phot.=JD 2448547.954
$P_{cycle}=162^d$

The comparison stars are:
(Photographic magnitudes)

Star	Pg
1	10.22
2	10.77
3	11.85
4	12.75
5	13.86
6	14.95
7	13.86

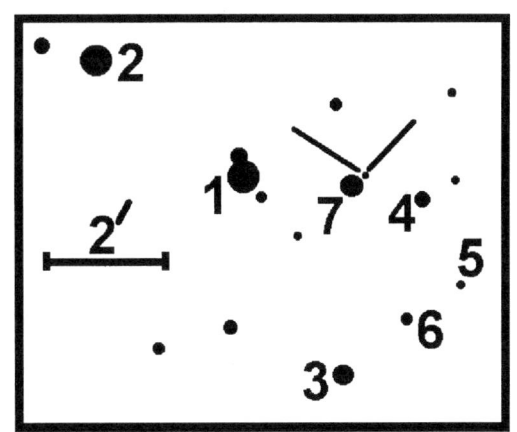

Visual binary system: in 5″, 305° there is a $13^m.86$ component.

References: [176, 350]

RU Peg

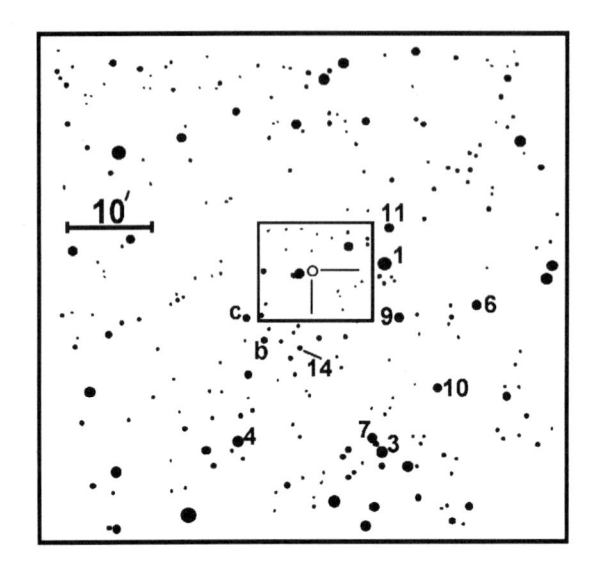

α (2000.0)=$22^h14^m02^s.53$
δ (2000.0)=$12°42'11''.1$
Type: DN UG XS
V=9.0–13.2
Spectrum: K2-3 V
P=$0^d.3575$
I.C.=JD 2447404.0918
$P_{cycle}=74^d.3$

The comparison stars are:

Star	V	B
a	10.31	11.56
b	11.10	12.33
c	11.94	13.55
d	12.52	13.20
e	12.70	13.41
f	13.41	14.13

(Photographic magnitudes)

Star	Pg	Star	Pg
1	8.45(5)	b	11.01(1)
2	8.70(5)	9	11.2(1)
3	9.62(6)	10	11.34(3)
4	9.76(5)	11	11.59(6)
a	10.41(7)	c	11.7(2)
6	10.52(9)	13	12.62(6)
7	10.63(7)	14	12.8(1)

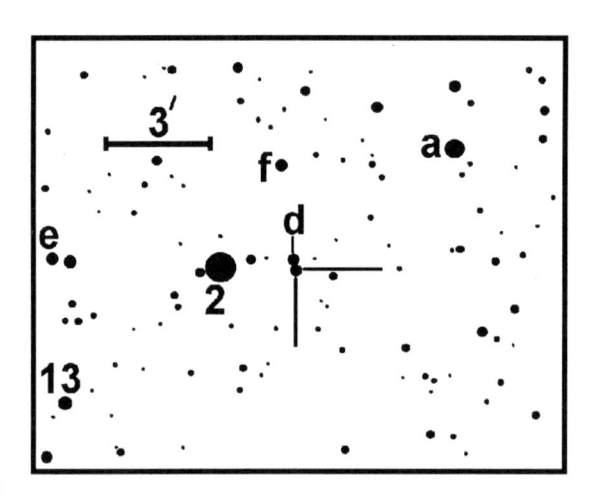

The southwestern component of a close pair. Light oscillations with $P \sim 11.6$ s and 50 s were observed.

References: [119, 157, 176, 331, 364, 402]

IP Peg

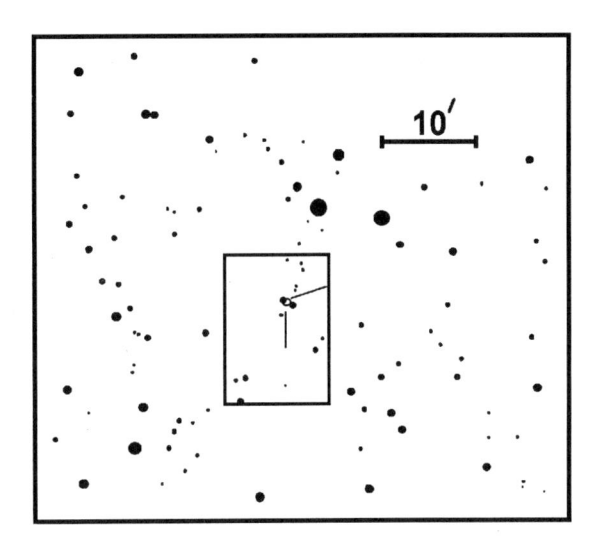

α (2000.0)=$23^h23^m08^s.7$
δ (2000.0)=$18°24'58''$
Type: DN UG
B=12.0–18.6
Spectrum: M4 V
P=0^d.15820764
Min.phot.=JD 2445933.4094

The comparison stars are:

Star	a	b
V	13.80	13.62
B	14.82	14.15
U	15.86	14.25

Possibly there is a third body with the period 4.7 yr. *Eclipsing* binary.

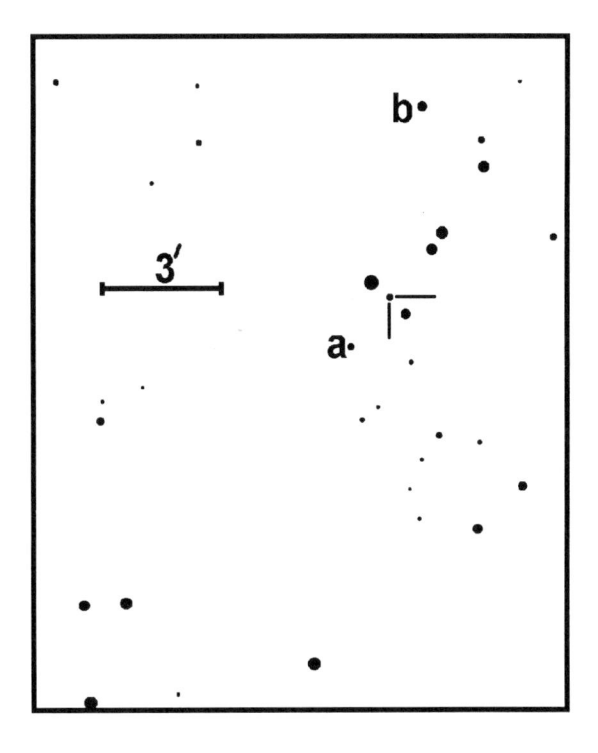

References: [176, 317]

UV Per

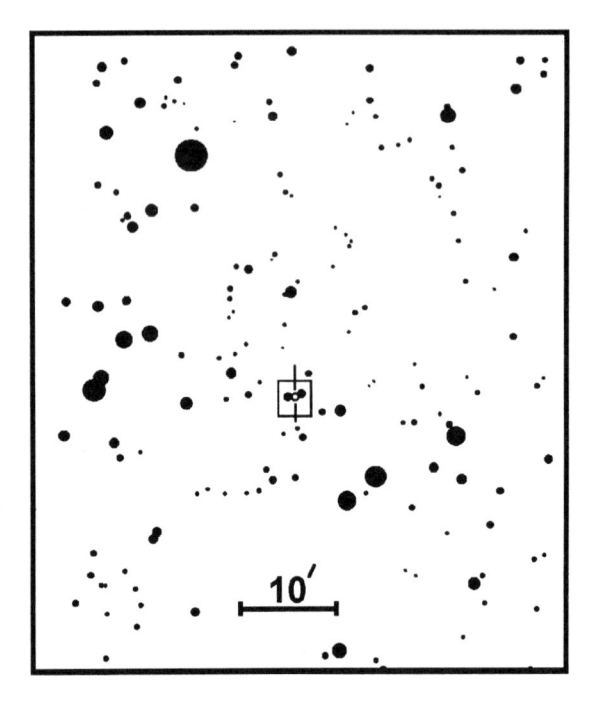

α (2000.0)=$02^h10^m07^s.4$
δ (2000.0)=$57°11'19''$
Type: DN SU
B=12–17
P=$0^d.0840$:
P_{cycle}=320^d :

The comparison stars are:

Star	V	B
a	10.58	10.89
b	10.83	10.94
c	11.72	11.46
d	13.75	14.15
e	14.09	14.03
f	14.47	14.26

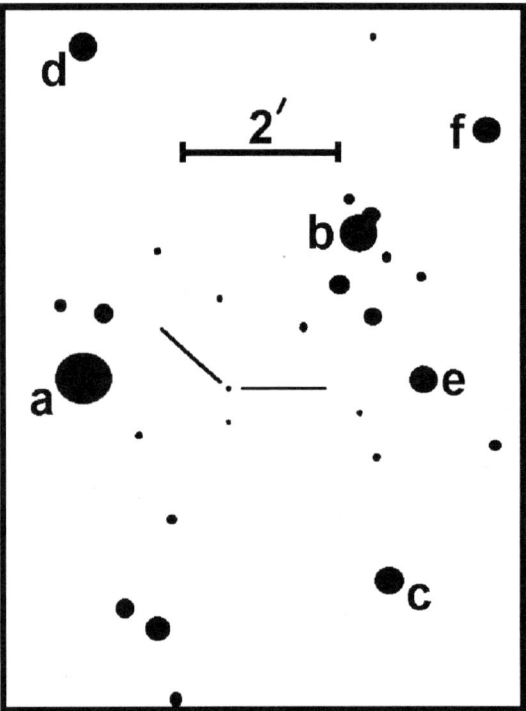

References: [6, 48, 52, 157, 176]

KT Per

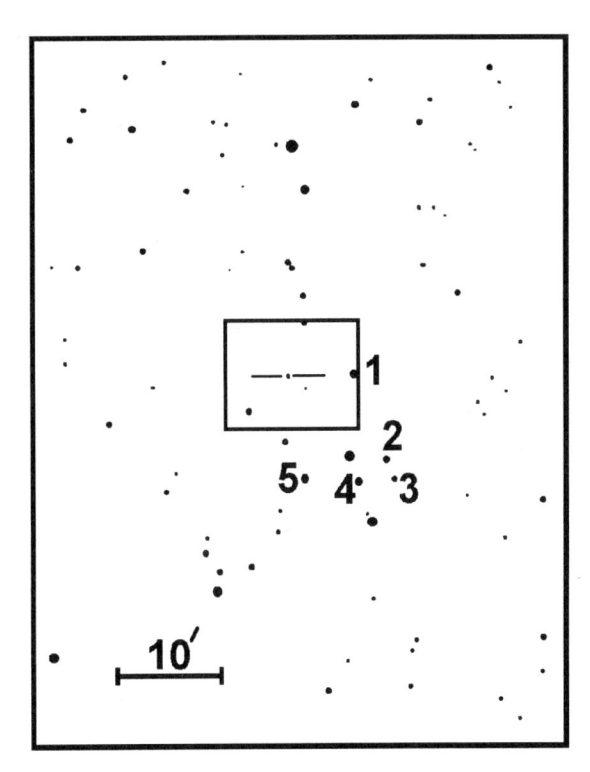

α (2000.0)=$01^h37^m08^s.65$
δ (2000.0)=$50°57'20''.7$
Type: DN ZC
V=11.5–15.4
Spectrum: G7–K2 V
$P=0^d.1633$:
$P_{cycle}=26^d$

The comparison stars are:
(Photovisual magnitudes)

Star	Vis
1	10.2
2	10.7
3	11.3
4	12.0
5	12.6
6	13.2
7	13.7

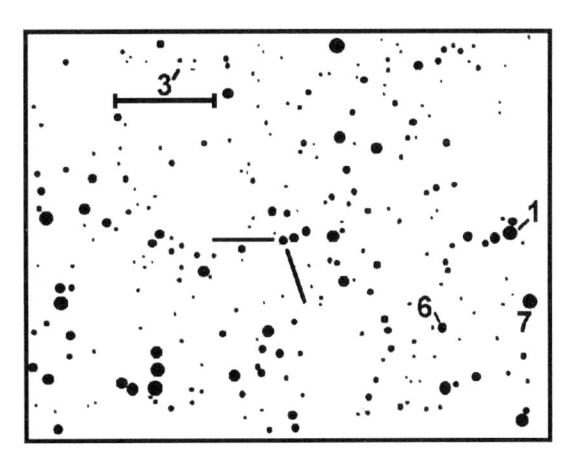

Quasiperiodic oscillations during outbursts with $P \sim 22.9 - 29.5, 82, 90, 147$ s, Δm up to $\sim 0^m.01$ were observed. X-ray source.

References: [52, 82, 176, 244, 284, 372, 402]

TY Psc

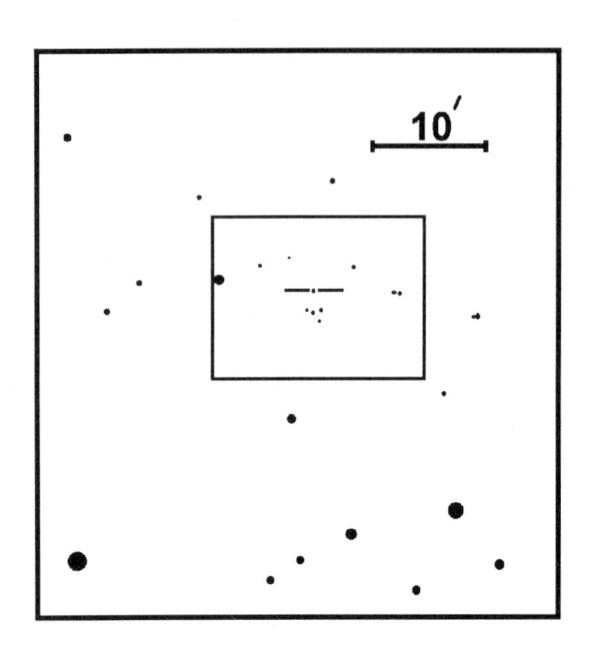

α (2000.0)=$01^h25^m39^s.35$
δ (2000.0)=$32°23'09''.8$
Type: DN SU
Pg=10.8–16.3
P=$0^d.06824$
P_{cycle}=$41^d.5$

The comparison stars are:
(Visual magnitudes)

Star	Vis
1	12.3
2	13.2
3	14.1
4	14.6

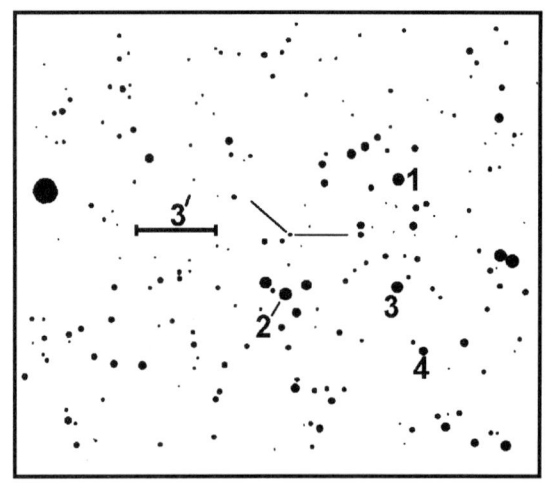

Supermaxima cycle is $\sim 370^d$. Superhump period is $P_{hump} = 0^d.080$ during long extended outbursts. Ellipsoidal variations with the amplitude $A \sim 0^m.11$ in optical light are observed.

References: [48, 52, 176, 211, 215, 258, 337, 351]

AY Psc

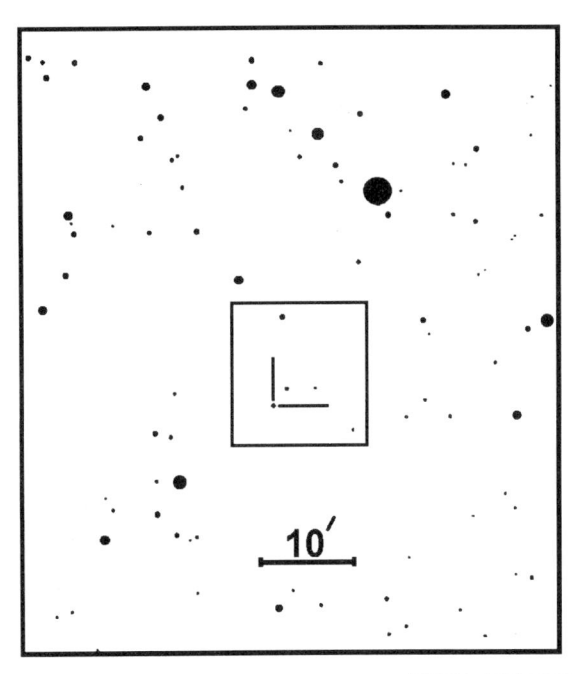

α (2000.0)=$01^h36^m55^s.7$
δ (2000.0)=$07°16'25''$
Type: NL UX
B=15.2–19
P=$0^d.2173209$
Min.phot.=JD 2447623.3463

The comparison stars are:

Star	V	B
1	13.03	13.53
2	14.27	14.93
3	14.29	14.92
4	14.65	15.36
5	...	15.55
6	14.88	15.77
7	...	15.90
8	15.61	16.38
9	15.17	16.19
10	15.78	16.52

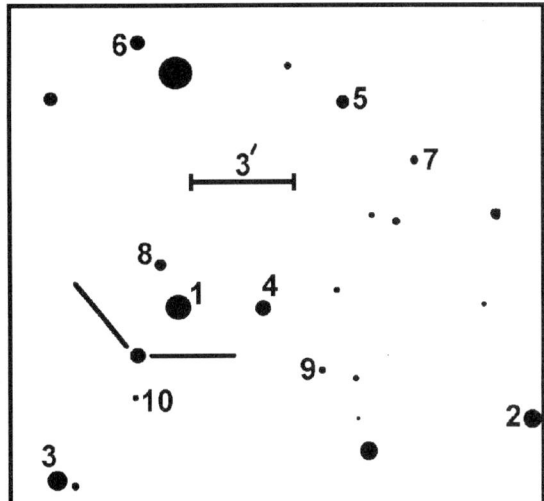

Eclipsing binary. The depth of eclipse is $1^m.7$ B. Oscillations with $A \sim 0^m.1$ in several minutes were observed.

References: [174]

TY PsA

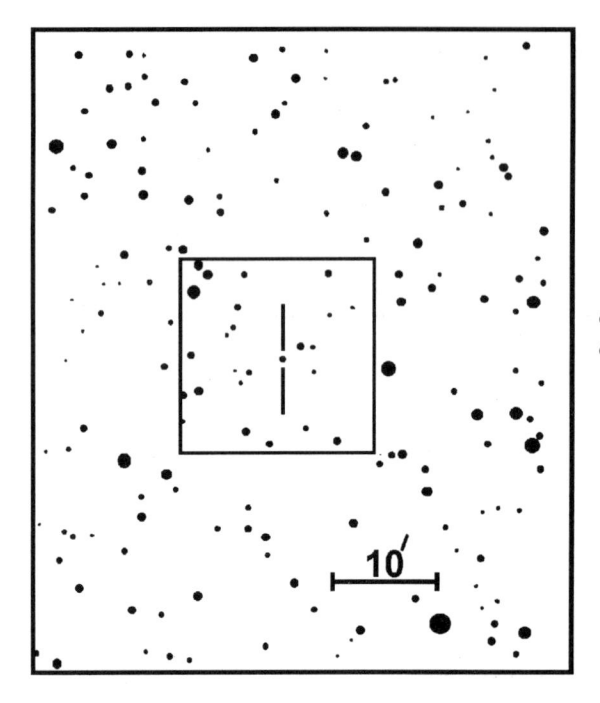

α (2000.0)=$22^h 49^m 39^s.0$
δ (2000.0)=$-27°06'52''$
Type: DN SU
B=12–17
P=$0^d.0840$:

The comparison star a has

U=13.76
B=13.83
V=13.34
R=13.03
I=12.73

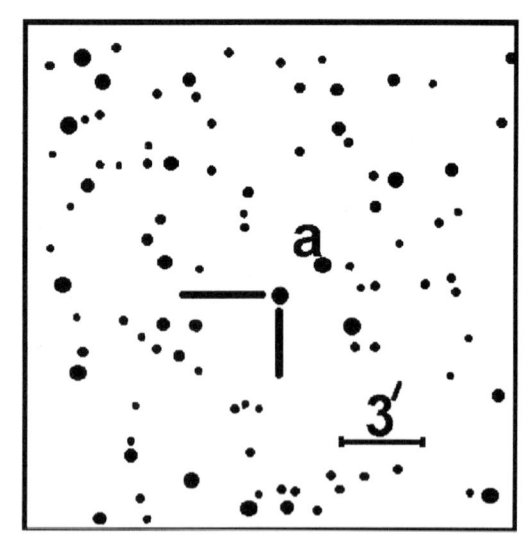

Superhump period is $0^d.08409$,
Max.= HJD 2445148.860(3). QPOs, 27–29 s and 248 s, were observed. Probably only the hot spot undergoes eclipses while the white dwarf and central parts of the disk remain visible.

References: [176, 386]

BV Pup

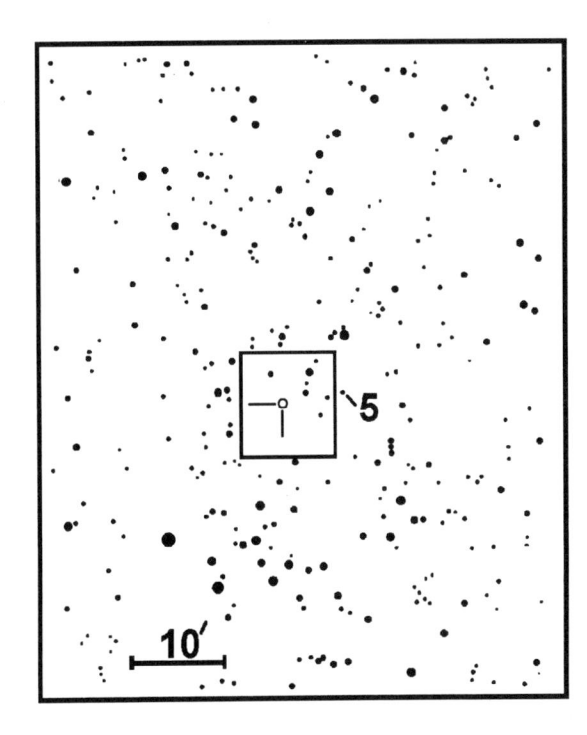

α (2000.0)=$07^h49^m05^s.1$
δ (2000.0)=$-23°34'00''$
Type: DN UG
V=13.1–15.7
Spectrum: M1 V
$P=0^d.167$ or $0^d.225$
$P_{cycle}=18^d.1$

Comparison stars are:
(Visual magnitudes)

Star	Vis
1	12.1
2	14.1
3	14.1
4	14.4
5	10.1

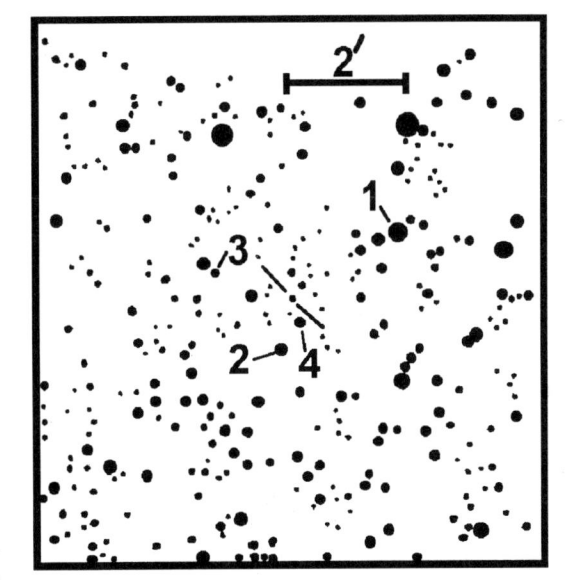

Ellipsoidal variations with the amplitude $\Delta V \sim 0.17$ are observed. Till now no optical eclipses have been registered. Flickering up to $0^m.5$ is observed. Possibly it is an X-ray source.

References: [6, 48, 176, 337, 362, 364]

RZ Sge

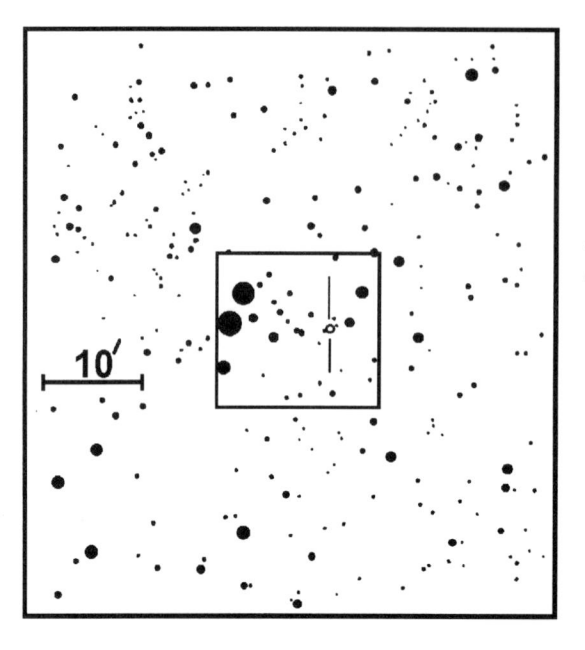

α (2000.0)=$20^h03^m18^s.56$
δ (2000.0)=$17°02'52''.5$
Type: DN SU
B=12.2–17.4
P=$0^d.0686$
$P_{cycle}=64^d$

The comparison stars are:
(Photovisual magnitudes)

Star	Vis
1	12.1
2	12.7
3	13.3
4	14.0
5	14.5

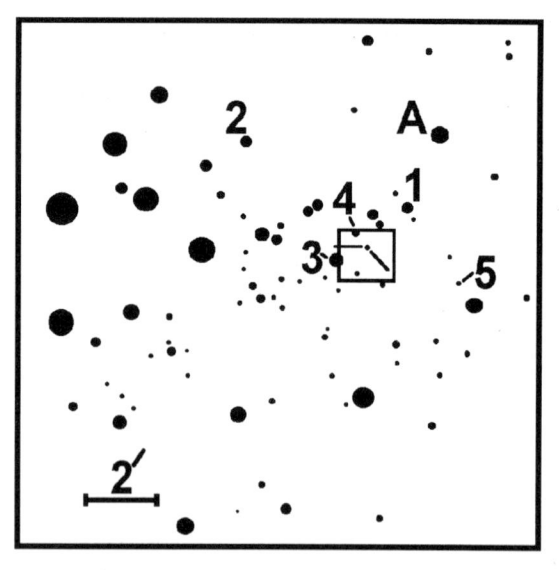

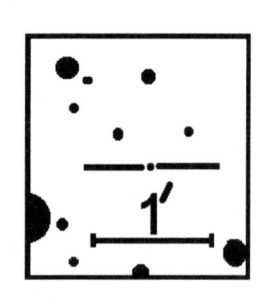

A = BD+16°4115. Superhump period is $P_{hump} = 0^d.07035$. Supermaxima cycle is ~ 266^d.

References: [30, 48, 176, 364, 374, 392]

WY Sge

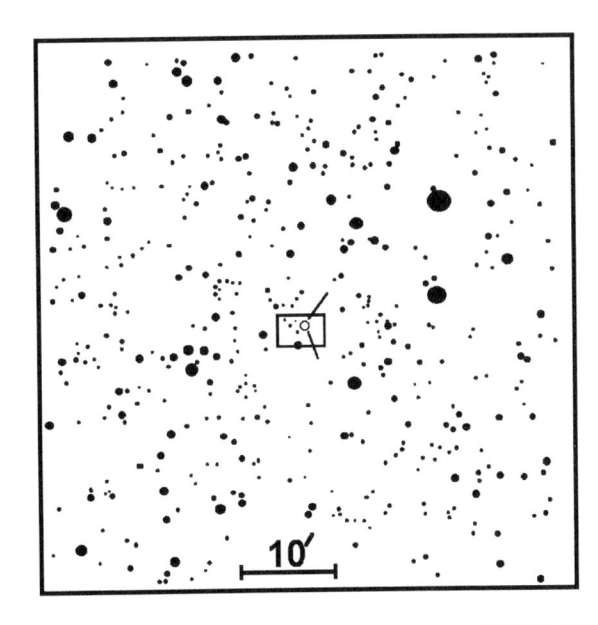

α (2000.0)=$19^h 32^m 43^s.80$
δ (2000.0)=$17°44'54''.4$
Type: N DN?
B=5.4:->21
P=$0^d.1536342$
Min.phot.=JD 2445137.8993

The comparison star a has

B=16.56 Sp A3-5

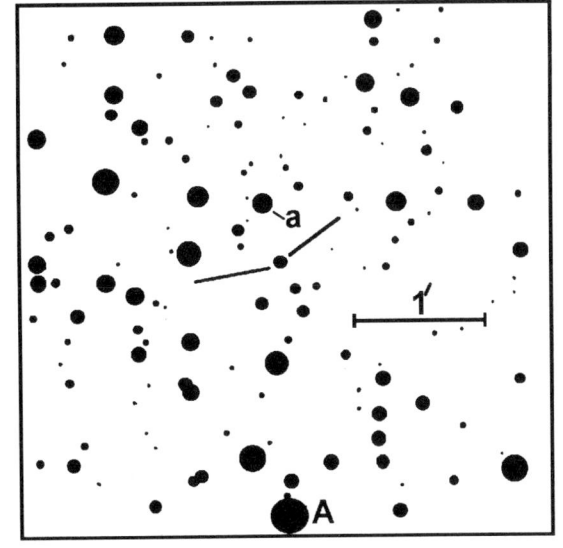

A = BD+17°3997. Nova Sge 1783, Max = JD 2372494. *Eclipsing* binary, brightness outside eclipses is $\sim 18^m.7$, the depth of the eclipses is $\sim 1^m.5 - 2^m$. Outbursts up to $\sim 1^m.6$ were observed. There is rapid flickering of light.

References: [43, 107, 176, 315]

WZ Sge

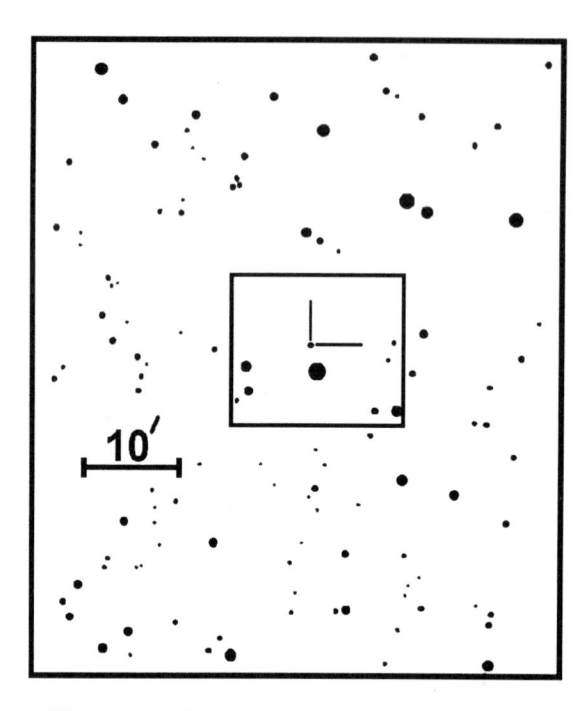

α (2000.0)=$20^h07^m36^s.1$
δ (2000.0)=$17°42'15''$
Type: DN SU CP XS
B=7.0–15.5
P=$0^d.0566878455$
Min.phot.=JD 2437547.72845
P_{cycle}=11900^d

The comparison stars are:

Star	V	B	U
A	9.72	10.75	11.51
B	11.84	12.23	12.45
C	8.76	8.92	9.05

Photographic and visual magnitudes

Star	Pg	Star	Vis
1	13.13	6	11.8
2	13.82	7	12.8
3	14.14	8	13.1
4	14.94	9	13.3
5	15.33	10	14.3

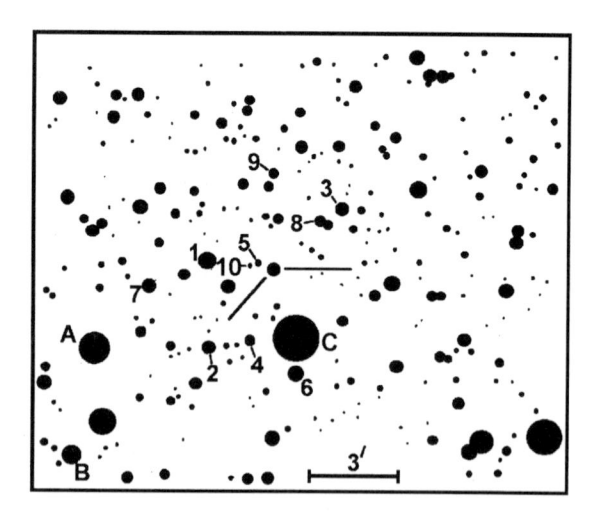

A = BD+17°4228, C = BD+17°4225. The system is the eastern component of a close pair, B = $15^m.76$. Sometimes Algol-like eclipses are observed with depth A ~ $0^m.2$. There are rapid oscillations with P ~ 27 − 29 s and A < $0^m.01$.

References: [6, 48, 107, 176, 181, 221, 256, 283, 364, 402]

V 4140 Sgr

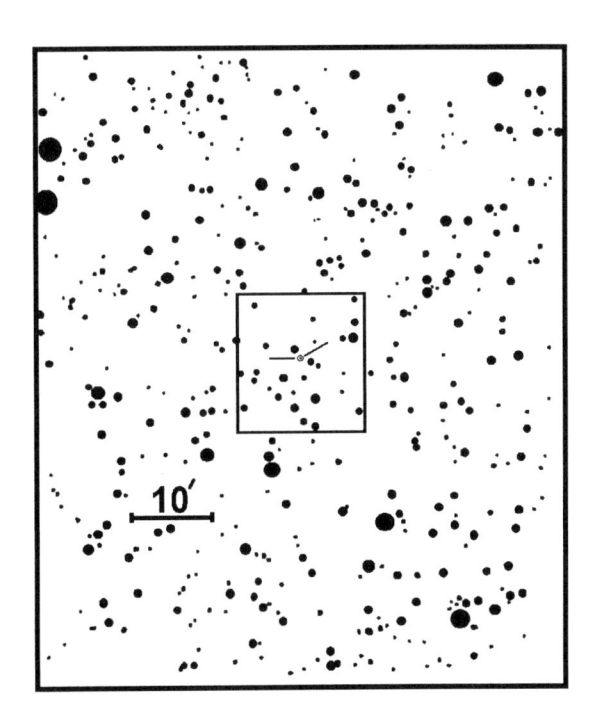

α (2000.0)=$19^h 58^m 49^s.76$
δ (2000.0)=$-38°56'13''.3$
Type: UG SU?
Pg=15.5–19.0
$P=0^d.0614296516$
$Min.phot.$=JD 2446261.67135

Eclipsing binary. The depth of eclipses is $1^m.4$, their duration is 502 s.

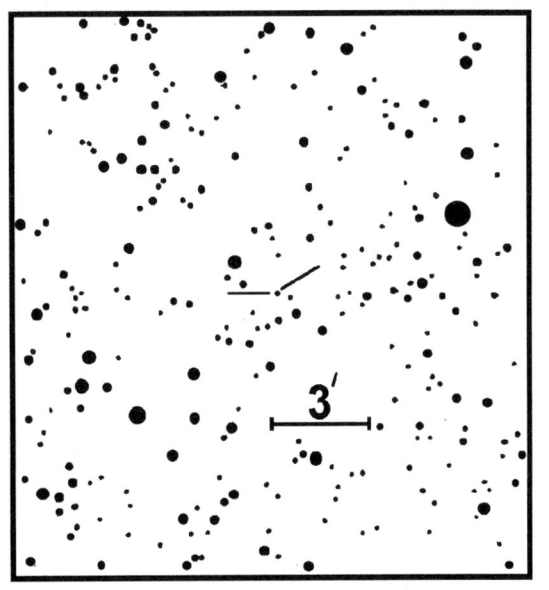

References: [142, 166, 176, 235, 370]

UZ Ser

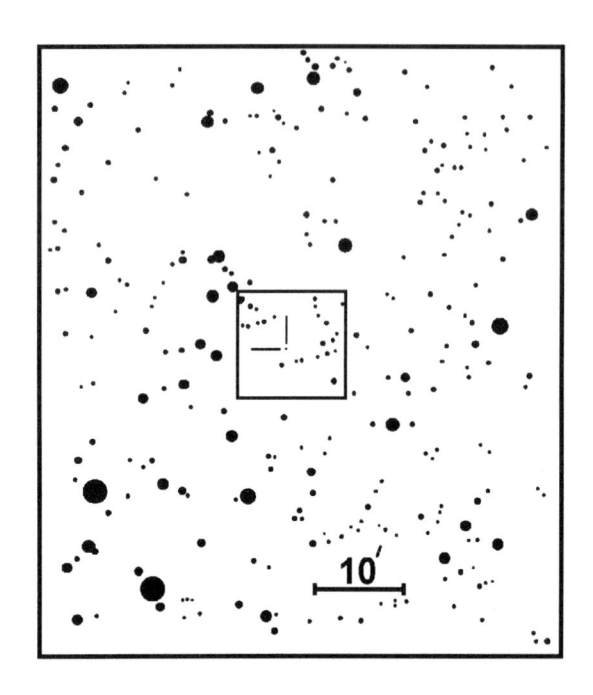

α (2000.0)=$18^h11^m24^s.0$
δ (2000.0)=$-14°55'33''$
Type: DN UG
Pg=12.0–16.7
P=$0^d.1730$
$Min.phot.$=JD 2446622.6480
P_{cycle}=$26^d.4$

The comparison stars are:

Star	a	b
V	11.81	12.31
B	13.02	12.58
U	13.99	12.80

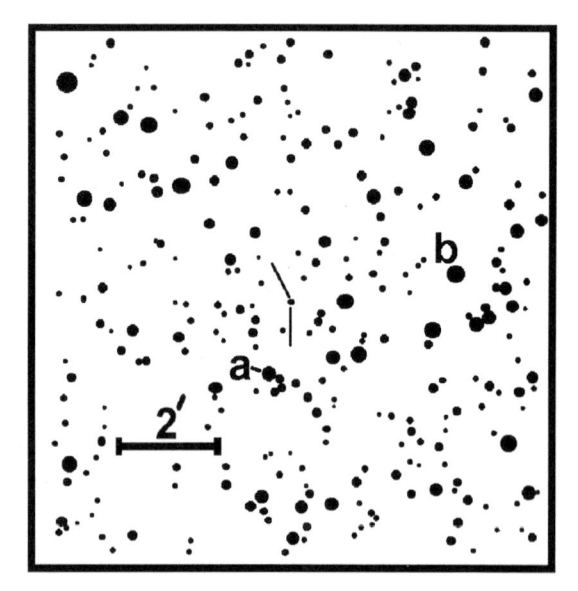

UBV photometric observations are published in [362], spectroscopic observations are in [110, 146]. Hard X-ray source [81].

References: [48, 81, 110, 146, 176, 351, 360, 362, 364]

LX Ser

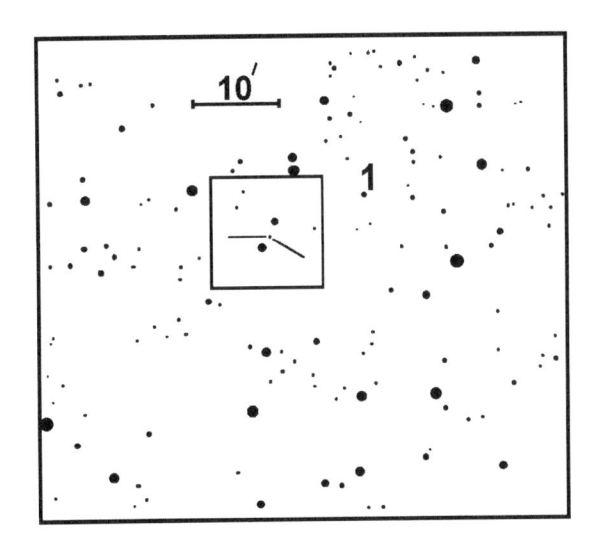

α (2000.0)=$15^h38^m00^s.18$
δ (2000.0)=$18°52'02''.3$
Type: UG VY SW
B=13.3–17.4
Spectrum: later than M2
$P=0^d.158432441$
Min.phot.=JD 2444293.0240

The comparison stars are:
(Photovisual magnitudes)

Star	Vis
1	13.2
2	14.2

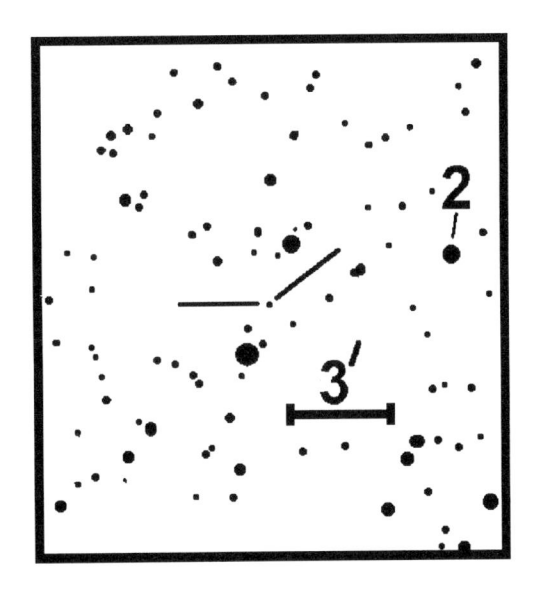

Eclipsing binary, the depth of eclipses is $\sim 2^m.5$. Comparatively frequent outbursts.

References: [176, 326, 395, 396]

KK Tel

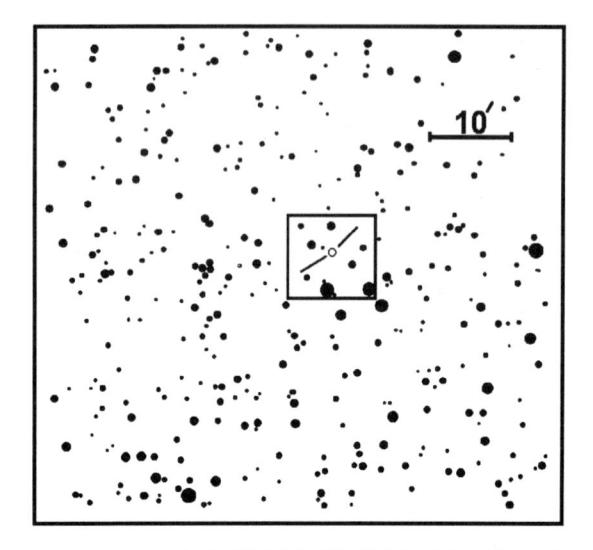

α (2000.0)=$20^h 28^m 38^s.5$
δ (2000.0)=$-52°18'46''.5$
Type: DN UG
B=13.5–19.7:
P=$0^d.084$:

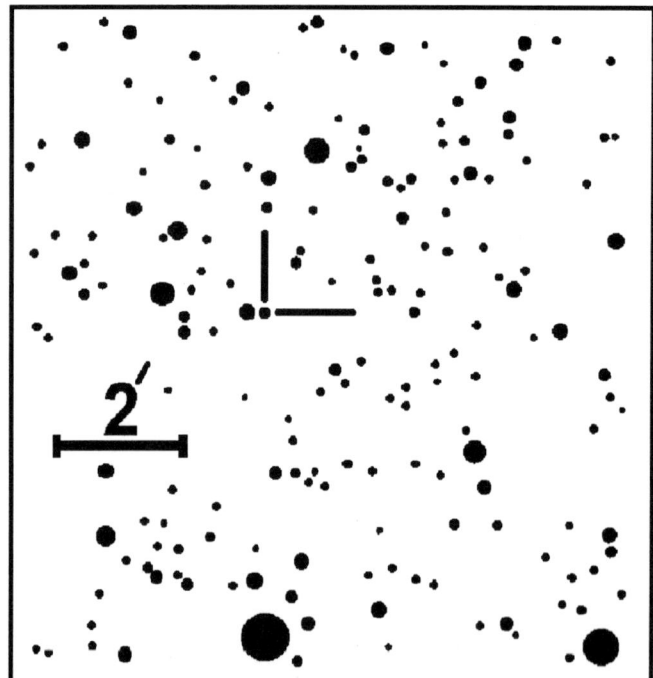

References: [176, 360, 364]

EK TrA

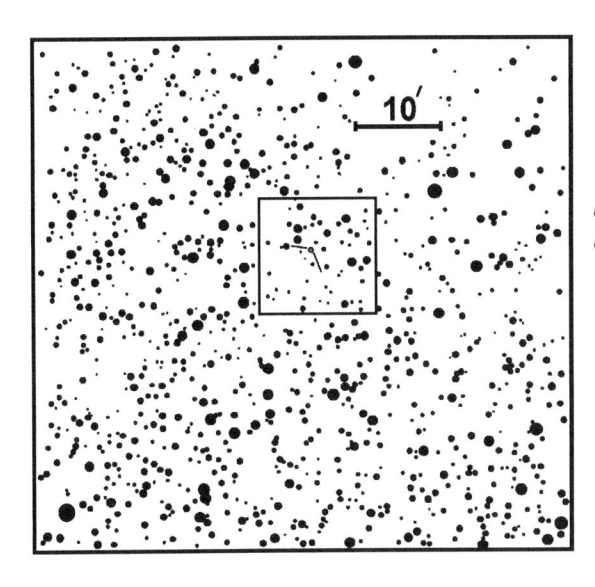

α (2000.0)=$15^h14^m01^s.9$
δ (2000.0)=$-65°05'42''$
Type: DN SU XS
V=10.4->17
P=$0^d.0635$
P_{cycle}=$50^d - 231^d$

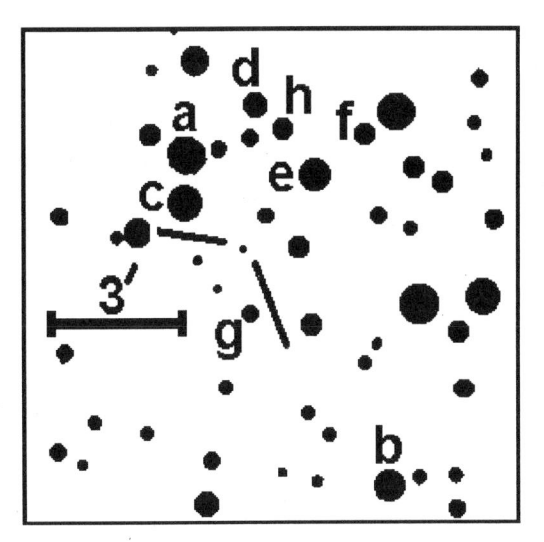

The comparison stars are:

Star	V	B
a	10.50	10.61
b	10.14	11.95
c	11.72	12.11
d	12.16	12.65
e	12.58	13.82
f	12.86	13.68
g	13.16	14.36
h	13.64	14.04

The nothern component of a close pair. The mean cycle between supermaxima is $442^d.1 - 551^d.5$. Superhumps ($\Delta V = 0^m.26$) with the period of $P_{hump} = 0^d.06492$ are observed.

References: [26, 48, 176, 364]

SU UMa

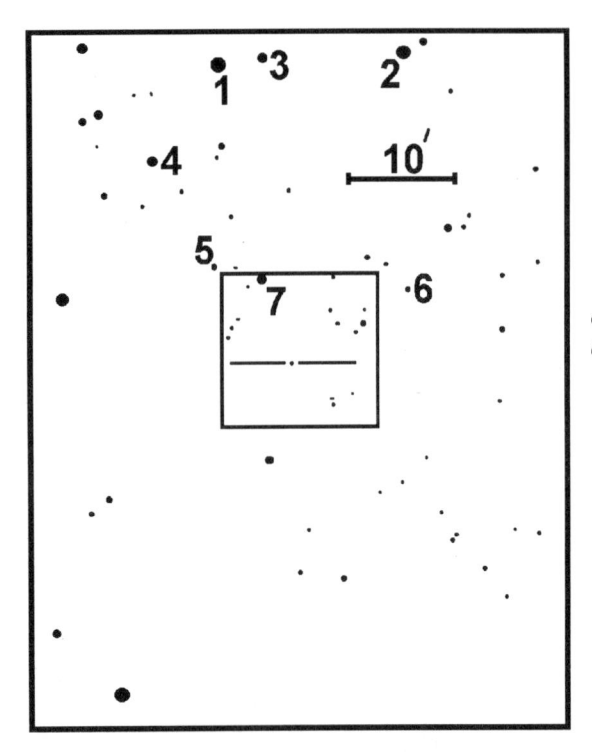

α (2000.0)=$08^h12^m28^s.9$
δ (2000.0)=$62°36'23''$
Type: UG SU XS
V=10.8–14.96
P=$0^d.076351$
$P_{cycle}=19^d$

The comparison stars are:
(Photographic magnitudes)

Star	Pg
1	8.48 ± 0.10
2	8.56 ± 0.13
3	9.38 ± 0.11
4	9.46 ± 0.09
5	10.57 ± 0.10
6	10.99 ± 0.11
7	11.48 ± 0.07
8	11.76 ± 0.12
9	12.31 ± 0.09
10	13.00 ± 0.05
11	13.03 ± 0.08

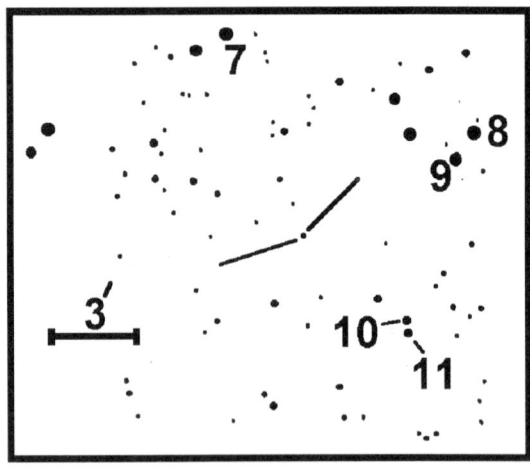

Three different mean values of the mean cycle of supermaxima change each other being active at different time intervals: $P_1 = 161^d.5$, $P_2 = 203^d.9$, $P_3 = 248^d.1$. The system is surrounded by a faint halo, 28″ diameter, of soft X-ray sources. Possible period in X-ray range: 33.93 s and 1.00434 s. $P_s = 0^d.07882$, A~ $0^m.2$.

References: [48, 119, 176]

BC UMa

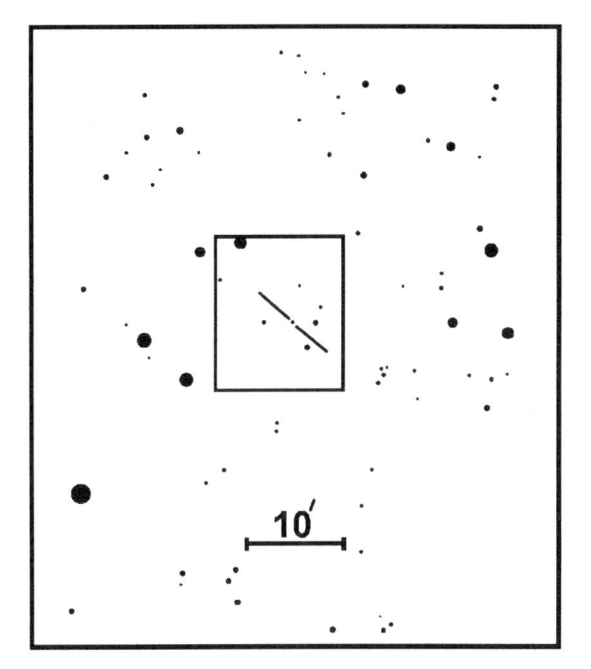

α (2000.0)=$11^h52^m15^s.6$
δ (2000.0)=$49°14'50''$
Type: DN UG
B=10.9–18.3
Spectrum: later than M5
$P=0^d.063$

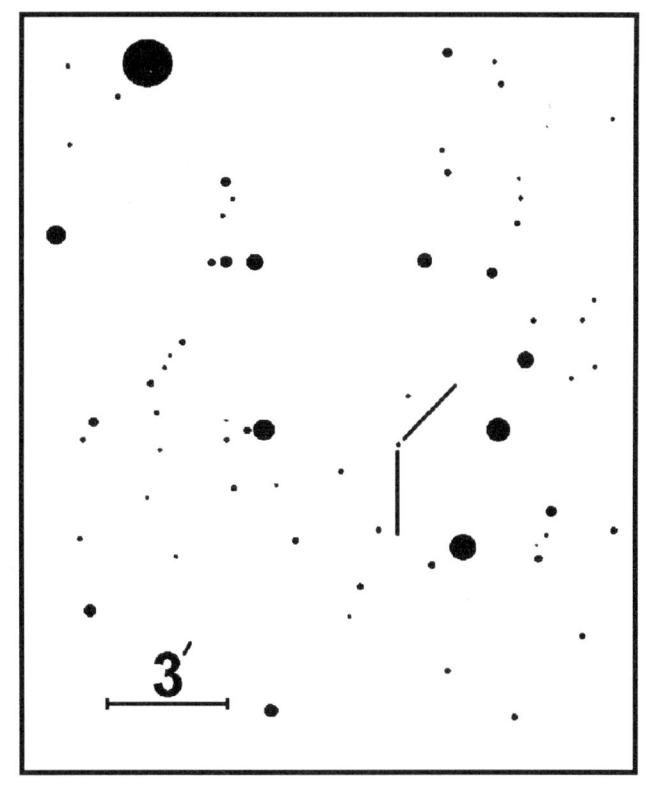

References: [21, 176]

BZ UMa

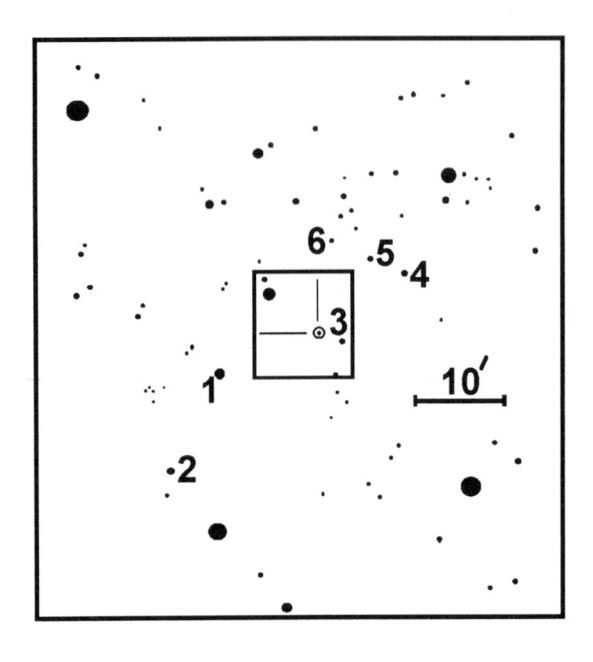

α (2000.0)=$08^h53^m44^s.6$
δ (2000.0)=$57°48'40''$
Type: DN
B=10.5–15.3
Spectrum: M5.5 V
$P=0^d.06798$
$P_{cycle}=97^d?$

The comparison stars are:
(Visual magnitudes)

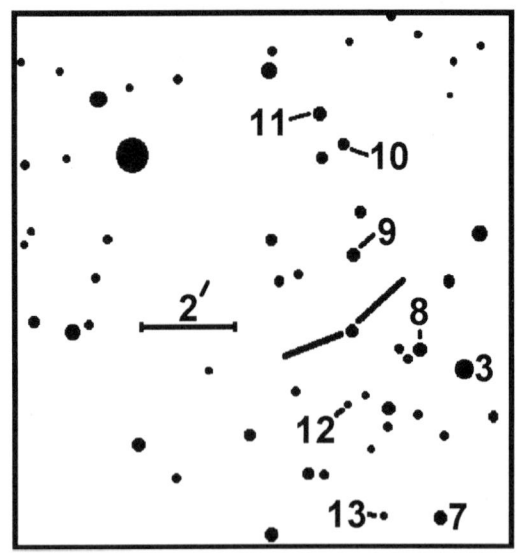

Star	Vis
1	10.8
2	11.3
3	11.7
4	11.9
5	12.1
6	12.3
7	12.6
8	13.7
9	15.4
10	15.4
11	15.8
12	16.5
13	17.3

Ellipsoidal variations with the amplitude $V \sim 0.12$ mag are observed.

References: [52, 109, 176, 206, 337]

CH UMa

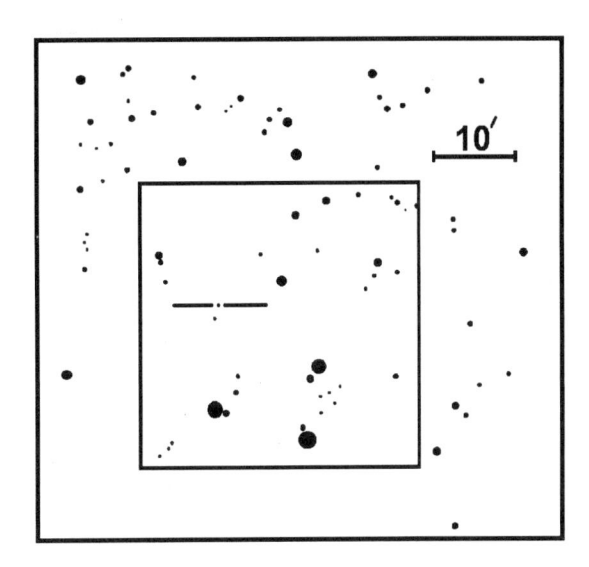

α (2000.0)=$10^h07^m00^s.73$
δ (2000.0)=$67°32'47''.5$
Type: DN UG
Pg=10.6–16.0
Spectrum: M0 V
$P=0^d.3448$
I.C.=JD 2446864.670
$P_{cycle}=204^d$

The comparison stars are:
(Photographic magnitudes)

Star	Pg
1	10.67
2	12.19
3	12.70
4	13.66
5	14.94
6	15.75

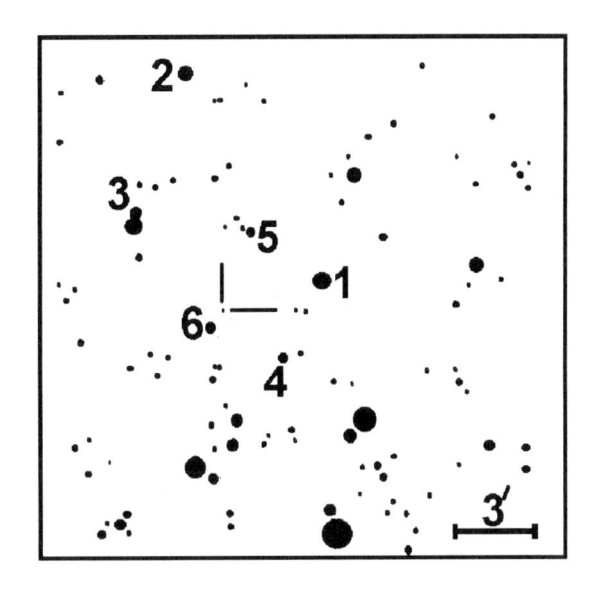

X-ray source 1E 1003+67, $L_x/L_{opt} \sim 0.04$ (2–10 keV).

References: [32, 48, 53, 126, 176, 341]

CI UMa

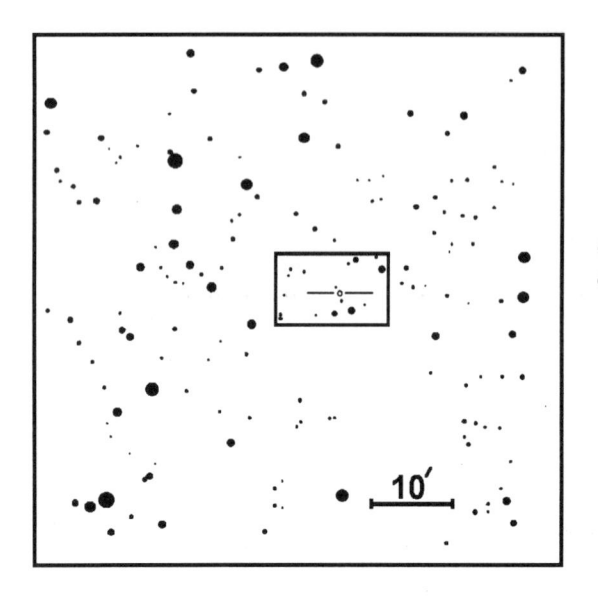

α (2000.0)=$10^h 18^m 12^s.5$
δ (2000.0)=$71°55'44''$
Type: DN UG
Pg=13.8–18.8
$P=0^d.0604$
$P_{cycle}=34^d$

The comparison stars are:
(Photographic magnitudes)

Star	Pg
1	12.65
2	14.15
3	14.46
4	15.36
5	15.91
6	16.49
7	16.95
8	var?
9	17.14
10	17.45

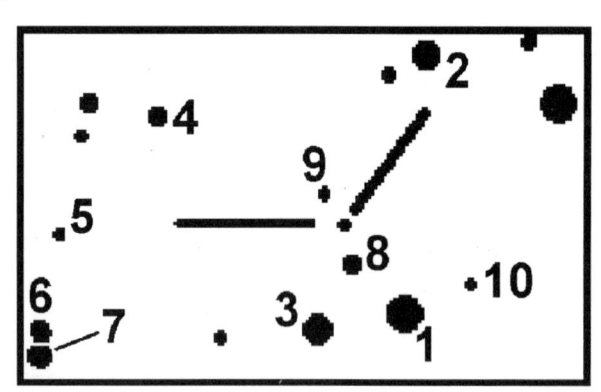

References: [128, 176]

CY UMa

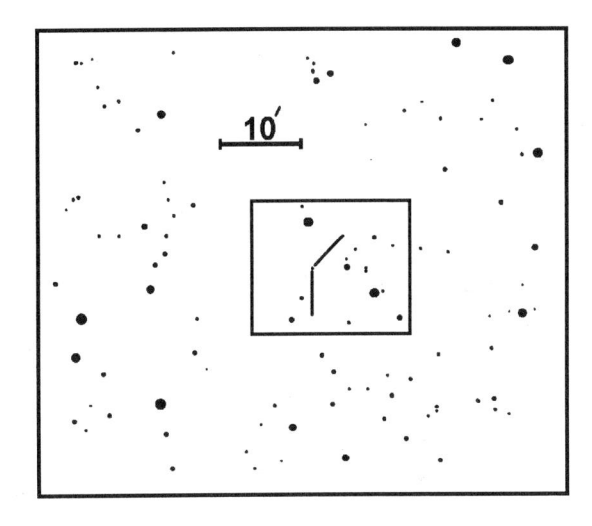

α (2000.0)=$10^h56^m58^s.6$
δ (2000.0)=$49°40'59''$
Type: UG SU
Pg=11.9–17
P=$0^d.0593$
P_{cycle}=$64^d - 165^d$

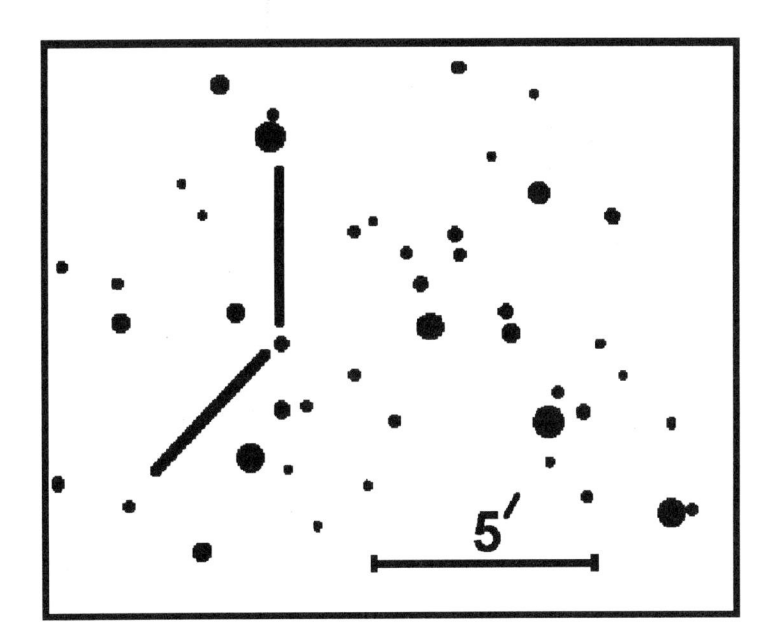

Superhump period is $P_{hump} \simeq 0^d.0593$.

References: [127, 173, 176]

DV UMa

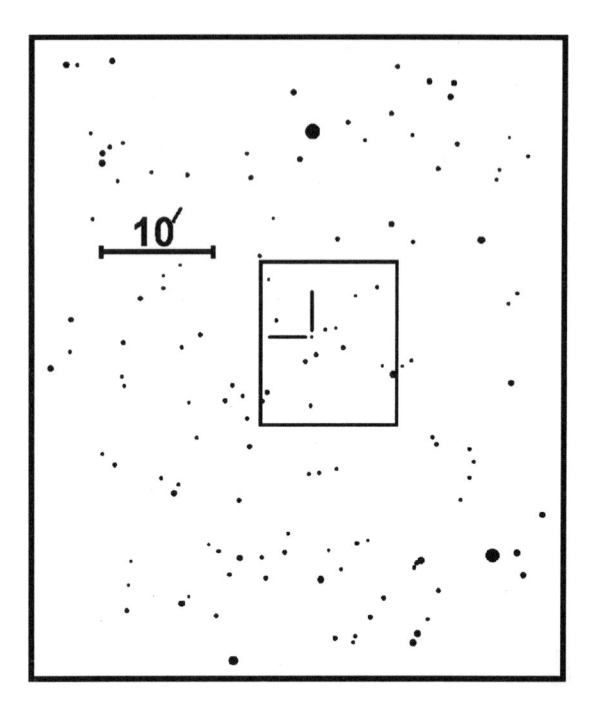

α (2000.0)=$09^h46^m36^s.9$
δ (2000.0)=$44°46'42''$
Type: UG SU
B=14.0-19.8
Spectrum: M4.5 V
$P=0^d.08597$
Min.phot.=JD 2446854.7451

US 943. *Eclipsing* binary.
The depth of eclipse is $1^m.5V$.

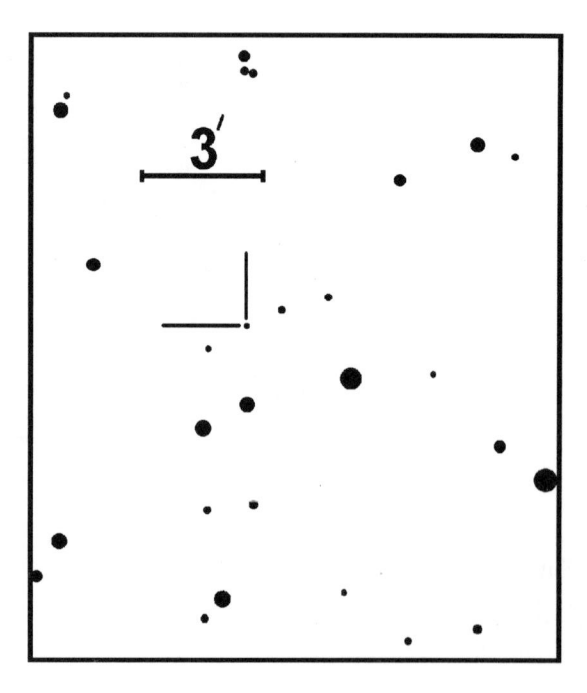

References: [159, 176, 348]

CU Vel

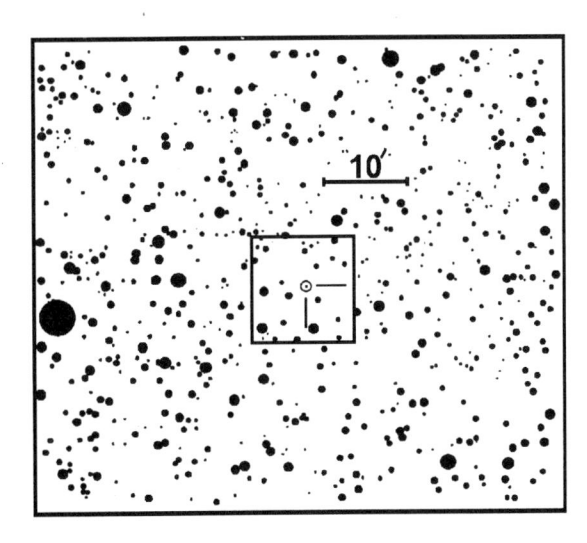

α (2000.0)=$08^h58^m32^s.96$
δ (2000.0)=$-41°47'51''.9$
Type: DN SU
V=10.0–15.5:
P=$0^d.0773$:
P_{cycle}=$164^d.7$

The comparison stars are:

Star	V	B	U
e	10.78	11.09	11.16
k	11.45	11.79	11.91
m	12.05	12.58	12.53
q	12.41	13.47	13.83
s	12.53	13.99	14.27
t	12.88	13.80	14.41
u	13.31	13.88	13.88
y	13.52	14.17	14.32
n	13.69	14.23	14.39
x	13.94	14.69	15.15
z	15.08	16.19	16.58

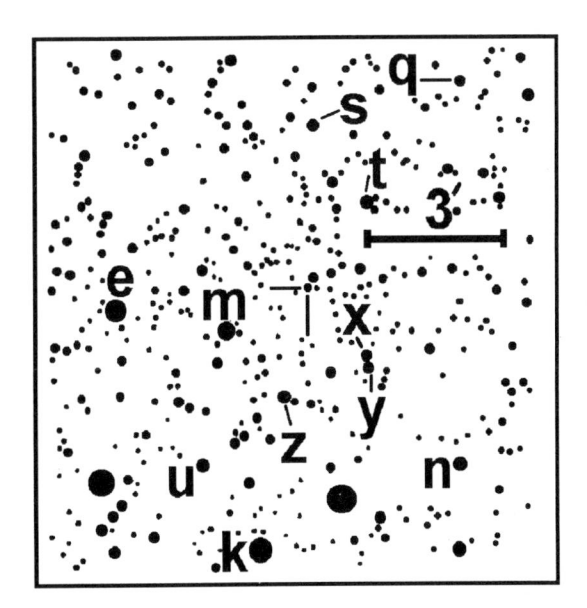

The period of superhumps is $P_{hump} = 0^d.0799$, $A \sim 0^m.4$.

References: [26, 48, 176, 204, 364]

TW Vir

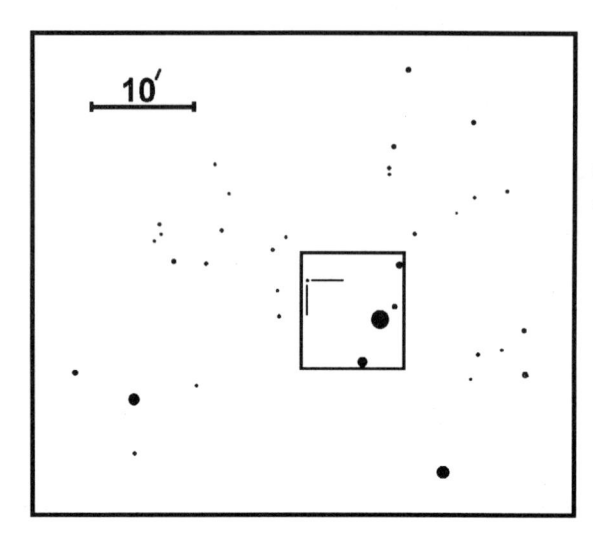

α (2000.0)=$11^h 45^m 20^s.4$
δ (2000.0)=$-04°26'04''$
Type: DN UG XS
V=11.2-16.4
Spectrum: M2-4 V
$P_{orb}=0^d.18267$
I.C.=JD 2445088.712
$P_{cycle}=32^d$

The comparison stars are:

Star	a	b
V	8.86	10.41
B	9.90	10.94
U	10.70	10.98

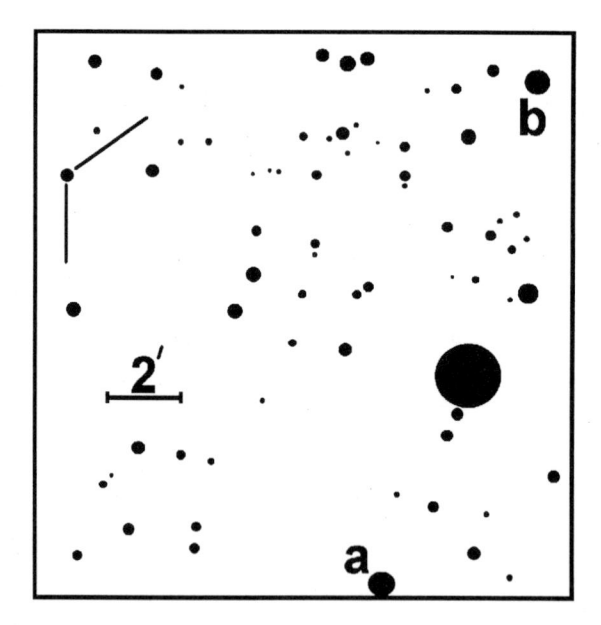

P_{orb} is the spectroscopic period, $K = 88\pm5$ km s^{-1}. Variability of ellipsoidal character in H light with $P = 0.5 P_{orb}$, $A_H \sim 0^m.13$ is observed. Superior conjunction for the broad Hα emission is $T = $ HJD 2445088.712(2). X-ray source.

References: [176, 307, 360, 364, 402]

SW Vul

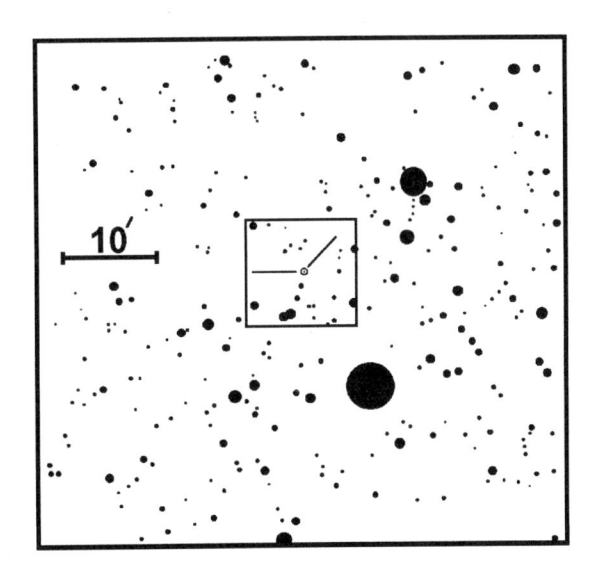

α (2000.0)=$20^h00^m05^s.14$
δ (2000.0)=$22°56'10''.7$
Type: NL?
Pg=15.0–18
P_{cycle}=15^d :

The comparison stars are:
(Photographic magnitudes)

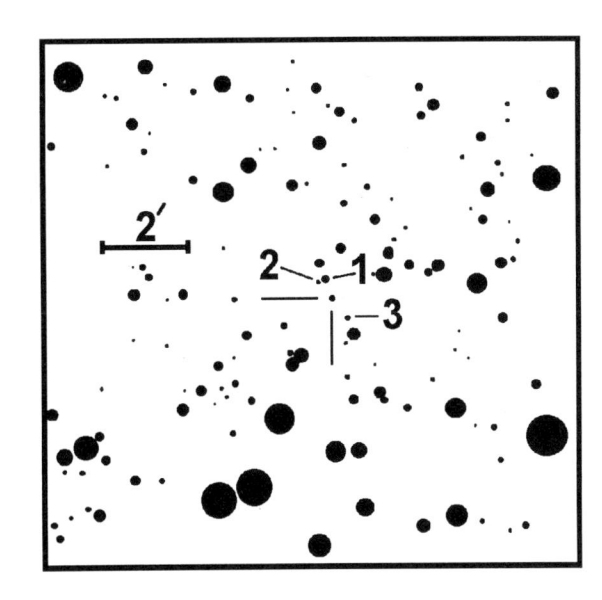

Star	Pg
1	15.40
2	16.05
3	16.37

After a brighter outburst (15^m) of 1923, the brightness of maxima dropped progressively: $15^m.6$ in 1947 and $16^m.0$ in 1961–91.

References: [52, 176, 349]

VW Vul

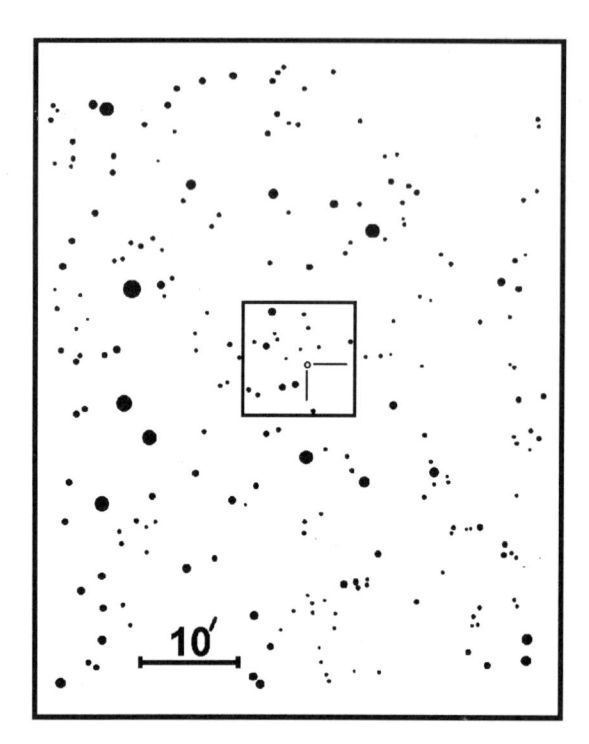

α (2000.0)=$20^h 59^m 56^s.8$
δ (2000.0)=$25°42'09''$
Type: DN UG?
B=13.1–16.3
$P_{orb}=0^d.0731$
I.C.=JD 2445163.767
$P_{cycle}=30^d$

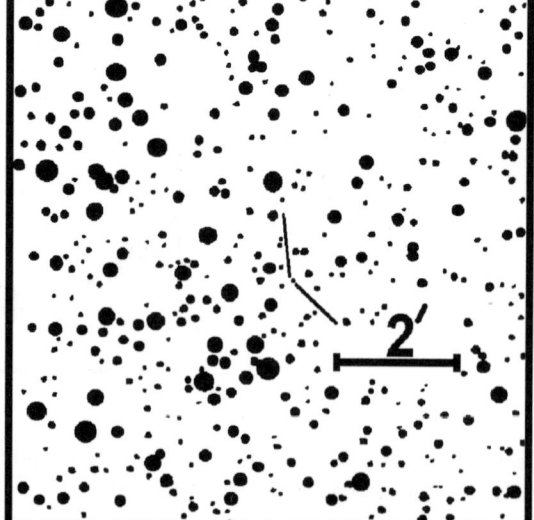

P_{orb} is the spectroscopic period, $K \sim 100$: km s^{-1}.

References: [176, 307, 308, 364]

PQ And/N And 1988

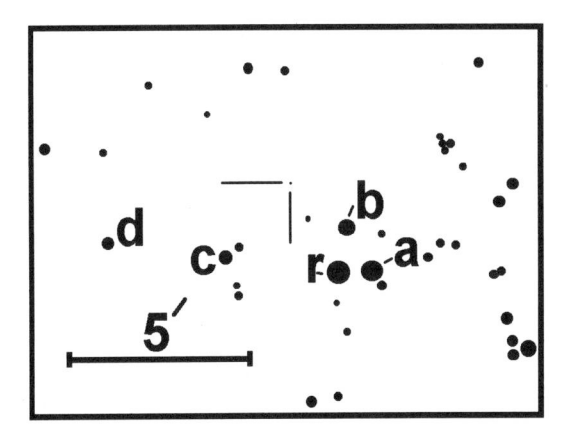

α (2000.0)=$02^h29^m29^s.5$
δ (2000.0)=$40°02'37''.8$
Type: RN
V=10.1–19
P_{cycle}~6000^d :

The comparison stars are:

Star	B
a	11.50
b	13.00
c	14.05
d	14.50
r	10.87

References: [102, 280]

PX And/PG 0027+260

α (2000.0)=$00^h30^m05^s.86$
δ (2000.0)=$26°17'25''.9$
Type: NL, VY?, SW
V=15.0–17.0
P=$0^d.146350$
Min.phot.=JD 2447783.798

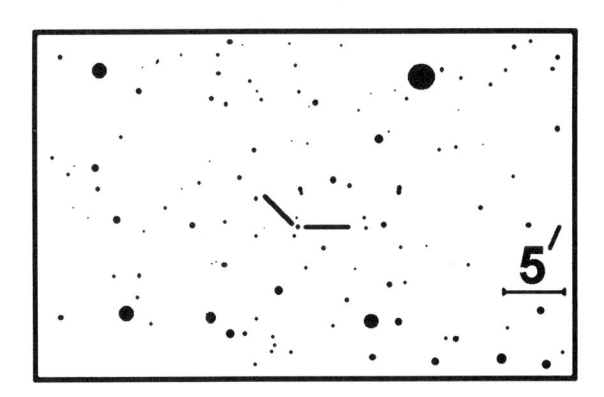

PX And is an *eclipsing* binary system, eclipse depth is $\Delta V \sim 0.6 - 1.3$, $D = 0^p.18$. The light of the system shows rapid variations $\sim 0^m.2$.

References: [102, 281, 318, 339]

HV And

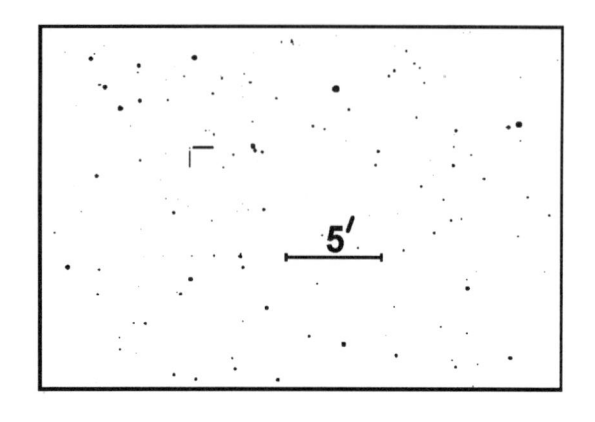

α (2000.0)=$00^h40^m55^s.41$
δ (2000.0)=$43°24'58''.4$
Type: NL
Pg=15.3–16.6
P=$0^d.055994$
Min.phot.=JD 2445991.506

References: [102, 318]

UU Aqr

α (2000.0)=$22^h09^m05^s.7$
δ (2000.0)=$-03°46'18''.5$
Type: NL?
Pg=9.6–16.8
P=$0^d.163579089$
Min.phot.=JD 2446347.2667

The comparison stars are:

Star	V	B	U
1	11.44	12.07	12.11
2	11.65	12.21	12.36

Eclipsing star.

References: [102, 305]

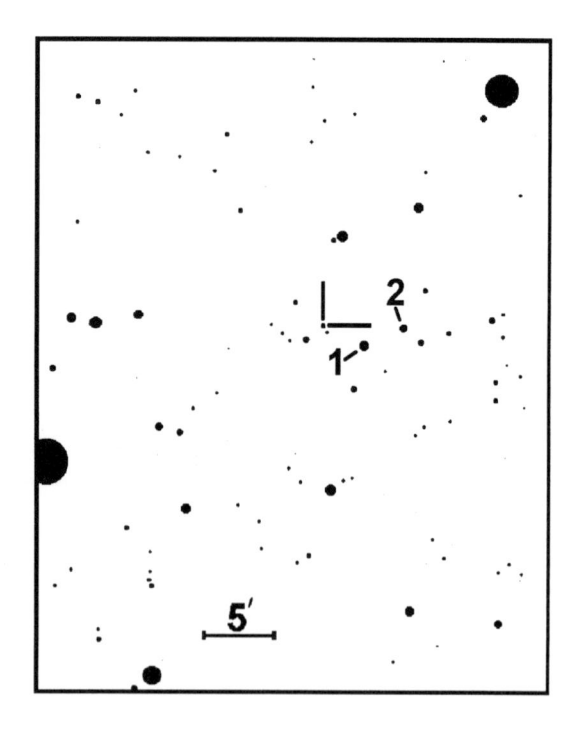

HL Aqr

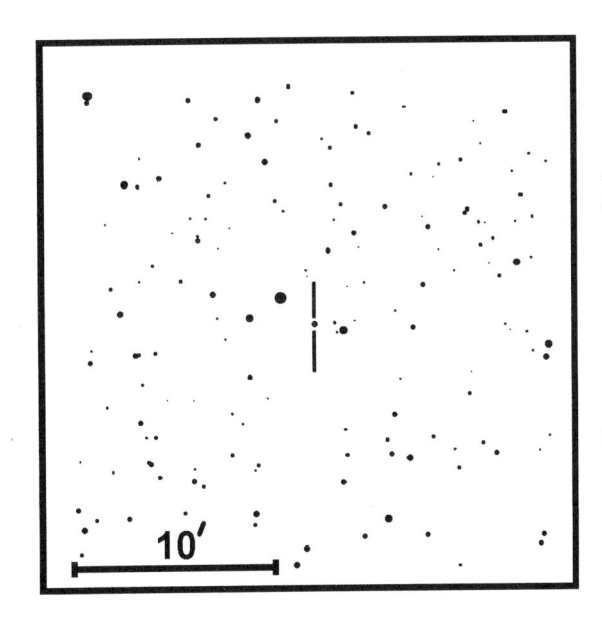

α (2000.0)=$22^h20^m27^s.03$
δ (2000.0)=$02°00'53''.2$
Type: NL UX
B=13.3–13.6
$P=0^d.1356$
$I.C.$=JD 2446309.484

References: [102]

V 794 Aql

α (2000.0)=$20^h17^m34^s.9$
δ (2000.0)=$-03°39'50''.5$
Type: NL VY XS
B=13.7–20.2
$P=0^d.23$:

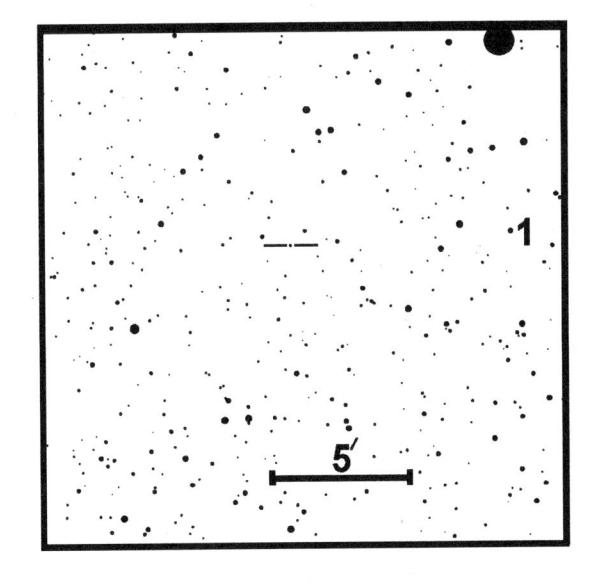

The comparison star *1* has

V= 13.10
B= 13.25
U= 13.26

$F_x = 4.5 \times 10^{-12}$ erg cm^{-2}s^{-1}
in the range 0.15–4.5 keV

References: [102, 305, 318]

VY Aqr

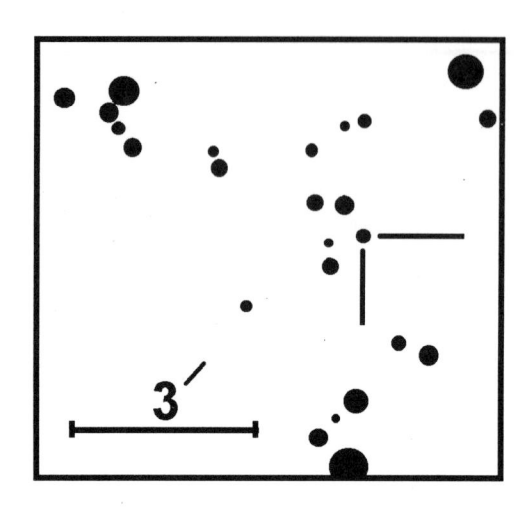

α (2000.0)=$21^h 12^m 09^s.25$
δ (2000.0)=$-08°49'37''.3$
   Type: DN, SU?, RN
   Pg=8.0–17.5
   $P=0^d.0632(2)$

Outbursts occured in 1907, 1962.
The period of superhumps is
$P_{hump} = 0^d.06445$.

References: [183]

TV Crv

α (2000.0)=$12^h 20^m 24^s.7$
δ (2000.0)=$-18°27'01''$
   Type: DN, SU?
   V=12–19
   $P=0^d.08:$

At the finding chart some comparison stars used for differential photometry are marked.

References: [102, 191]

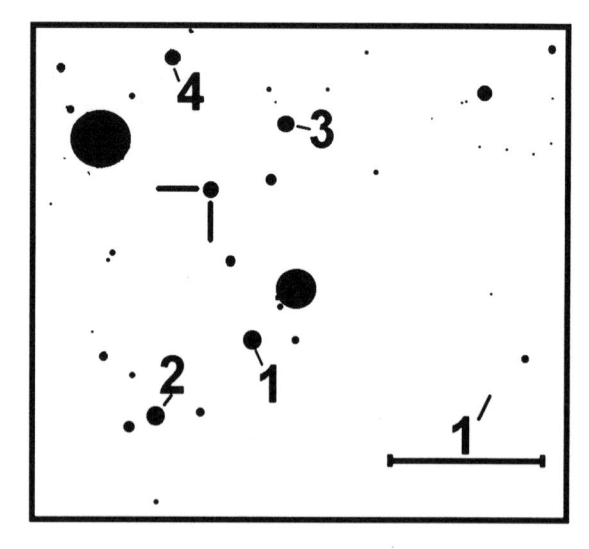

V 1315 Aql

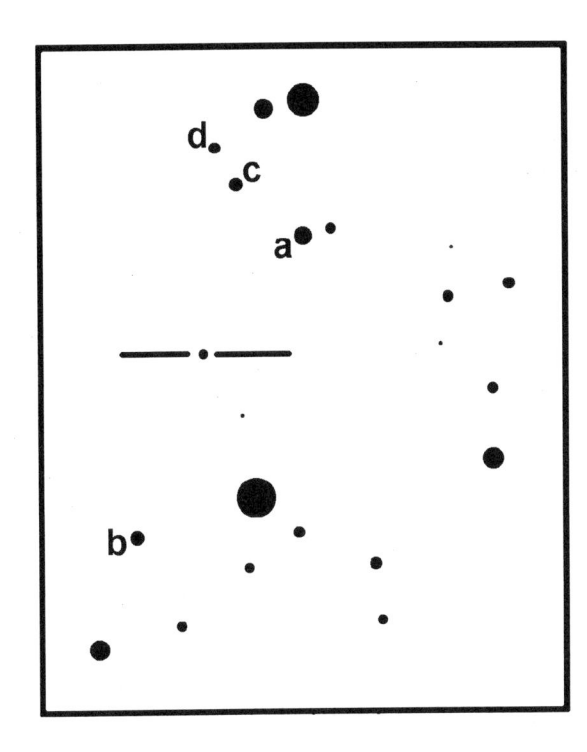

α (2000.0)=$19^h13^m54^s.59$
δ (2000.0)=$12°18'02''.3$
  Type: NL UX SW
  Pg=14.3–19
  $P=0^d.13968994$
  Min.phot.=JD 2445902.70088

Comparison stars are:

Star	Pg
a	13.28
b	14.47
c	15.28
d	16.26

The depth of the eclipse is $1^m.9V$, duration 50 min.

References: [102]

WX Ari

α (2000.0)=$02^h47^m36^s.30$
δ (2000.0)=$10°35'37''.7$
  Type: NL UX SW
  V=14.8–15.4
  $P=0^d.13934$
  I.C.=JD 2447840.7683

References: [102]

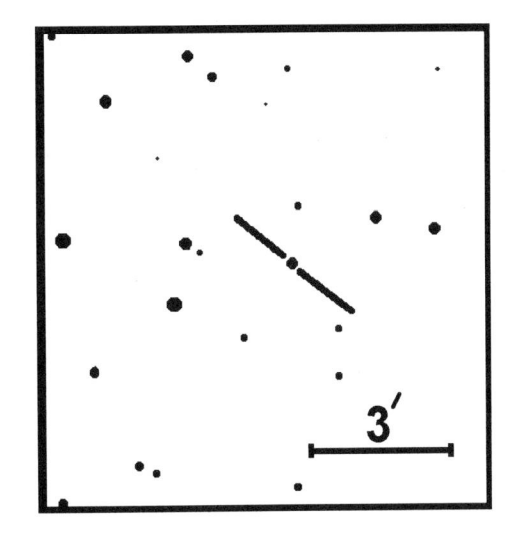

T Aur

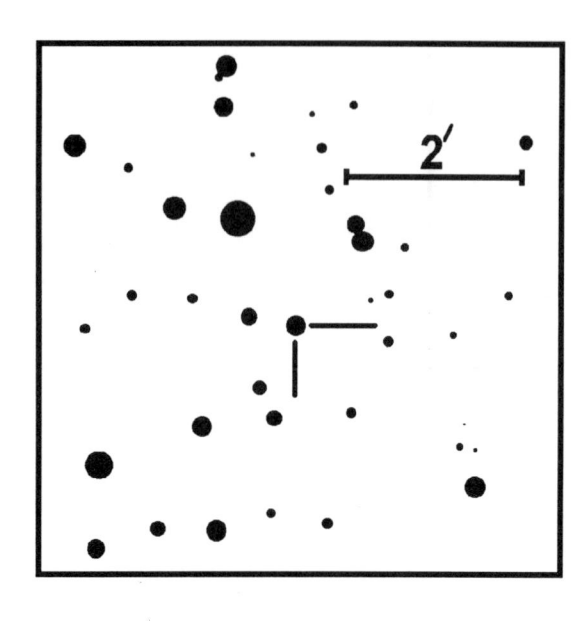

α (2000.0)=$05^h31^m59^s.3$
δ (2000.0)=$30°26'44''.6$
Type: NB
B=4.1–15.5
P=$0^d.20437851$
Min.phot.=JD 2437614.0088

There are brightness fluctuations between the eclipses. Depths of minima are $0^m.10$–$0^m.28$.

References: [107]

QZ Aur

α (2000.0)=$05^h28^m34^s.15$
δ (2000.0)=$33°18'21''.4$
Type: NA
Pg=6:–18:
P=$0^d.3575$

References: [107, 291]

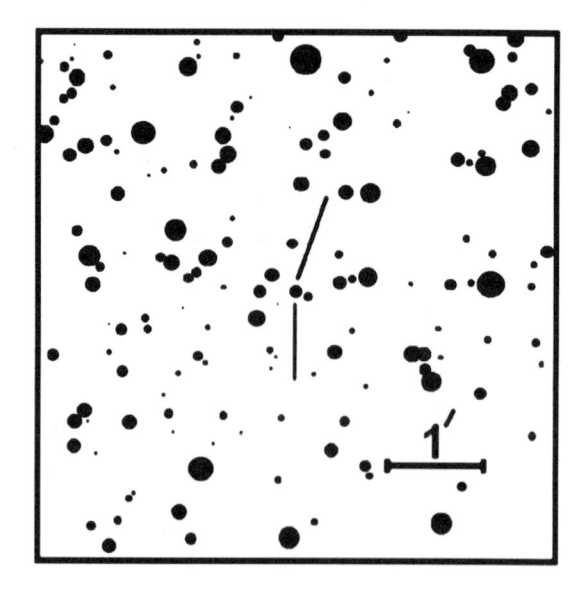

V 363 Aur

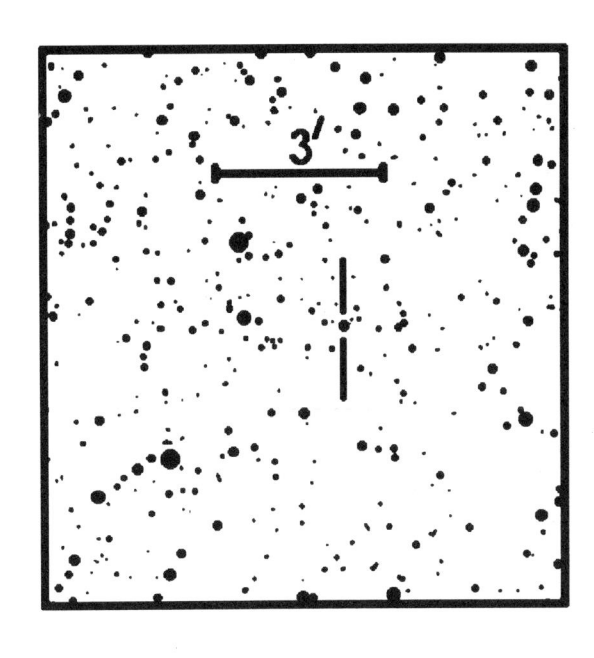

α (2000.0) = $05^h 33^m 33^s.51$
δ (2000.0) = $36°59'31''.7$
Type: NL UX
Spectrum: K0 V
B=14.4–15.5
P=$0^d.3122425$
Min.phot.=JD 2444557.9495

References: [102]

AC Cnc

α (2000.0) = $08^h 44^m 27^s.17$
δ (2000.0) = $12°52'32''.2$
Type: NL UX
Spectrum: G8-K2 V
V=13.8–15.4
P=$0^d.30047768$
Min.phot.=JD 2444290.309

Eclipsing star. A probable cycle of 11 days was found. The brightness outside eclipses varies in the range 13.8–15.2B. The constant stars *a, c, g* were used for differential photometry.

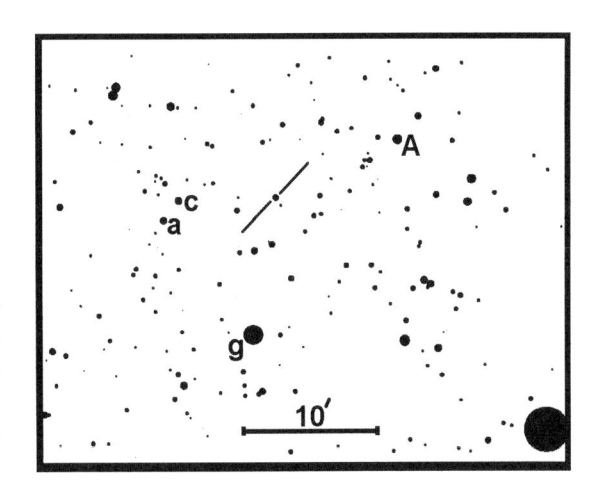

The comparison star *A* has

V	B–V	U–B
11.31	+0.33	+0.09

References: [102]

V 425 Cas

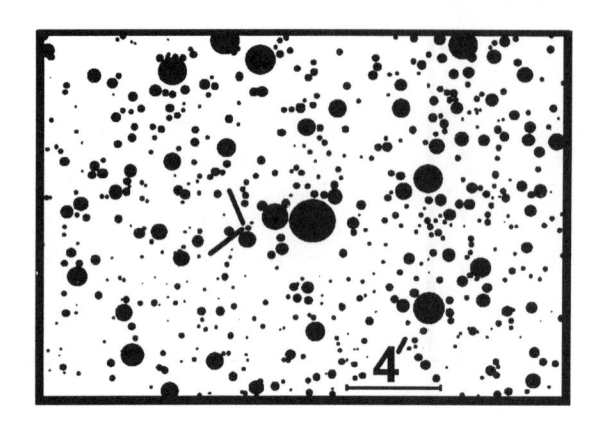

α (2000.0)=$23^h03^m46^s.61$
δ (2000.0)=$53°17'15''.1$
Type: NL VY
B=14.0–18
P=$0^d.14964$

References: [102, 318]

V 394 CrA

α (2000.0)=$18^h00^m26^s.0$
δ (2000.0)=$-39°00'35''.0$
Type: NR
Pg=7.2–20.2
P=$0^d.7577$:

Outbursts were observed in 1987 ($T = 5^d.5$) and 1949 ($T = 10^d$).

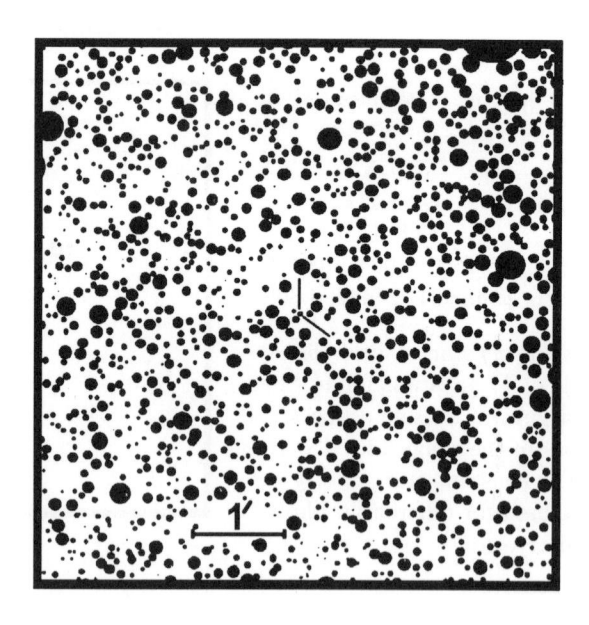

References: [107]

V 751 Cyg

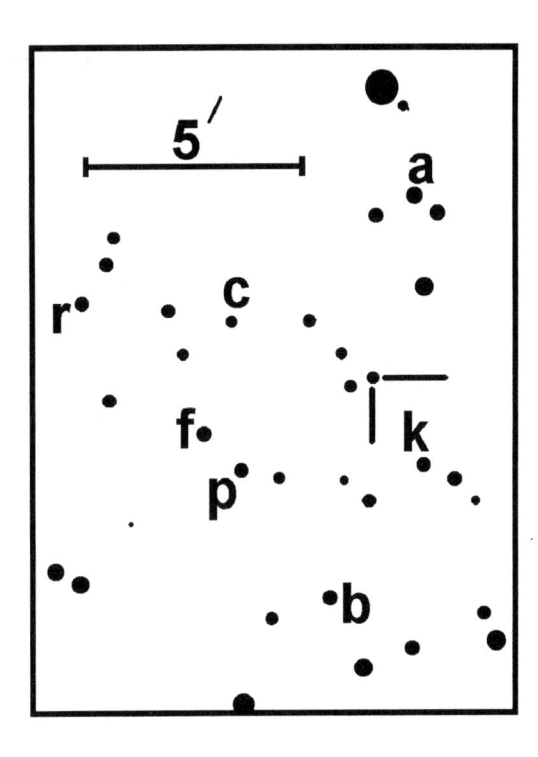

α (2000.0)=$20^h 52^m 12^s.76$
δ (2000.0)=$44°19'25''.8$
  Type: NL, VY
    V=13.2–16:
    P=$0^d.25$:

The comparison stars are:

Star	Pg	Vis
a	13.44	12.98
b	14.00	13.37
c	16.60	14.9:
f	14.50	13.70
k	14.82	13.44
p	15.32	14.22
r	15.74	14.51

In the region of the nebula IC 5070.

References: [102, 208]

HX Peg/PG 2337+123

α (2000.0)=$23^h 40^m 24^s.0$
δ (2000.0)=$12°37'39''$
  Type: DN
    Vis=13.04–16.63
    P=$0^d.2009$

Oscillations by $0^m.1$ within minutes.

References: [102, 133]

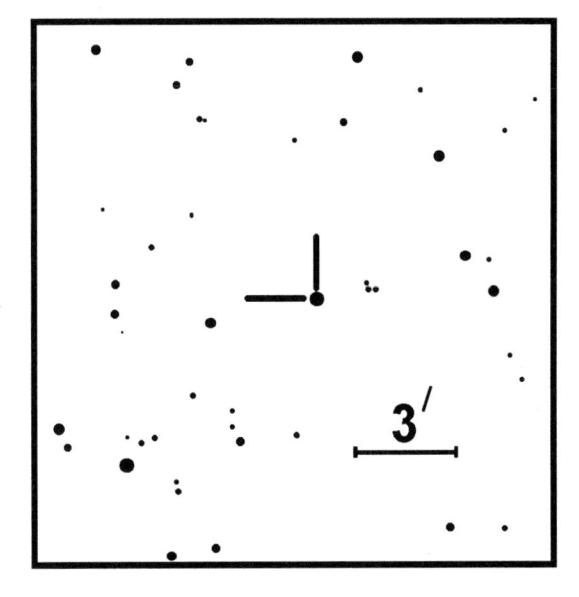

V 1668 Cyg

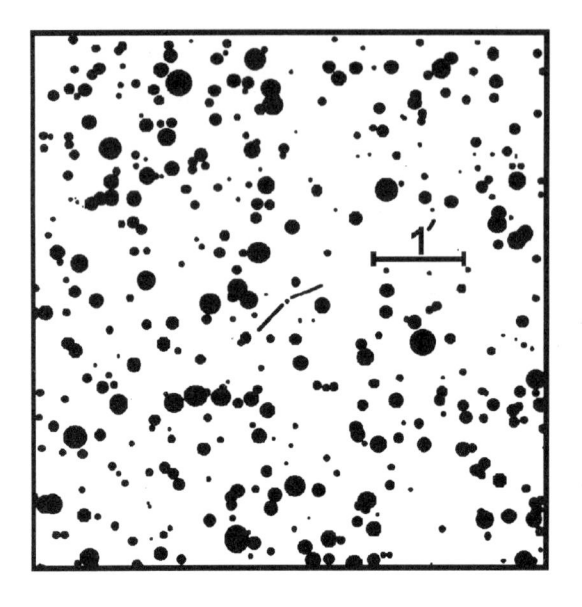

α (2000.0)=$21^h 42^m 35^s.37$
δ (2000.0)=$44°01'54''.8$
Type: NA
V=6.0–21:
P=$0^d.1384$
Min.phot.=JD 2447679.848

Sine-shaped variations of brightness with $\Delta m = 0^m.15$ and $P = 0^d.4392$ were noted on the descending branch after JD 2443765. On 29 October 1978 rapid variability with $A \sim 0.4$ mag in 1^h was observed.

References: [107]

V 1776 Cyg

α (2000.0)=$20^h 23^m 30^s.62$
δ (2000.0)=$46°31'29''.3$
Type: NL UX SW
V=16.24–17.16
P=$0^d.1647386$
Min.phot.=JD 2447048.7925

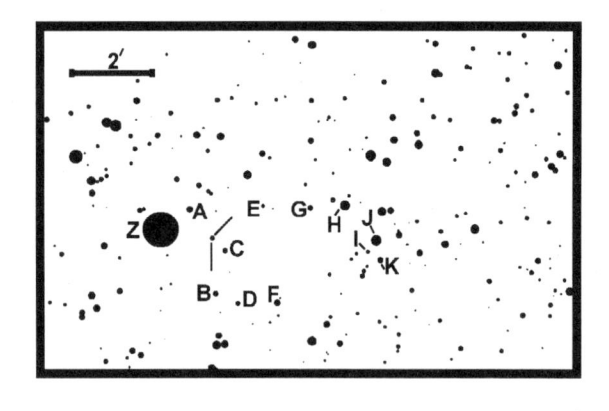

The comparison stars are:

Star	V	B–V	Star	V	B–V	Star	V	B–V
A	15.00(3)	0.70(6)	E	16.01(5)	0.84(9)	I	15.50(5)	1.7(2)
B	15.02(3)	1.10(6)	F	14.16(3)	0.86(5)	J	12.76(3)	0.78(7)
C	16.15(3)	1.08(7)	G	15.41(3)	0.84(5)	K	15.15(3)	0.84(9)
D	15.22(3)	0.99(5)	H	12.16(3)	0.56(4)	Z	16.42(4)	...

References: [102, 120]

V 1974 Cyg

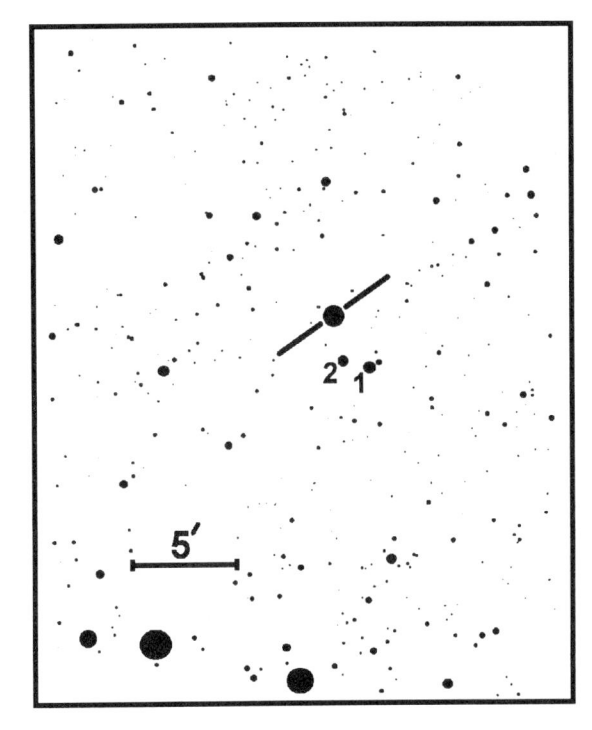

α (2000.0)=$20^h30^m31^s.7$
δ (2000.0)=$52°37'52''.9$
  Type: NA
  Pg=4.4–<16
  P=$0^d.081$

Oscillations by $0^m.01$ in 10 min were observed. Strong neon emissions. $T = 42^d$ (in V filter). A possible period is $0^d.814$. The constant stars 1, 2 were used for differential photometry.

References: [102]

HR Del

α (2000.0)=$20^h42^m20^s.36$
δ (2000.0)=$19°09'39''.8$
  Type: NB XS
  V=3.3–12

The system was noted as a radiosource. The constant star a was used as a comparison star.

References: [318]

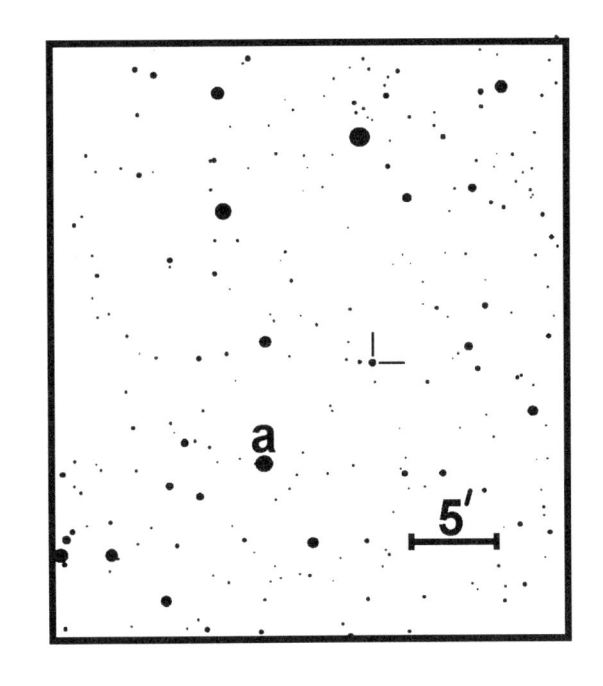

ES Dra

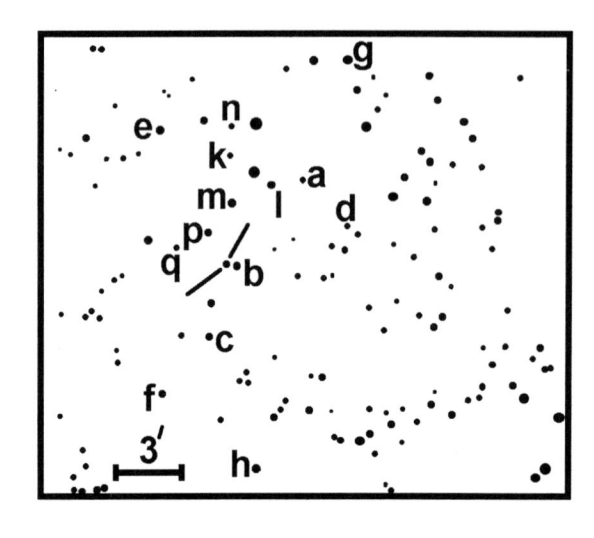

α (2000.0)=$15^h25^m31^s.97$
δ (2000.0)=$62°00'59''.1$
Type: DN?
Pg=13.9–16.3
P=$0^d.1766$

The comparison stars are:

Star	Pg	Star	Pg
a	13.86(7)	h	15.3(1)
b	13.95(5)	k	15.36(9)
c	14.22(7)	l	15.7(1)
d	14.45(8)	m	16.0(1)
e	14.9(1)	n	16.0(1)
f	15.0(1)	p	16.5(1)
g	15.1(1)	q	16.48(9)

References: [8, 102]

RZ Gru

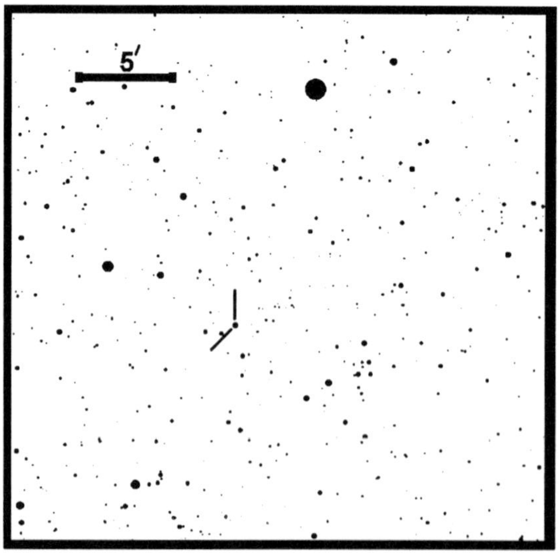

α (2000.0)=$22^h47^m12^s.15$
δ (2000.0)=$-42°44'39''.4$
Type: NL UX
Pg=11.5–13
P=$0^d.4170$:
P_{cycle}~103^d :

References: [102]

CF Gru

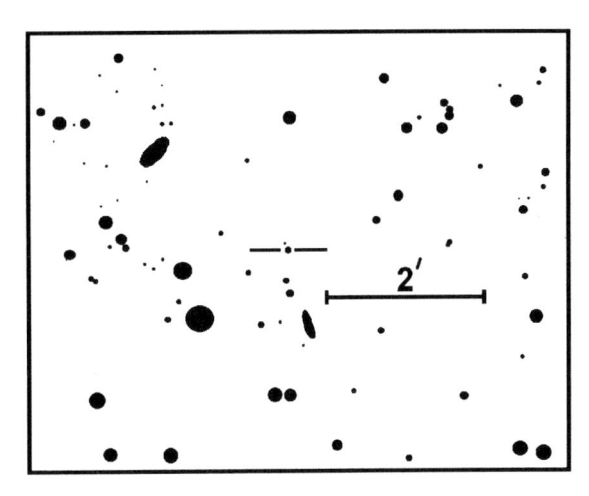

α (2000.0)=$21^h41^m23^s.15$
δ (2000.0)=$-45°04'32''.0$
Type: DN
V=19.9–20.2
P=$0^d.064$:

References: [102, 144]

V 825 Her

α (2000.0)=$17^h18^m37^s.06$
δ (2000.0)=$41°15'50''.2$
Type: NL
V=14.0–14.3
P=$0^d.206$

The comparison stars are:

Star	V	B
b	12.95	13.58
c	13.16	13.81
d	13.22	14.00
e	13.43	12.33
f	14.22	15.20
p	11.35	12.10
M	10.85	11.35

$0^m.1 - 0^m.2$ flickering was observed.

References: [102, 318]

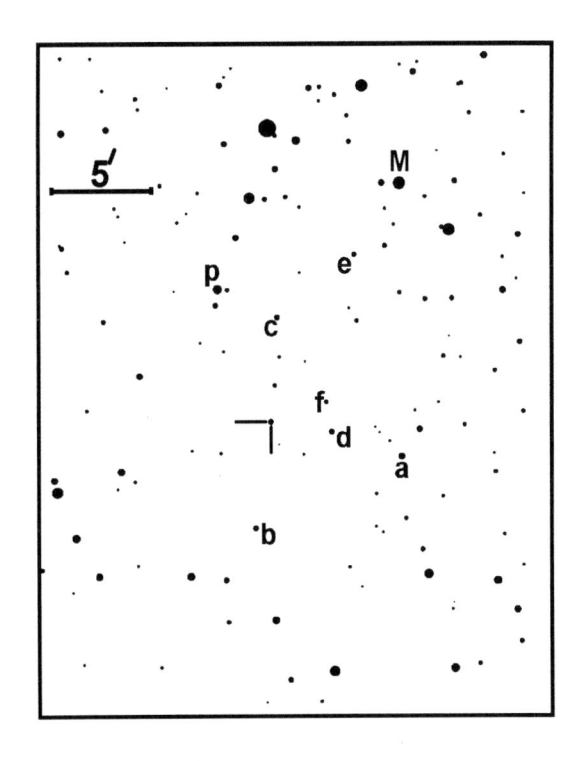

LY Hya

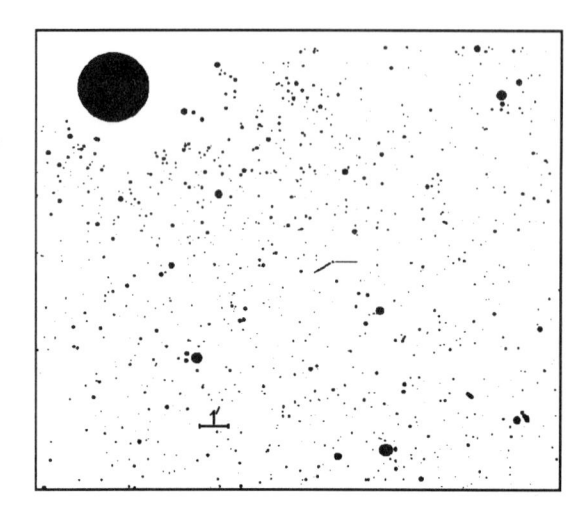

α (2000.0)=$13^h31^m53^s.93$
δ (2000.0)=$-29°40'59''.9$
Type: DN
V=17.4–18.4
P=$0^d.0748$

Possible period $0^d.159$.

References: [102, 318]

DI Lac

α (2000.0)=$22^h35^m48^s.52$
δ (2000.0)=$52°42'59''.1$
Type: NA
B=4.3–15.1
P=$0^d.543773$

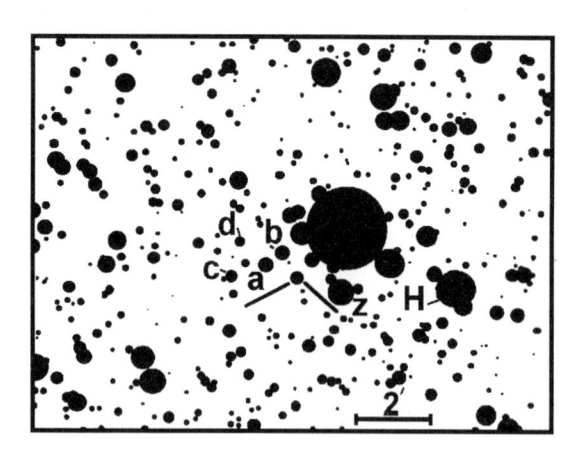

At the finding chart, some comparison stars used for differential photometry are marked.

References: [102, 318]

RZ Leo

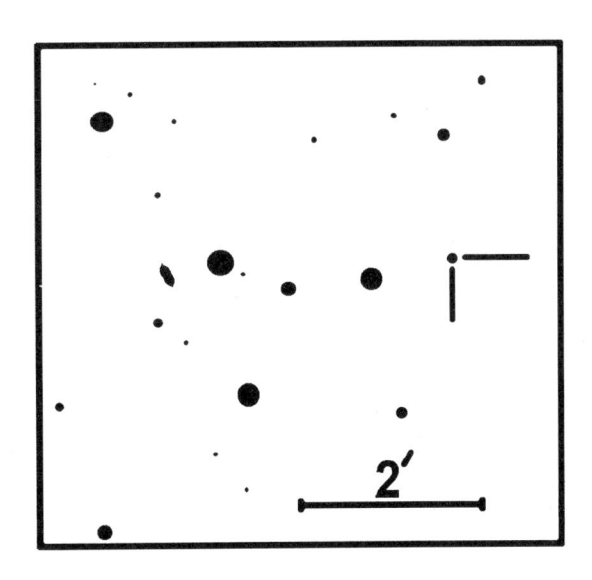

α (2000.0)=$11^h37^m22^s.9$
δ (2000.0)=$01°48'57''$
Type: DN SU?
Pg=11.5–19
$P=0^d.073$

References: [102]

DO Leo

α (2000.0)=$10^h40^m51^s.3$
δ (2000.0)=$15°11'35''$
Type: NL
B=16.0–17.0
Min.phot.=JD 2447225.75578

Eclipsing binary, the depth of minimum is $1^m.5$.

References: [102, 132]

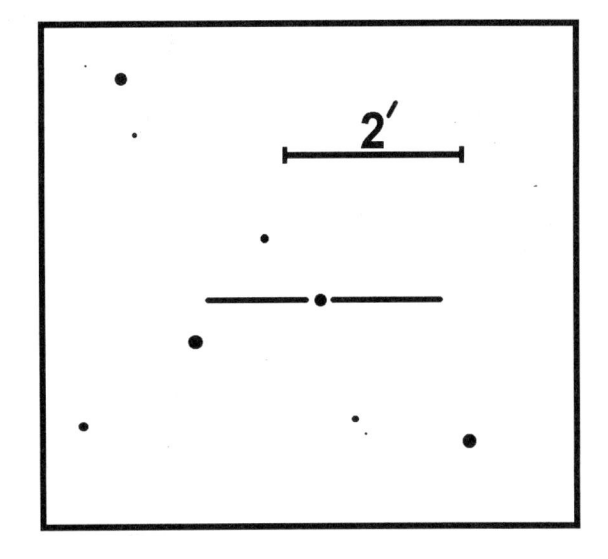

BH Lyn

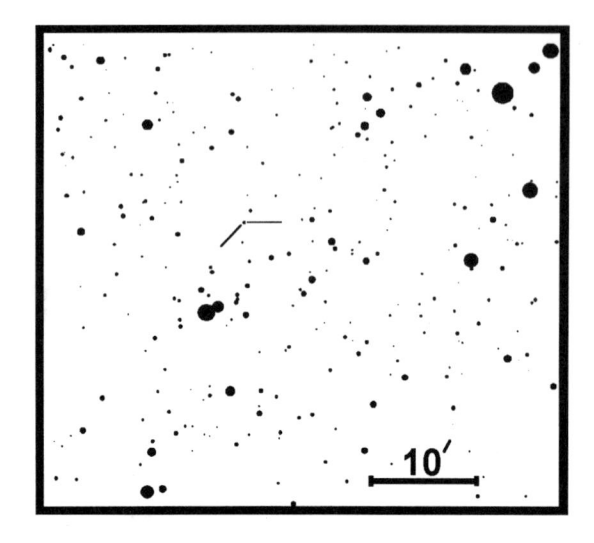

α (2000.0)=$08^h 22^m 35^s.7$
δ (2000.0)=$51°05'23''$
Type: NL VY SW
$Pg(B)$=13.7–16.3
$P=0^d.1558749$
Min.phot.=JD 2447180.3364

Eclipsing binary. The depth of the minimum is $1^m.3$, the duration of the eclipse is 42 min.

References: [102]

BK Lyn

α (2000.0)=$09^h 20^m 11^s.27$
δ (2000.0)=$33°56'41''.2$
Type: NL SU?
B=14.1–14.9
$P=0^d.078567$
Max.phot.=JD 2448956.899

In 1992 $P = 0^d.078533$.

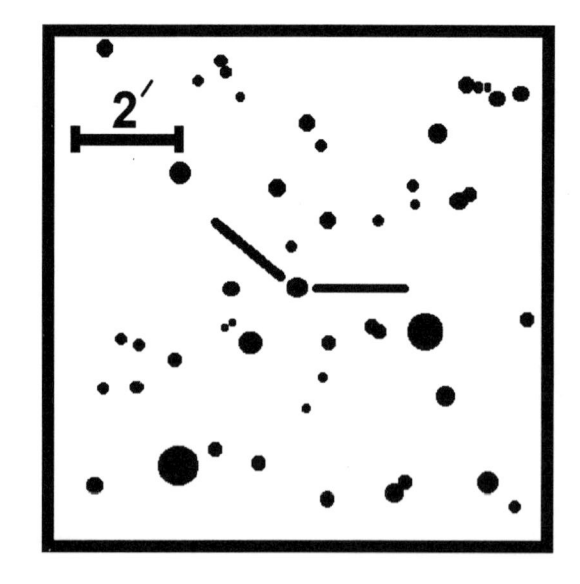

References: [102]

MV Lyr

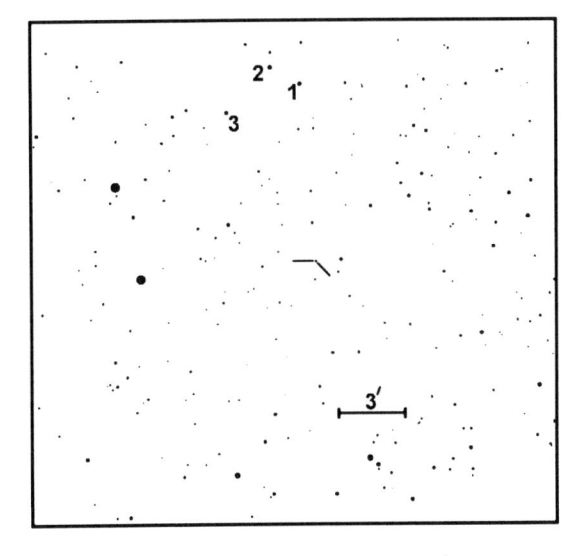

α (2000.0) =$19^h07^m17^s$.1
δ (2000.0) =$44°01'10''$.9
Type: NL VY XS
Pg=12.05–19
Spectrum: M5 V
$P=0^d$.1379
I.C.=JD 2444453.755

The comparison stars are:

Star	V	B	U
1	11.09	11.50	11.44
2	10.96	12.00	12.81
3	11.43	12.03	12.16

In 1979–1989 the "off" stage was observed. Rapid variability with $A \sim 0^m.4$ in 5 minutes. The $3^d.8$ period is a beat period betwen the photometric period ($0^d.1379$) and the spectroscopic one ($0^d.1336$) with $A \sim 0^m.15$. Soft X-ray source.

References: [102, 289]

V 345 Pav

α (2000.0) =$19^h35^m42^s$.8
δ (2000.0) =$-59°08'22''$
Type: NL UX
V=13.4–14.0
Spectrum: later M2 V
$P=0^d$.198096
Min.phot.=JD 2447769.2801

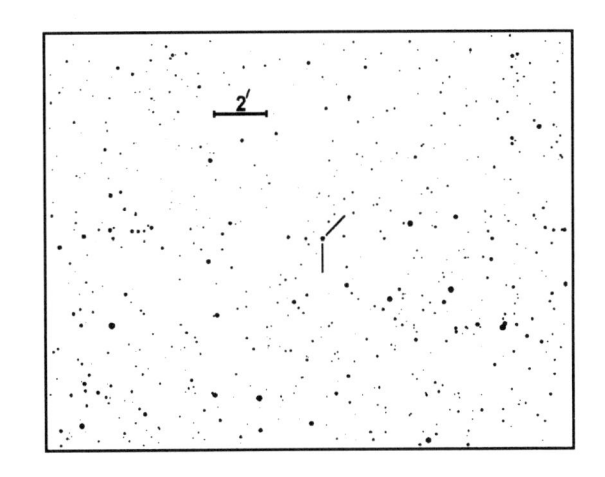

Eclipsing binary. The duration of eclipses is $0^d.18$.

References: [54]

V 1193 Ori

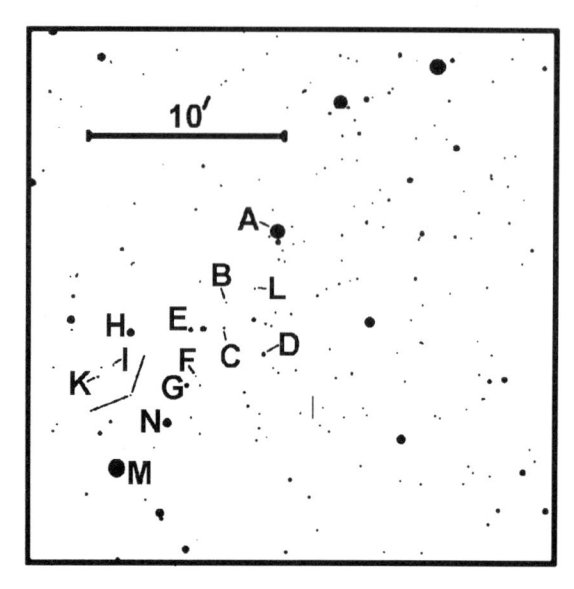

α (2000.0)=$05^h16^m26^s.3$
δ (2000.0)=$-00°12'13''.6$
Type: NL UX
V=13.95–14.28
$P=0^d.165$:

Rapid brightness oscillation was observed.

References: [102]

The comparison stars are:

Star	V	Star	V	Star	V	Star	V
A	8.93(1)	E	12.38(1)	H	11.07(1)	L	14.95(1)
B	14.68(1)	F	14.83(1)	I	13.84(1)	M	8.46(1)
C	13.91(1)	G	11.89(1)	K	13.92(1)	N	10.38(1)
D	12.32(1)						

V Per

α (2000.0)=$02^h01^m53^s.82$
δ (2000.0)=$56°44'03''.9$
Type: NL
V=4–18.5
$P=0^d.10712$
Min.phot.=JD 2447445.9322

Eclipsing binary. The duration of eclipses is 11 minutes.

References: [107]

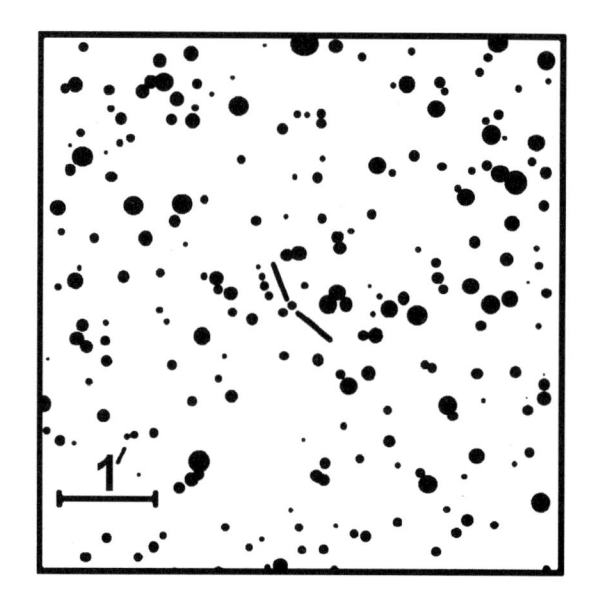

RS Oph

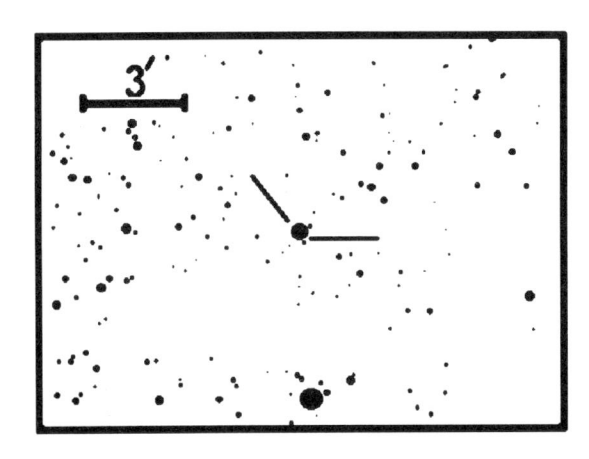

α (2000.0) = $17^h 50^m 13^s.0$
δ (2000.0) = $-06°42'28''.2$
  Type: N
Spectrum: M0–2 III
  V=4.3–12.5
  $P=230^d$

Outbursts in 1989, 1933, 1958, 1967, 1985 were observed.

References: [102, 402]

V 380 Oph

α (2000.0) = $17^h 52^m 40^s.3$
δ (2000.0) = $06°04'51''$
  Type: NL VY?
  Pg=14.2–17.2
  $P=0^d.16$
  $I.C.$=JD 2445464.928

In 1976 the possible orbital period was $0^d.15003$, 14.6–14.8Pg. "Off" stage: 14–15Pg, "On" stage: (JD 2444000–44150): 16.0–17.2Pg.
The constant star a was used as a comparison star.

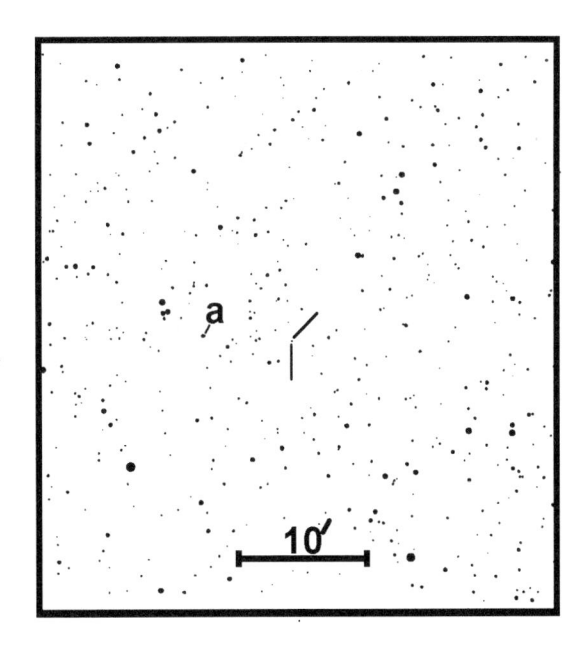

References: [102]

V 442 Oph

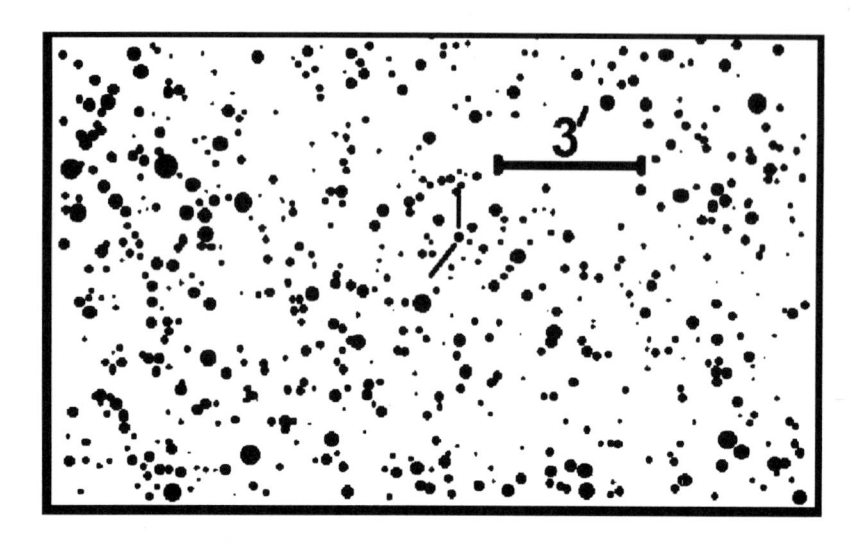

α (2000.0)=$17^h32^m14^s.0$ $Pg(B)$=13.8–15.5
δ (2000.0)=$-16°15'22''.4$ $P=0^d.1406$
Type: NL VY?

Flickering was observed.

References: [102]

V 841 Oph

α (2000.0)=$16^h59^m30^s.30$
δ (2000.0)=$-12°53'26''.8$
Type: NB
Vis=4.3–13.5
$P=0^d.60423$:

Maximum of an outburst was observed in JD 2396146.

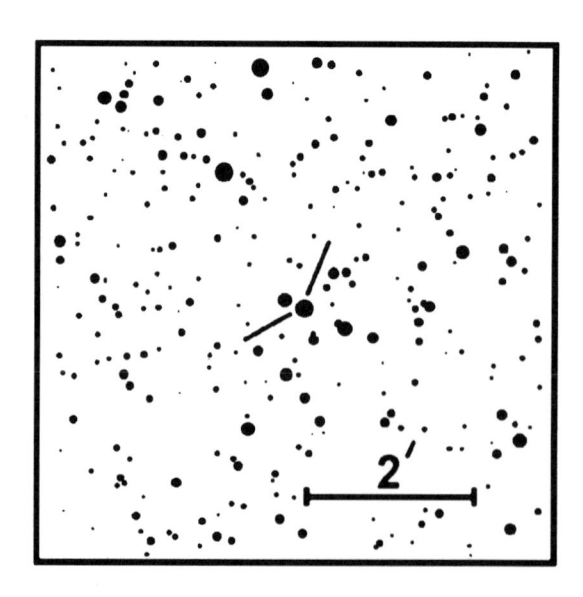

References: [107]

FY Per

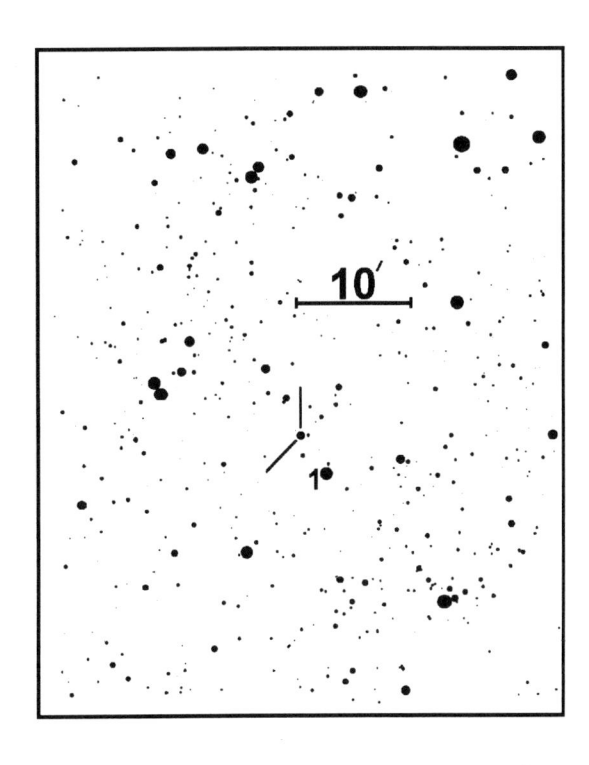

α (2000.0)=$04^h 45^m 47^s.5$
δ (2000.0)=$50°48'02''.3$
Type: NL?
Pg=11.0–14.5
P=$0^d.0648479$:
Min.phot.=JD 2447494.4403:

The comparison star *1* has

V=10.42
B=11.00
U=11.10

According to [246], there are no periodic variations. Possible Herbig Ae/Be star.

References: [102, 246]

RR Pic

α (2000.0) =$06^h 35^m 36^s.5$
δ (2000.0) =$-62°38'23''.7$
Type: NB XS
Spectrum: later G8
Vis=1.2–12.52
P=$0^d.1450255$
Max.phot.=JD 2438815.379

Variations with $A \sim 0^m.036$ and $P \sim$15.4–15.6 min are observed. There are rapid oscillations with cycles 20–40 s.

References: [102]

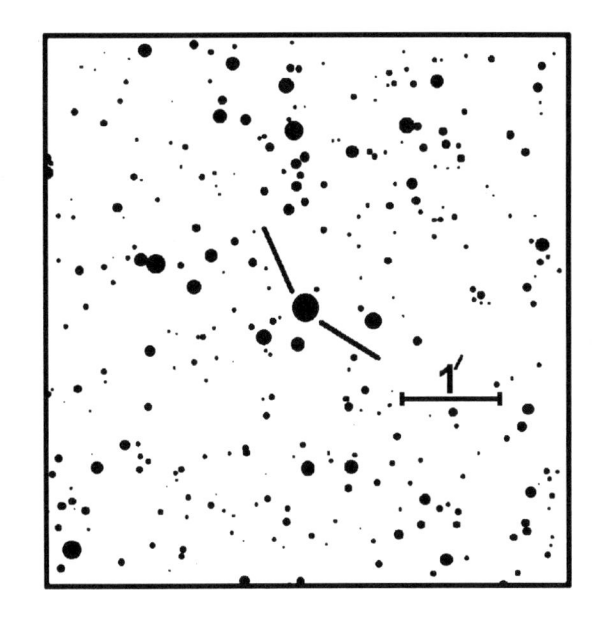

T Pyx

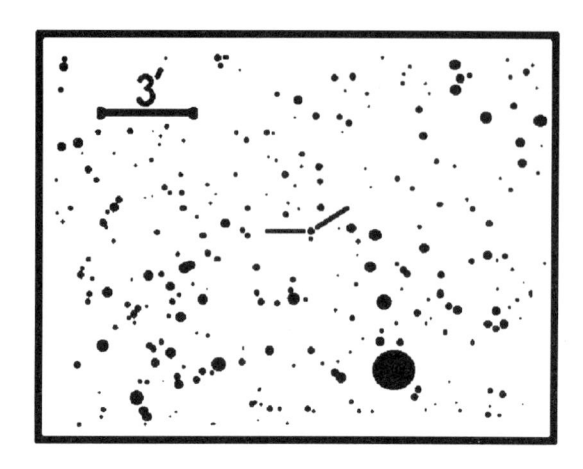

α (2000.0) $=09^h04^m41^s.57$
δ (2000.0) $=-32°22'47''.1$
Type: NR
Spectrum: M5 V
$Pg(B)=7.0-15.8$
$P=0^d.07616$

Rapid oscillations to $0^m.1$ are observed.

References: [107]

V Sge

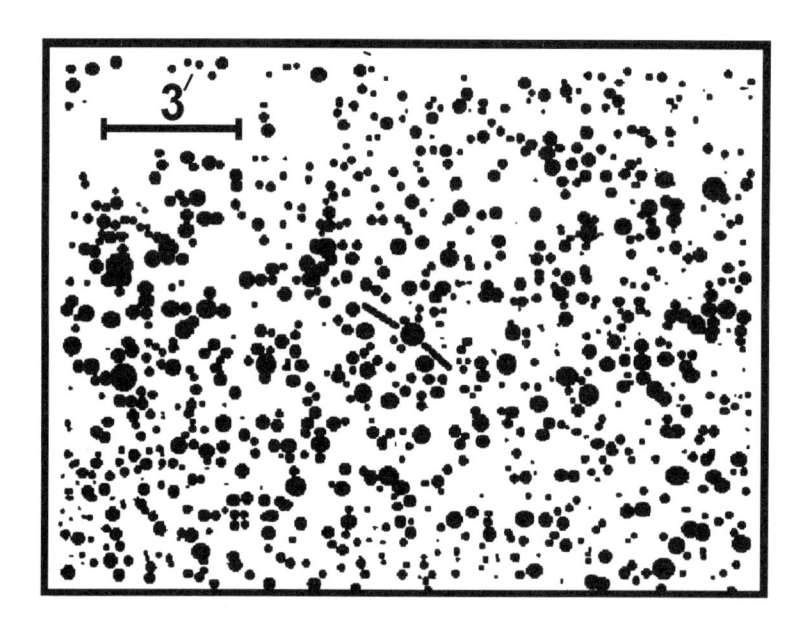

α (2000.0) $=20^h20^m14^s.84$
δ (2000.0) $=21°06'10''.2$
Type: NL XS
Spectrum: F6–G0 V

$V=8.6-13.9$
$P=0^d.5141980$
Min.phot.=JD 2437889.9136

Eclipsing binary. The depth of the eclipse is $0^m.6 - 1^m.3$. Rapid oscillation to $0^m.5$ were observed.

References: [102]

U Sco

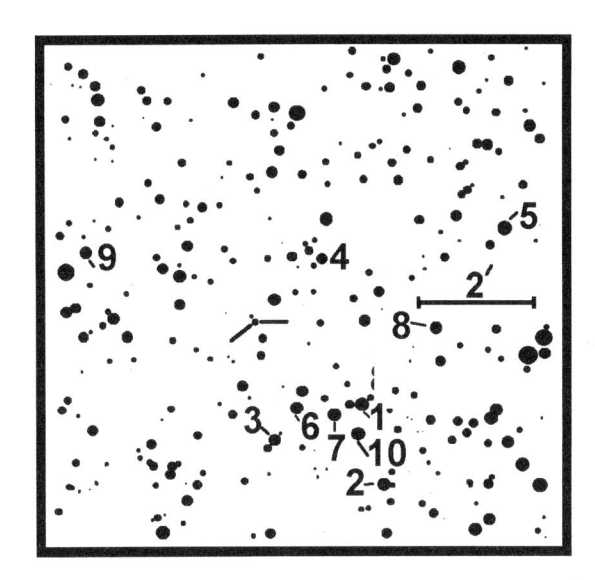

α (2000.0) $=16^h 22^m 30^s.2$
δ (2000.0) $=-17°52'42''.2$
  Type: NR
Spectrum: F5 V
  Vis=8.7–20.5
  $P=1^d.234518$
Min.phot.=JD 2447717.606

Short-period variability: $Min=$ JD $2446979.3153+0^d.053125 \cdot E$ was found in [58]. Rapid flickering to $0^m.32$ was observed.

References: [58, 107]

The comparison stars are:

Star	V	B	U	Star	V	B	U
1	14.71	15.26	15.00	6	15.45	16.18	16.23
2	14.75	15.33	15.07	7	15.53	16.27	15.35
3	14.82	15.49	15.29	8	15.76	16.52	...
4	14.85	16.56	...	9	15.80	16.03	16.00
5	15.15	15.52	...	10	...	16.90	...

RT Ser

α (2000.0) $=17^h 39^m 51^s.94$
δ (2000.0) $=-11°56'38''.1$
  Type: NC?
Spectrum: M6 V
  Pg(B)=10.5–16
  $P=3497^d$

Possibly symbiotic.

References: [107]

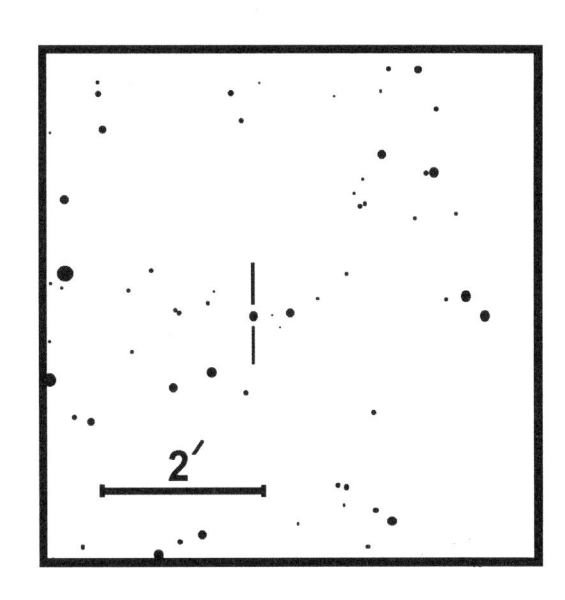

CT Ser

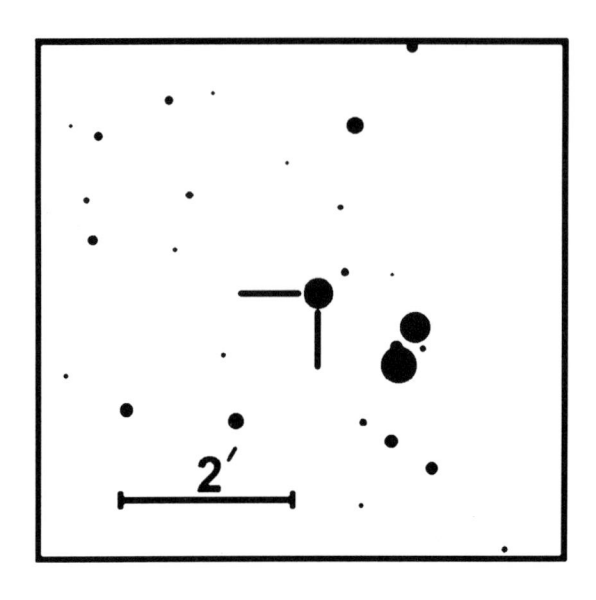

α (2000.0)=$15^h 45^m 39^s.0$
δ (2000.0)=$14°22'33''.1$
Type: N
Pg=6:–16.6
P=$0^d.1950$

References: [107]

SW Sex

α (2000.0)=$10^h 15^m 09^s.42$
δ (2000.0)=$-03°08'34''.6$
Type: NL UX SW
B=14.8–16.7
P=$0^d.1349384$
Min.phot.=JD 2444339.65087

Eclipsing binary. The depth of Min is $\sim 2^m$ B. A high exitation-spectrum star.

References: [102, 132]

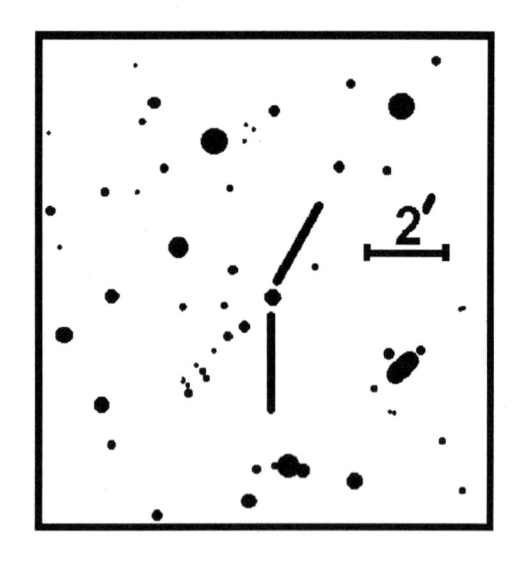

RW Tri

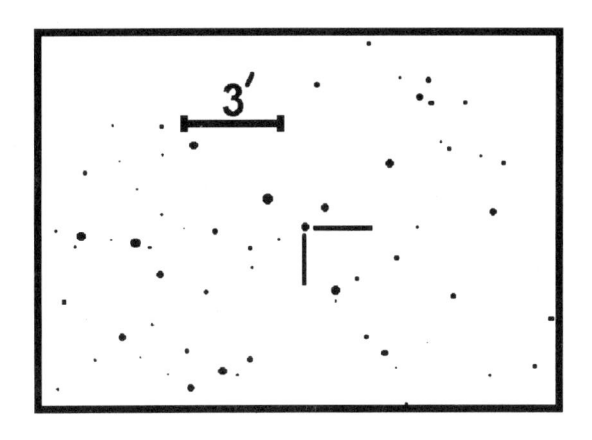

α (2000.0) =$02^h25^m36^s.01$
δ (2000.0) =$28°05'51''.4$
Type: NL UX
Spectrum: M0 V
V=12.5–15.6
P=$0^d.231883297$
Min.phot.=JD 2441129.36487

Eclipsing binary. The depth of the minimum is $1^m.1 - 2^m.1$, its duration is $0^p.05 - 0^p.07$. Strong filckering of light outside eclipses.

References: [102]

UX UMa

α (2000.0) =$13^h36^m41^s.3$
δ (2000.0) =$51°54'48''$
Type: NL UX XS
V=12.6–14.2
Spectrum: K4–M6 V
P=$0^d.19667126$
Min.phot.=JD 2443904.87775

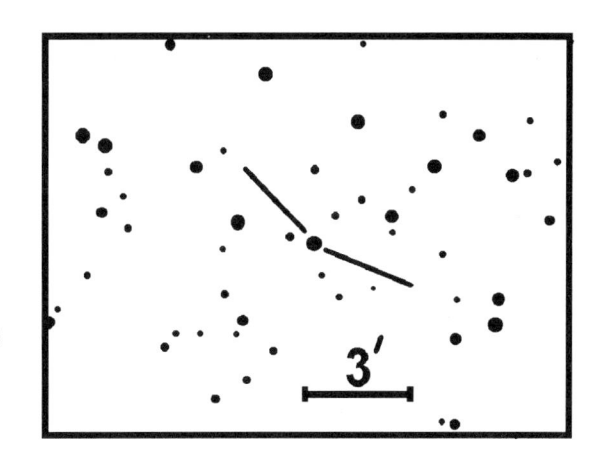

Eclipsing binary. The depth of eclipses is $0^m.7 - 1^m.3$ B, $0^m.7 - 1^m.1$ V.

References: [102]

EI UMa

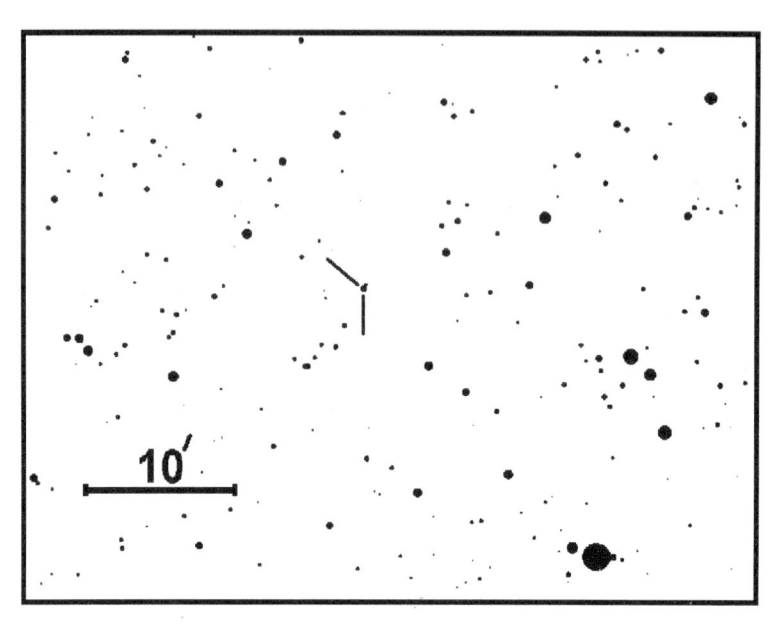

α (2000.0)=$08^h38^m22^s.07$ Pg=13.4–14.9
δ (2000.0) =$48°38'00''.8$ $P=0^d.26810$
Type: DN UG XS   Min.phot.=JD 2446149.834

References: [12, 102]

RW UMi

α (2000.0)=$16^h47^m55^s.6$
δ (2000.0) =$77°01'41''$
Type: NB
Vis=6–21
$P=0^d.081$:

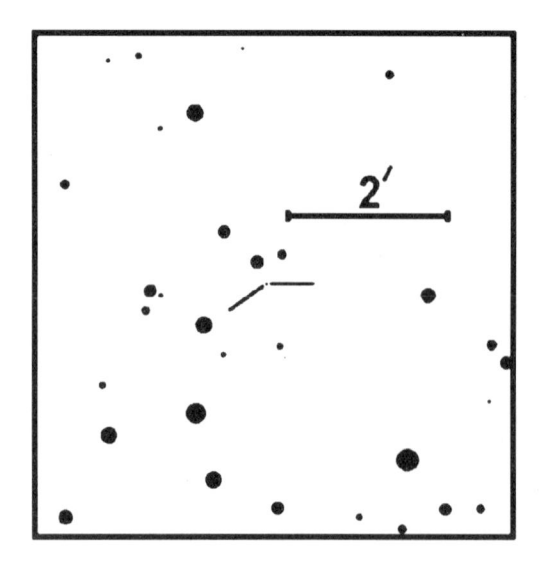

References: [107]

ER UMa/PG 0943+521

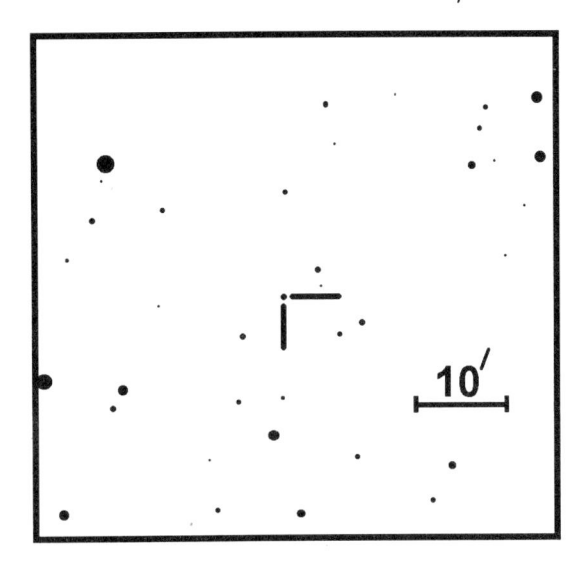

α (2000.0)=$09^h47^m11^s.91$
δ (2000.0)=$51°54'08''.0$
Type: NL, UX
V=12.5–15
P=$0^d.1997$

References: [102, 132]

HV Vir

α (2000.0)=$13^h21^m03^s.17$
δ (2000.0)=$01°53'28''.8$
Type: DN, SU
Pg=11–19
P=$0^d.05070$
Max.phot.=JD 2448737.3486

Outbursts in 1929, 1992.

References: [102, 219]

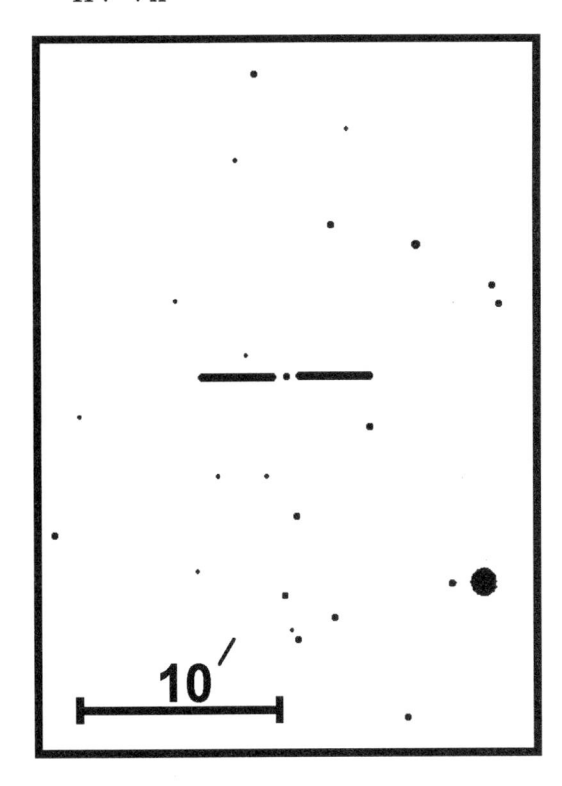

HS Vir

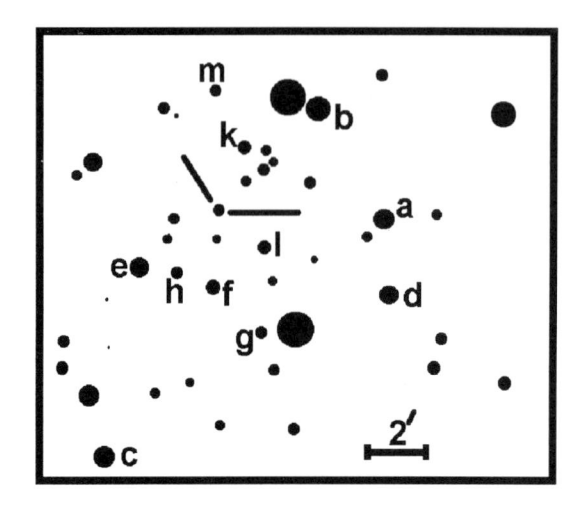

α (2000.0)=$13^h43^m38^s.51$
δ (2000.0)=$08°14'03''.9$
Type: DN UG
B=13.0–15.8
$P=0^d.0836$:
$P_{cycle}=75^d$

The comparison stars are:

Star	Pg	Star	Pg
a	12.52	g	15.12
b	12.63	h	15.62
c	13.12	k	16.28
d	14.47	l	16.37
e	14.49	m	16.95
f	14.55		

References: [102, 132, 249]

PW Vul

α (2000.0)=$19^h26^m05^s.01$
δ (2000.0)=$27°21'57''.9$
Type: NA
Vis=6.43–17:
$P=0^d.21372$
Min.phot.=JD 2446704.263

References: [107]

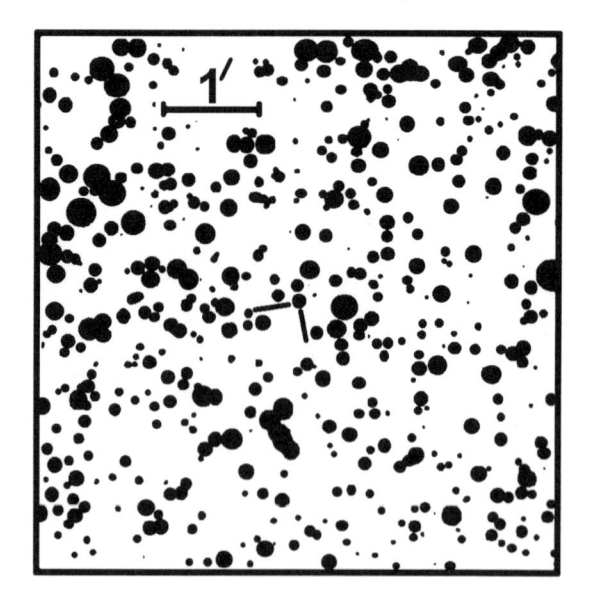

0903+41=PG 0859+415

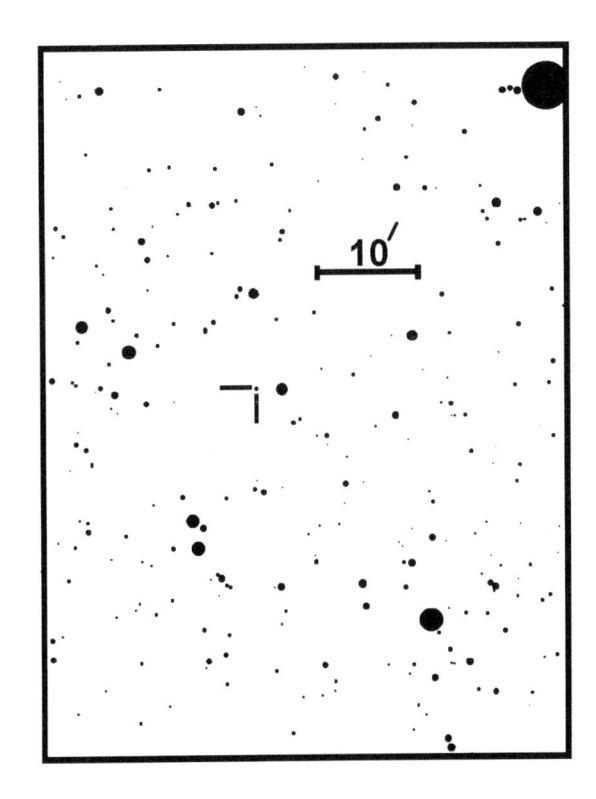

α (2000.0)=$09^h03^m08^s.96$
δ (2000.0)=$41°17'46''.9$
Type: NL UX SW
Vis~14.3
$P=0^d.152813$

References: [318]

2136+11=PG 2133+115

α (2000.0)=$21^h36^m19^s.20$
δ (2000.0)=$11°40'53''.6$
Type: NL VY?
Pg=14.7–17
$P=0^d.121$

The constant star a was used as a comparison star.

References: [318]

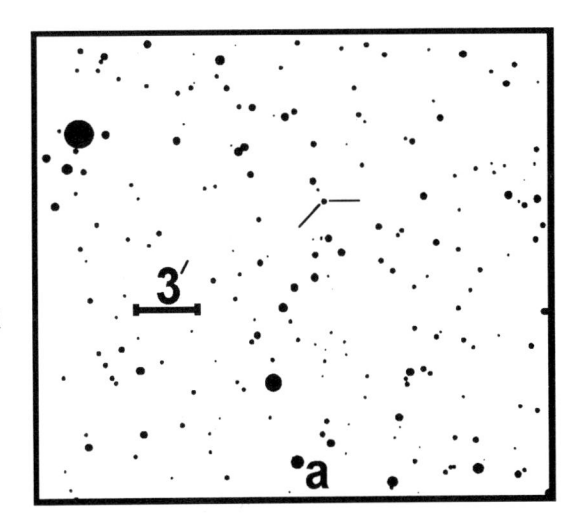

INTERMEDIATE POLARS

V 603 Aql/N Aql 1918

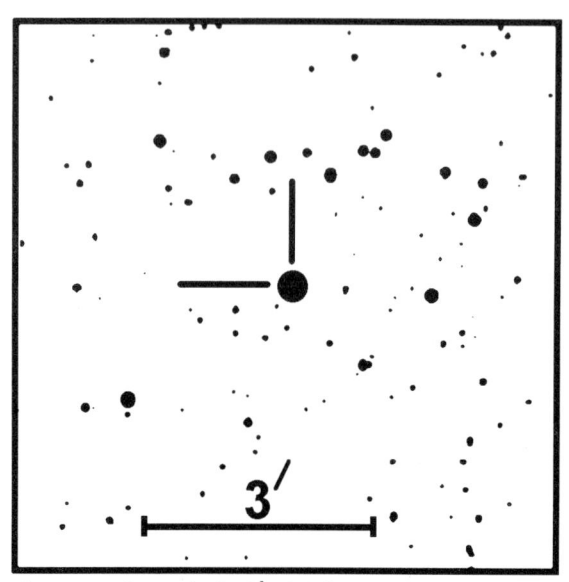

α (2000.0)=$18^h48^m54^s.5$
δ (2000.0)=$00°35'02''.9$
Type: IP, NA, XS
Spectrum: later than G8
V=−1.1 −12.03
P=$0^d.138154$
$Min.phot.$=JD 2444401.399

V 603 Aql is a fast Nova, one of the brightest *IPs* in the optics, UV and X-rays. This object exhibits the following periods: 63 min modulations corresponds to the spin period of the WD; the $3^h.3$ spectroscopic period, to the orbital motion; $3^h.5$ modulation of light corresponds to the period of dips; $15^m.6$ is a photometric period; 2^h48^m is a polarimetric period; and $2^d.5$ corresponds to the precession period of the accretion disc.

References: [40, 259, 402]

AE Aqr

α (2000.0)=$20^h40^m09^s.1$
δ (2000.0)=$-00°52'16''.3$
Type: IP, NL, XS
Spectrum: K4–5 V
V=9.8–12.5
T=JD 2445172.283(5)
P=$0^d.4116548(7)$

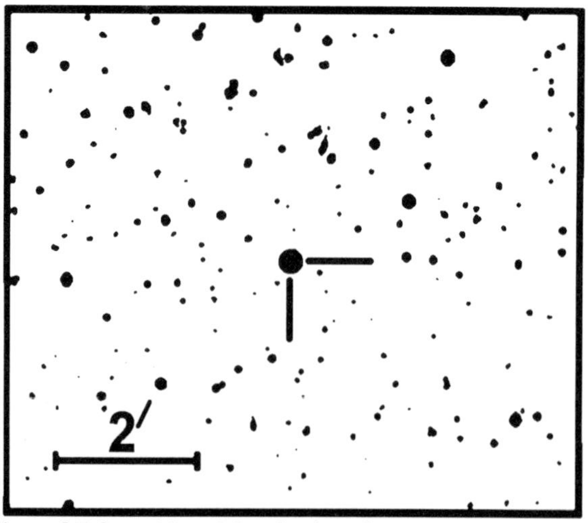

AE Aqr showed in 1938 unusual outbursts ($\Delta V \simeq 2$) in optical range. The smaller flares are more numerous and regular. It exhibits large amplitude flares, flickering, radio emission and linear polarization. Rapid, coherent oscillations (16.5 s, and 33.08 s), transient QPOs ~18 s, 36 s during flares.

References: [259, 402]

FO Aqr/H 2215–086

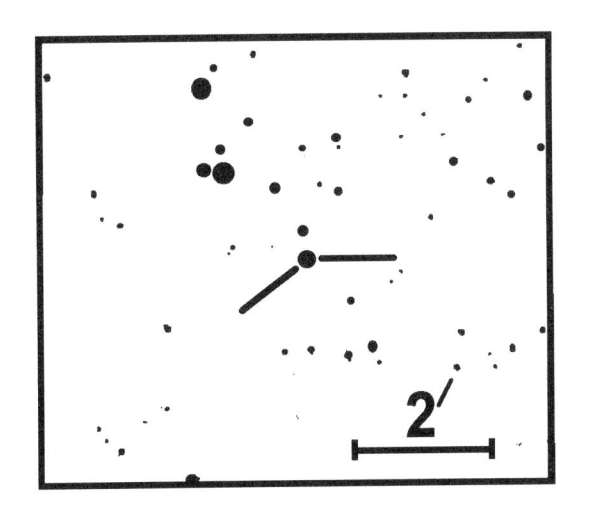

α (2000.0) $=22^h17^m55^s.5$
δ (2000.0) $=-08°21'05''.4$
Type: IP, NL, XS
Spectrum: K1–5 V
$Pg=13.0$–17.5
$P_1=0^d.16802(1)$
$Min.=$JD 2444782.879(4)
$P_2=0^d.20206$

References: [259, 402, 429]

TV Col/2A 0526–328

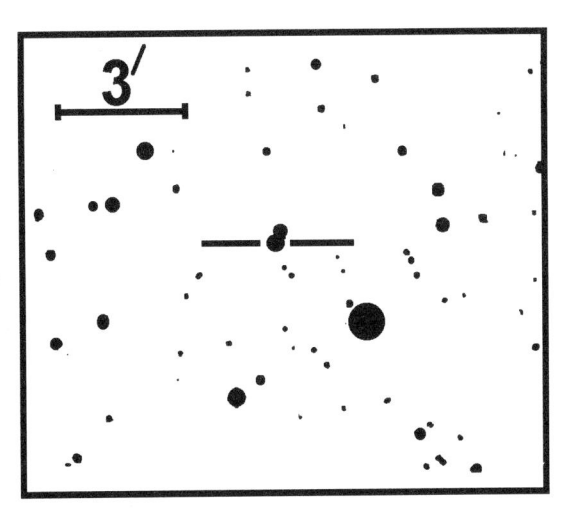

α (2000.0) $=05^h29^m25^s.5$
δ (2000.0) $=-32°49'05''.3$
Type: IP, NL, XS
Spectrum: K1–5 V
$V=13.6$–14.1
$P=0^d.2285529(2)$
$T_{ecl}=$JD 2447537.111(2)

TV Col contains a white dwarf emitting X-rays pulses with the 1911±5 s spin period, it also exhibits photometric modulations at $5^h.2$ and 4^d interpreted as the result of disk precession. The four reported outbursts of TV Col are similar in amplitude (∼2 mag) and duration (hours). The bright S-wave is present during the outbursts.

References: [102, 259]

TX Col/H 0542−407

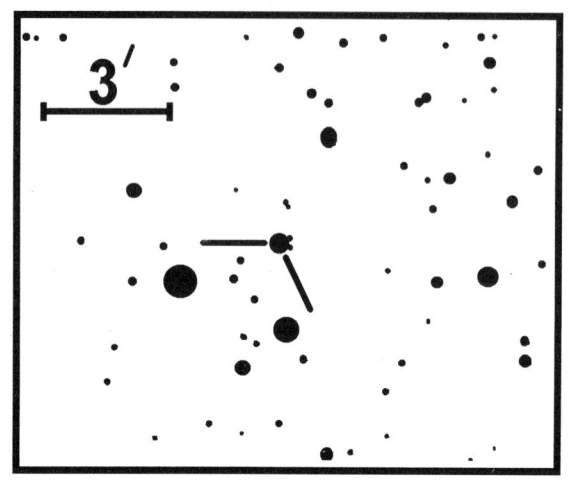

α (2000.0)=$05^h43^m20^s.3$
δ (2000.0)=$-41°01'56''.1$
  Type: IP, NL, XS
  V=15–16
  P=$0^d.2383$

From *EXOSAT* light curve the period of $P = 1920 \pm 30$ s was found, from optical light curve two other periods follow: 2106 and 1054 s.

References: [102, 259]

HL CMa/1E 0643−1648

α (2000.0)=$06^h45^m17^s.0$
δ (2000.0)=$-16°51'35''$
  Type: IP, DN, XS, ZC
  V=10–14.5
  P_{orb}=$0^d.2145(6)$
Min.phot.=JD 2445329.560(7)

The star "A" is Sirius. HL CMa is a dwarf nova with the time of recurrence $t_{rec} \sim 15^d$. The circular polarization is of the order of 0.34± 0.11%. $L_X/L_{opt} \sim$ 0.002 at optical maximum and $\sim$1 at optical minimum.

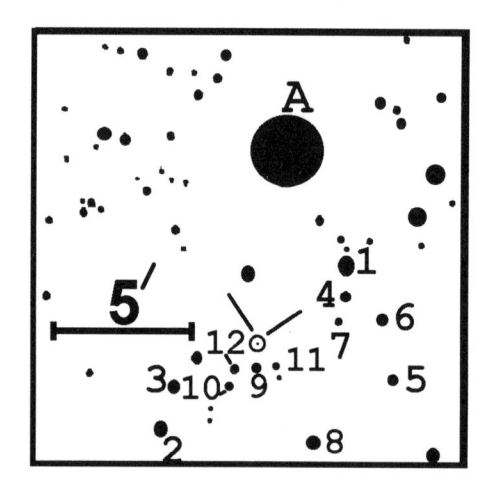

The comparison stars are [176]:

Star	Vis	Star	Vis	Star	Vis
1	8.5	5	11.6	9	13.7
2	8.7	6	11.8	10	14.1
3	9.2	7	13.0	11	14.4
4	10.4	8	13.5	12	14.5

References: [74, 176]

BG CMi/3A 0729+103

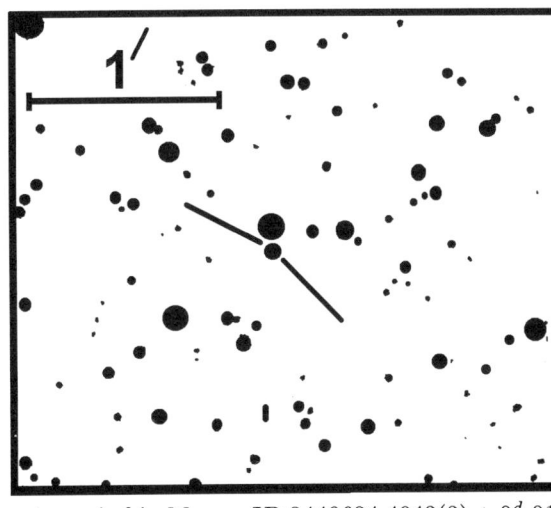

α (2000.0)=$07^h31^m29^s.0$
δ (2000.0)=$09°56'21''.8$
Type: IP, NL, XS
V=14.5–15.5
$P=0^d.1347486(3)$
Min.phot.=JD 2445020.384(6)

A hard X-ray source. The periods 847, 913, 1152 and 1140 s have been found. The 913 s period dominates at all wavelenghts. The ephemeris of the spin period is $Max = $ JD $2448624.4943(3) + 0^d.010572712(6) \cdot E - 5.9(4) \times 10^{-13} \cdot E^2 - 4.9(7) \cdot 10^{-19} \cdot E^3$.

References: [214, 259, 419]

DQ Her/N Her 1934

α (2000.0)=$18^h07^m30^s.2$
δ (2000.0)=$45°51'31''.9$
Type: IP, NA
V=1.4–18.0
Spectrum: M3 V
$P=0^d.193620796$
Min.phot.=JD 2434954.9443

The average light of DQ Her from 1954 till 1977 shows cyclic oscillations with $A \sim 0^m.6$. Its light curves are very unstable. There are some hints that the true spin period is 142 s but not 71 s because of a 142 s period in the circular polarization.

References: [40, 421]

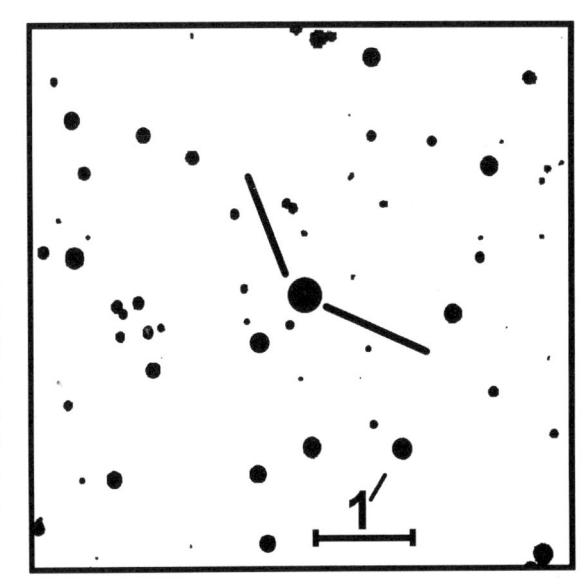

AL Com

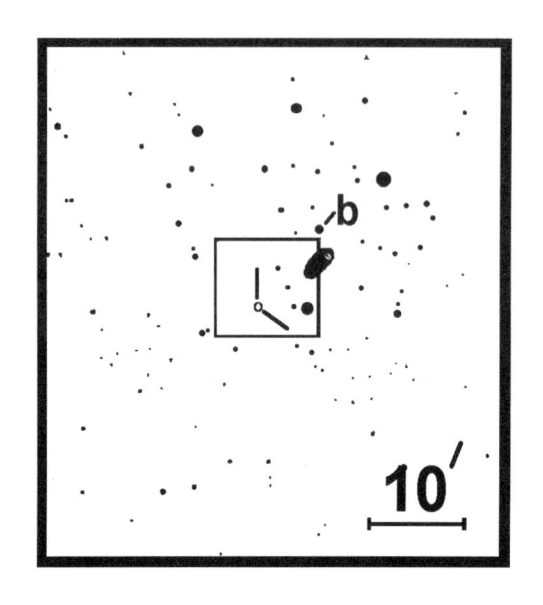

α (2000.0)=$12^h32^m25^s.6$
δ (2000.0) =$14°20'57''.5$
Type: IP?, DN, SU
V=12.8–20.8
$P=0^d.061$:
$P_{cycle}=325^d.4$

The comparison stars are [176]:

Star	a	b	c
U	12.51	13.32	14.87
B	12.40	12.77	14.37
V	11.85	11.96	13.51
Pg	12.21	12.63	14.23

Photografic magnitudes:

1	16.44
2	17.30
3	17.30
4	17.50
5	18.43
6	19.10
7	19.45

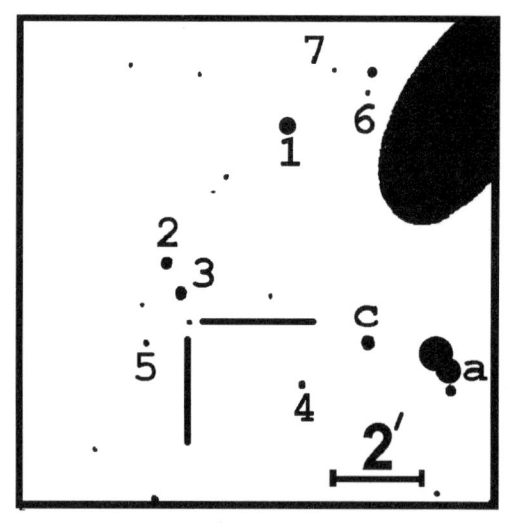

AL Com is situated within 8' to SE from the nucleus of the galaxy M 88 (NGC 4501). It is an unusual CV. It underwent an outburst to 14^m with $A \simeq 5$ in 1962. The photometric behavior of AL Com is complex and variable and the earlier believed 41 min period is not the orbital one. Abbott et al. [1] give the more probable orbital period, 87–90 min.

References: [1, 36, 176, 426]

SS Cyg

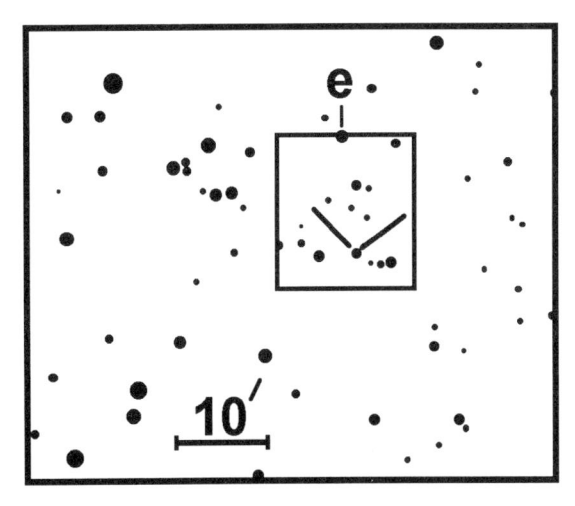

α (2000.0) $=21^h 42^m 43^s.0$
δ (2000.0) $=43°35'08''.7$
Type: DN, UG?, XS, SU?
$V=7.7-12.4$
Spectrum: K5–M5 V
$P=0^d.2751302(4)$
Min.phot.I=JD 2444841.9378
$P_{cycle}\sim 49^d.5$

The comparison stars are:

Star	V	B
a	8.52(2)	9.84(2)
b	9.62(2)	10.62(2)
c	9.86(2)	10.22(2)
d	10.29(2)	10.70(2)
e	10.86(2)	11.04(2)
f	10.92(2)	11.43(2)
g	11.01(2)	12.16(2)
h	11.44(2)	12.04(2)
k	11.91(2)	12.50(2)
l	12.31(3)	12.93(3)

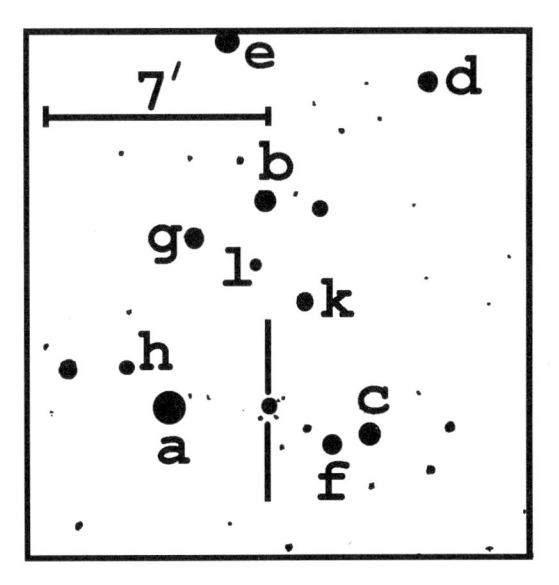

The system exhibits the following states: "mean" – 15-25 days before outbursts with $V = 11.43(1)$, "rise" – 1-9 days before outburst with $V=11.15(3)$, "decline" state – 5-10 days after outburst with $V = 11.66(1)$ and "quiet" state with $V=11.53(2)$ [365, 368]. The quasi-period oscillations probably exist. There are modulations in the soft X-rays from $7^s.4$ at the maximum to $10^s.4$ at the late "decline" state, in the hard X-rays there are $\sim 100^s$ modulations.

References: [48, 52, 152, 176, 190, 257, 381, 402]

DO Dra/3A 1148+719/2A 1150+72/E1140.8+7158

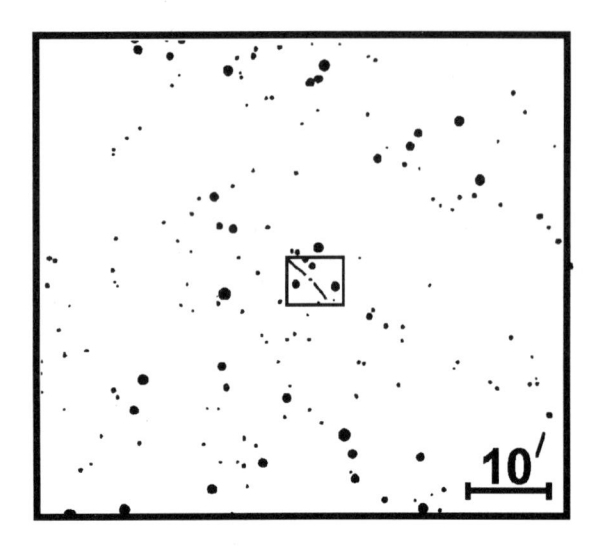

α (2000.0)=$11^h 43^m 38^s.5$
δ (2000.0)=$71°41'19''.2$
Type: IP, DN, XS
B=10.6–16.7
Spectrum: M3–5 V
$P=0^d.166 \pm 0^d.002$
$P_{cycle}=1034^d$

The comparison stars are:
(Visual magnitudes)

Star	Vis
1	14.8
2	10.2
3	11.4
4	11.4
5	11.1

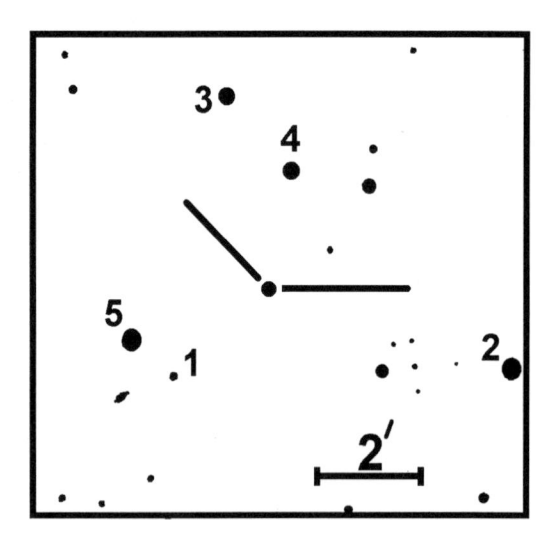

DO Dra exhibits stable optical 265 s and X-ray (529 and 550 s) pulsations. The spin period is 529 s. The primary is probably a two-pole accretor.

References: [156, 176, 259, 261, 279]

V 533 Her/N Her 1963

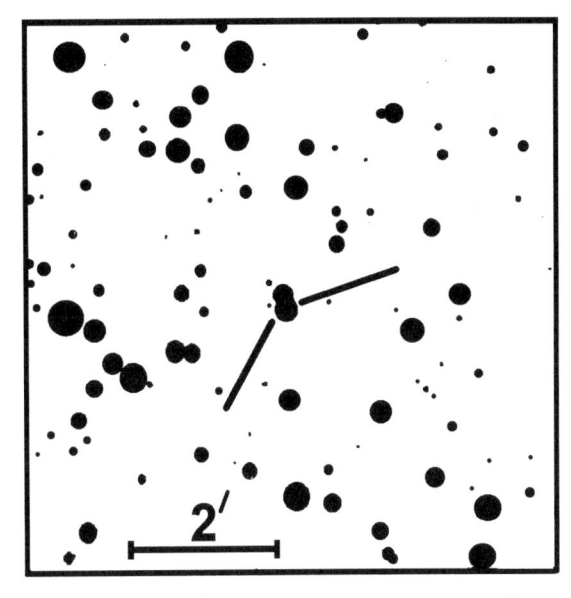

α (2000.0)=$18^h14^m20^s.3$
δ (2000.0)=$41°51'21''.3$
Type: IP, NA, CP
Pg=3.0–16.0
Spectrum: later than G9
$P=0^d.2097774$
Min.phot.=JD 2446594.870

This old nova shows the stable 63 s periodicity in the light curves. There are two arguments against including it into the *IPs*: a) the absence of pulsed X-rays; b) the disappearance of the pulsed signal in photometry in 1981. The ephemeris for pulse maximum is: P_{max} = JD 2443283.83167(3) + $0^d.00073649343(2) \cdot E$.

References: [40, 259]

V 795 Her/PG 1711+336

α (2000.0)=$17^h12^m56^s.7$
δ (2000.0)=$33°31'19''.$
Type: IP?, NL, SH
B=12.5–13.2
$P_{sp}=0^d.1082648$
$P_{phot}=0^d.1164863(3)$
Min.phot.=JD 2446584.205(2)

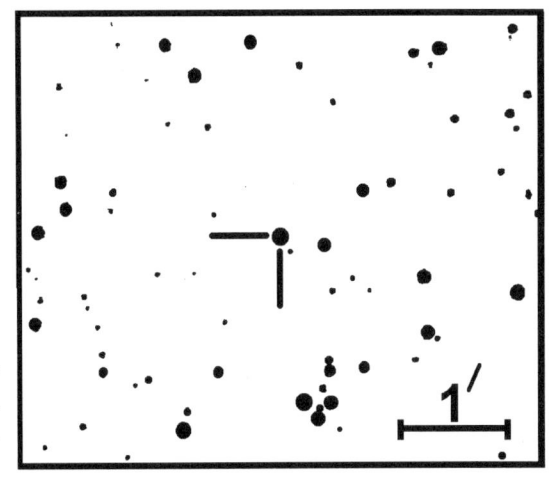

The system is perhaps in a permanent superoutburst. UV spectrum is not typical for an *IP*, X-ray emission was not detected by *EXOSAT*. A statistically significant period for resonance UV lines is 4.77 ± 0.12 hr. It is the beat period of the $2^h.6$ orbital period and the observed $\sim 1^h.7$ period of the white dwarf rotation.

References: [132, 134, 259, 439]

EX Hya

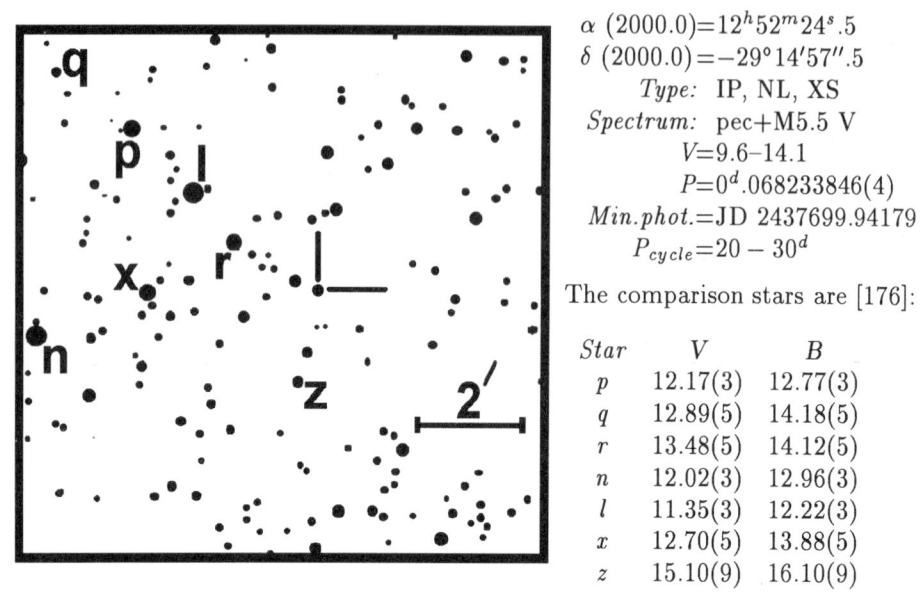

α (2000.0)=$12^h52^m24^s.5$
δ (2000.0)=$-29°14'57''.5$
Type: IP, NL, XS
Spectrum: pec+M5.5 V
V=9.6–14.1
P=$0^d.068233846(4)$
Min.phot.=JD 2437699.94179
P_{cycle}=20 – 30^d

The comparison stars are [176]:

Star	V	B
p	12.17(3)	12.77(3)
q	12.89(5)	14.18(5)
r	13.48(5)	14.12(5)
n	12.02(3)	12.96(3)
l	11.35(3)	12.22(3)
x	12.70(5)	13.88(5)
z	15.10(9)	16.10(9)

EX Hya is a unique *eclipsing* IP and dwarf nova with the recurrence time ~ 4^d and $\Delta m \sim 2.5$. Its orbital period is below the "period gap". It is the brightest IP in the hard X-rays in quiescence. The partial eclipse is also present in hard X-rays.

References: [176]

V 426 Oph

α (2000.0)=$18^h07^m51^s.8$
δ (2000.0)=$05°51'48''$
Type: IP?, DN, XS, ZC?
Spectrum: K2–4 V
V=11.5–13.4
P=$0^d.28531(1)$
Min.phot.=JD 2445526.7261(8)

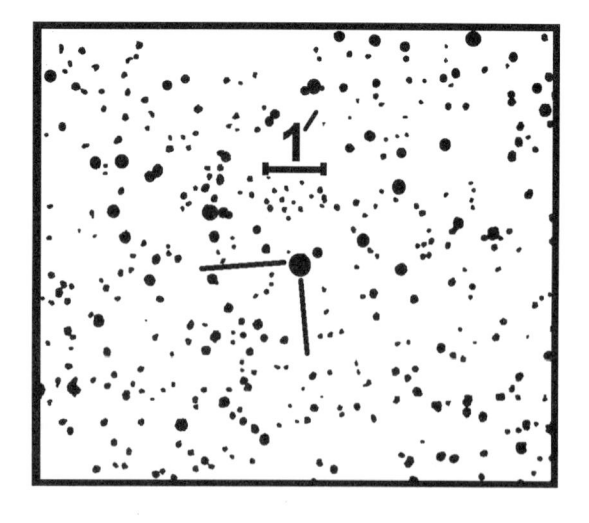

V426 Oph is possibly a *Z Cam* dwarf nova with the recurrence time of 22 days or an *IP* candidate, it has a 60 minute modulation in soft and hard X-rays and a set of X-ray periods: $1^h.25$, $2^h.5$, $4^h.5$ and $12^h.5$.

References: [102, 259]

V 2301 Oph/1H 1752+081

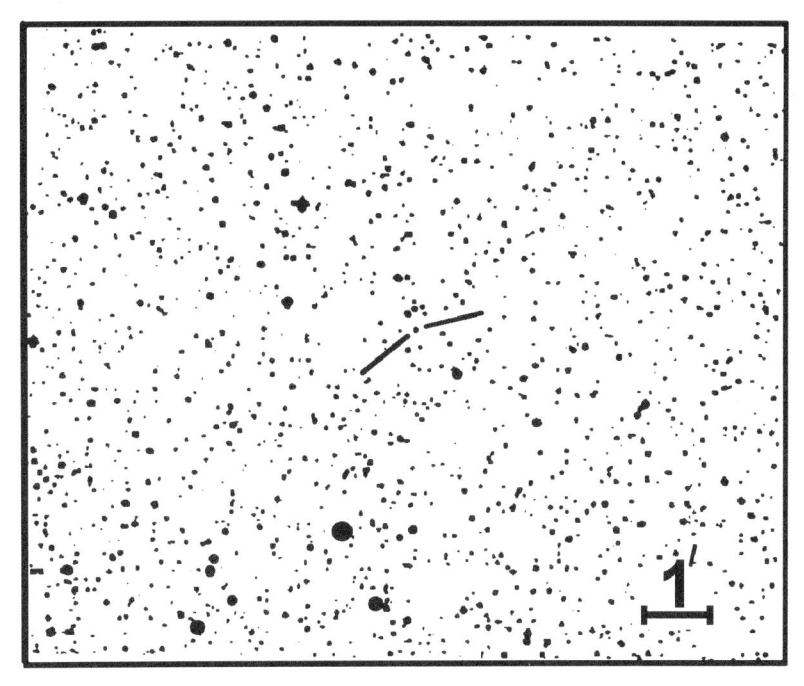

α (2000.0)=$18^h00^m35^s.6$ V=16.1–16.7
δ (2000.0)=$08°10'11''.5$ $P=0^d.07845003(5)$
Type: IP?, NL, XS   *Min.phot.*=JD 2441071.02001(2)
Spectrum: M5.5–6.5 V

References: [22, 102, 278, 320]

GK Per/N Per 1901

α (2000.0)=$03^h31^m11^s.8$
δ (2000.0)=$43°54'16''.8$
Type: IP, NA, XS, DN
V=0.2–14.0
Spectrum: K0 IV
$P=1^d.996803(7)$
Max.vr=JD 2444912.966(8)

GK Per shows strong fluctuations in quiescence, by $V \sim 0.1$ in a few minutes, ~ 0.3 in days, ~ 1.5 in dozens days. From 1948 it began to show dwarf nova outbursts with the recurrence time between $\sim$880 and 1240 days.

References: [40, 259]

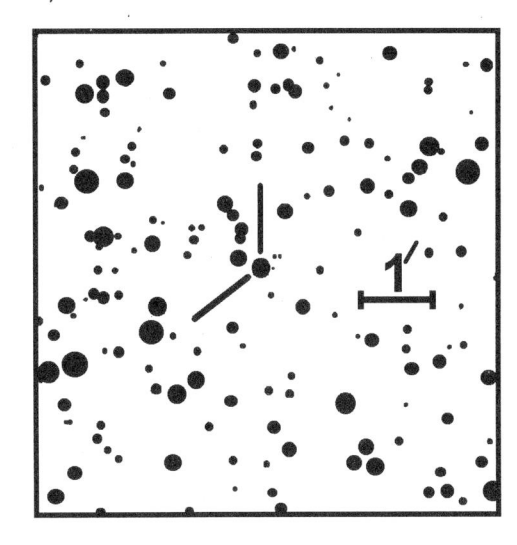

AO Psc/H 2252-035

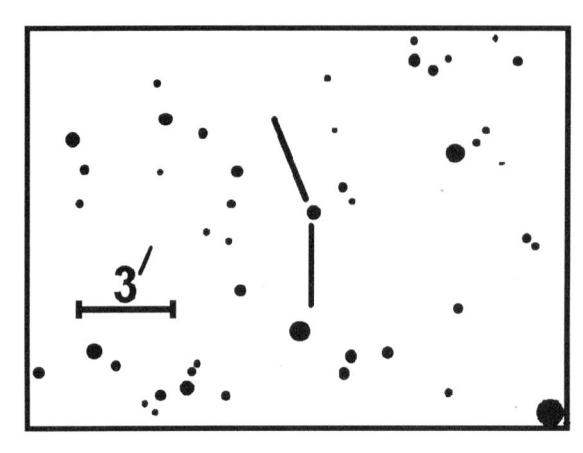

α (2000.0)=$22^h55^m18^s.0$
δ (2000.0)=$-03°10'41''.3$
 Type: IP, NL, XS
 V=13.3-14
 P=$0^d.149626(2)$
 Max.phot.=JD 2444846.143(3)

AO Psc has 859 s and 805 s X-ray pulsations and a 805 s period in UV. The 859 s period is the "beat" frequency.

References: [259, 402]

CP Pup/N Pup 1942

α (2000.0)=$08^h11^m46^s.0$
δ (2000.0)=$-35°21'05''.7$
 Type: IP?, AM?, NA,
 Type: XS, SH?
 B=0.2-15

CP Pup has two periods:
 spectroscopic:
 Min = HJD 2446812.5961(1) +
     + $0^d.06129(1) \cdot E$,
 and photometric:
 Min = HJD 2446812.0254(1) +
     +$0^d.06834(1) \cdot E$.

References: [40, 107, 438, 418]

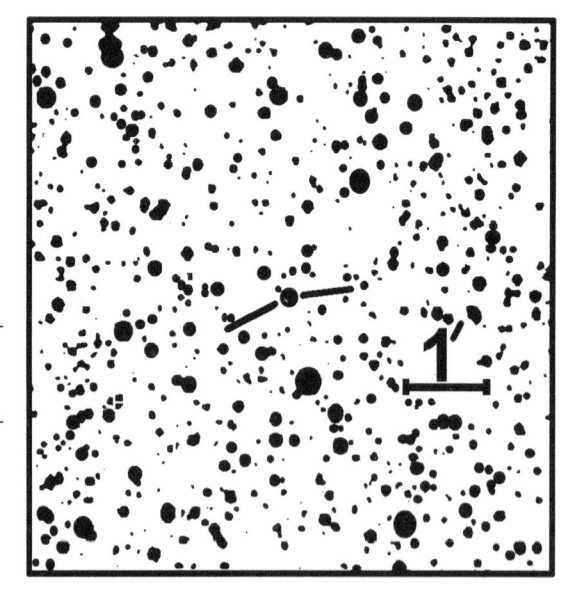

V 347 Pup/4U 0608−49/LB 1800

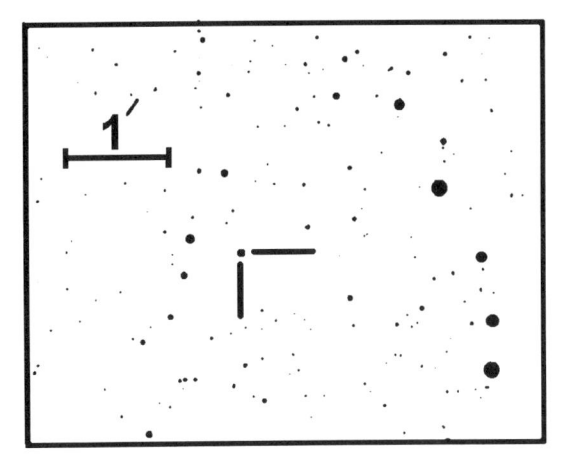

α (2000.0)=$06^h 10^m 33^s.6$
δ (2000.0)=$-48°44'26''.7$
Type: IP?, NL, XS
V=13.2–15.8
P=$0^d.23193(1)$
Min.phot.=JD 2446836.962(1)

V347 Pup is an *eclipsing* nova-like variable. It was discovered as the optical counterpart of the X-ray transient source 4U 0608−49 with circular polarization ≤ 0.5%.

References: [55, 102, 259, 428, 437]

V 348 Pup/1H 0709−360

α (2000.0)=$07^h 12^m 32^s.9$
δ (2000.0)=$-36°05'40''.4$
Type: IP?, NL, XR
V=15.0–17.0
P=$0^d.1018401(6)$
Min.phot.=JD 2447210.12091(7)

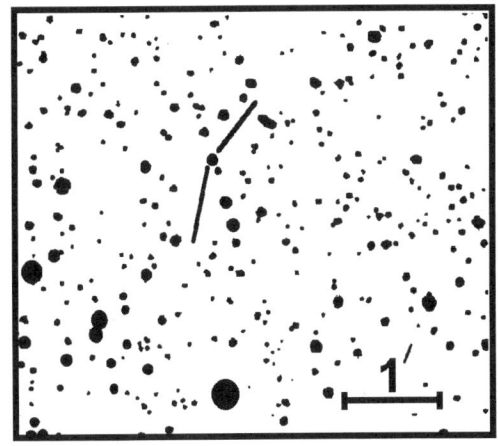

V348 Pup/1H 0709+360 is a faint X-ray source (*HEAO-1*), an *eclipsing* cataclysmic variable. Its orbital period is in the "period gap". There is no visible polarization, although strong He II lines are seen in spectra.

References: [102, 259]

TW Pic/H 0534–581

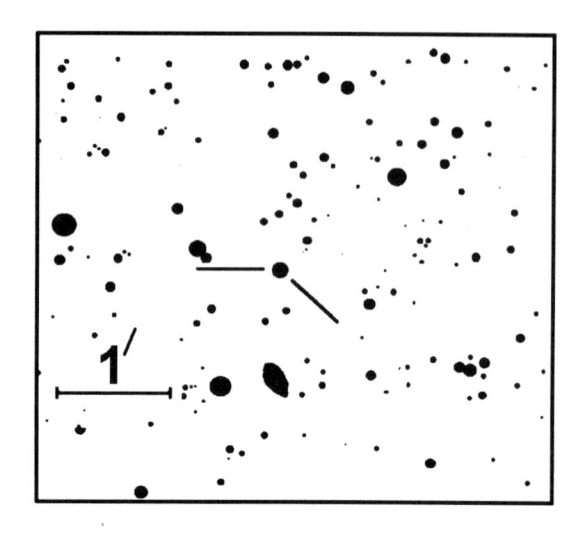

α (2000.0)=$05^h 34^m 50^s.8$
δ (2000.0)=$-58°01'41''.7$
Type: IP?, NL, XS
V=14.1–18.0
P=$0^d.2583$:
or $0^d.2708$:

The system exhibits 2707 and 1938 s, 2.1(1) and 6.5 ± 1.0 hr periods [56].

References: [56, 346]

KO Vel/1E 1013–477

α (2000.0)=$10^h 15^m 58^s.4$
δ (2000.0)=$-47°58'11''$
Type: IP?, NL, XS
V=16.6–19.0
P=$0^d.422(3)$

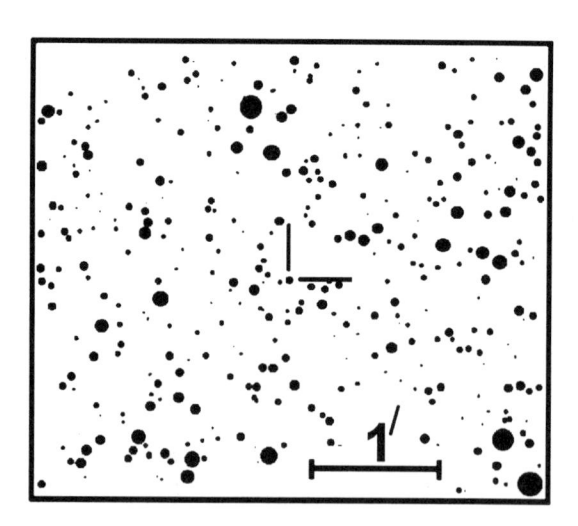

KO Vel is an X-ray source with a polarization of optical light of 1.0(3)%, its classification as *IP* or *CV* is debated [259]. The next periods were observed for KO Vel: $6^h.4$, 71^m, 68^m and 89^m [182].

References: [102, 182, 259]

V 1223 Sgr/3A 1851–312

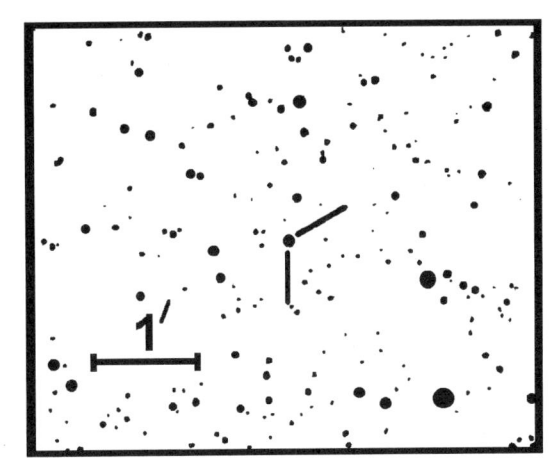

α (2000.0)=$18^h55^m02^s.3$
δ (2000.0)=$-31°09'49''.3$
Type: IP, NL, XS
V=12.3–17.0
P=$0^d.1402440(4)$
Max.phot.=JD 2444749.986(3)

V1223 Sgr/4U 1849–31 has optical pulsations with $P = 794$ s and the amplitude $\simeq$15–40%. It shows $P = 745$ s in X-rays.

References: [102]

VZ Pyx/H 0857–242

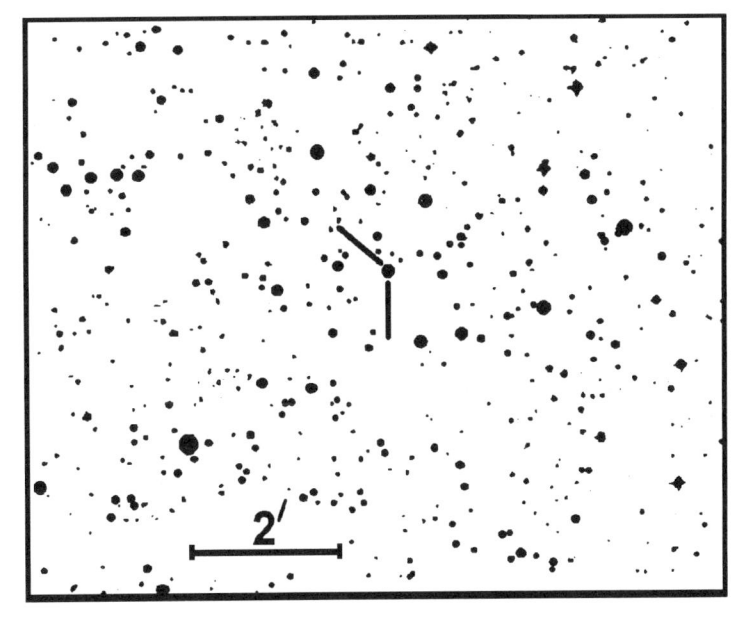

α (2000.0)=$08^h59^m19^s.9$ B=11.2–17
δ (2000.0)=$-24°28'55''.7$ P=$0^d.0742(4)$
Type: IP?, DN, XS   Max.vr=JD 2447241.175(5)

VZ Pyx shows superoutbursts of the SU UMa type, has bright broad H I, He I, He II emission lines, X-ray pulsations. Remillard et al. [278] found the following periods: 6400 ± 24 s (radial velocity), 2924.64 ± 7.56 s (V and I bands), 290.12 ± 30.6 s and 1920 ± 10.8 s (V band).

References: [102, 278]

V 1062 Tau/H 0459+246

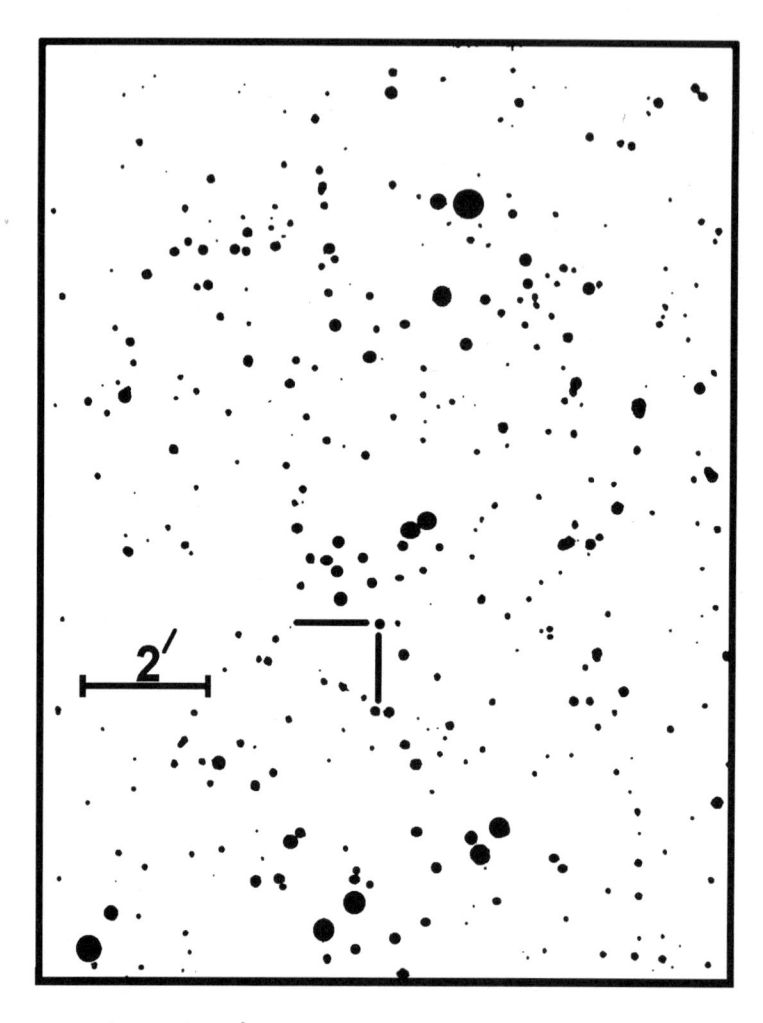

α (2000.0)=$05^h02^m27^s.5$ B=15.1–16.1
δ (2000.0)=$24°45'22''.1$ $P=0^d.4147 \pm 0.0031$
   Type: IP?, XS T=JD 2447494.586±0.010
   Spectrum: later than K

V 1062 Tau/H 0459+246 has bright broad H I, Hc I, He II emission lines, TiO absorptions, X-ray pulsations with a 3726 ± 36 s period, optical pulsations with a period $9^h.95(7)$ and infrared pulsations with periods $1^h.054(5)$ and $9^h.95(7)$.

References: [278]

SW UMa

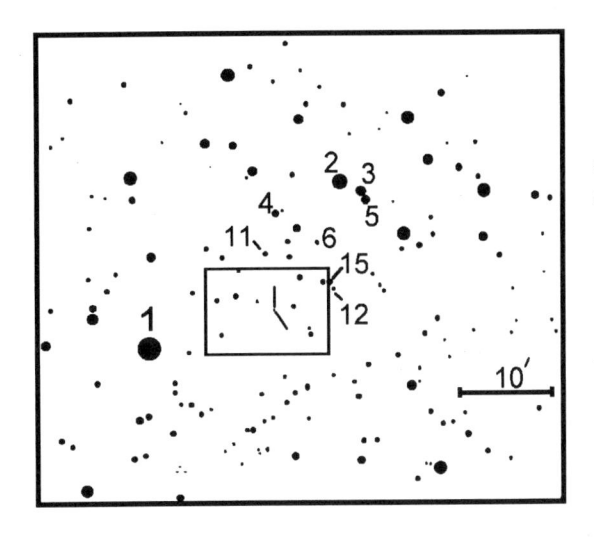

α (2000.0)=$10^h36^m42^s.9$
δ (2000.0) =$53°28'37''$
 Type: IP?, DN, XS, SU
 V=9.7–17.0
 Spectrum: later than M2
 P=$0^d.0568162(7)$
 Min.phot.I=JD 2448706.7253(6)

The comparison stars are:
(Visual magnitudes)

Star	Vis	Star	Vis
1	5.7	9	12.3
2	8.7	10	13.5
3	9.6	11	13.9
4	11.7	12	14.5
5	11.8	13	14.6
6	12.3	14	15.5
7	12.3	15	15.6
8	12.3		

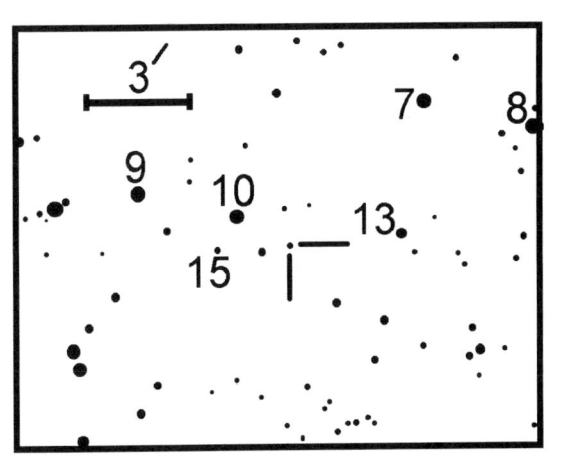

An eruption with amplitude $\Delta m \geq 7$ was in 1986 March. Every 460 days (on average) SW UMa rises by 6 mag or more. The superhumps began from 8 March 1986. Max_{hump} = JD 2446502.654(1) + 0^d.05833(6) · E, A ~ 11%. In X-rays and optics there is a 15.9 min periodicity – the white dwarf rotation period? There are 5 minute and 22 second QPOs.

References: [48, 109, 176, 285, 312, 402]

1H 0551−819

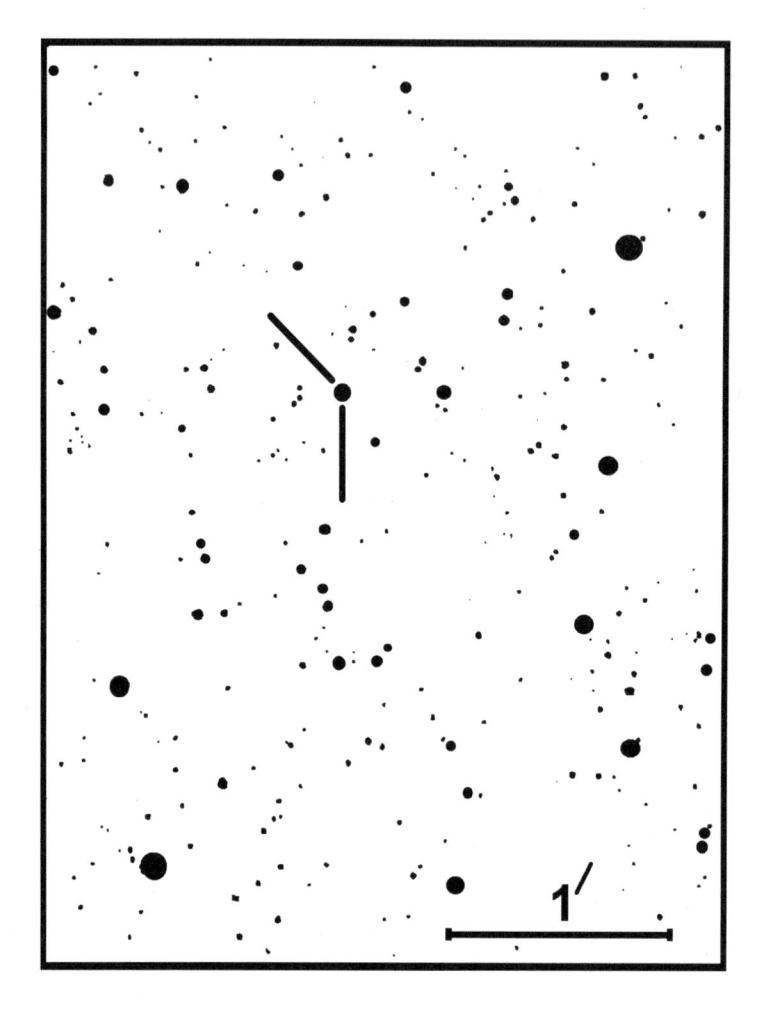

α (2000.0)=$06^h11^m12^s.1$ $V\sim17$
δ (2000.0)=$-81°49'17''.2$ $P=0^d.139(5)$
Type: IP Max=JD 2447737.554(3)

1H 0551−819 was discovered in the *HEAO-1* Survey [57]. It is a non-eclipsing novalike *CV*. The star is a 3-acrsec double, the easternmost brighter component is a *CV*, nearby optical contaminant component is a K0 star with $V \sim 14.034$, $B-V = 0.73$, $U - B = 0.41$. Max refers to the time of maximum redshift. Flickering is observed throughout the light curve of the star from 600 to 2400 s.

References: [57]

PG Gem/RE J0751+14

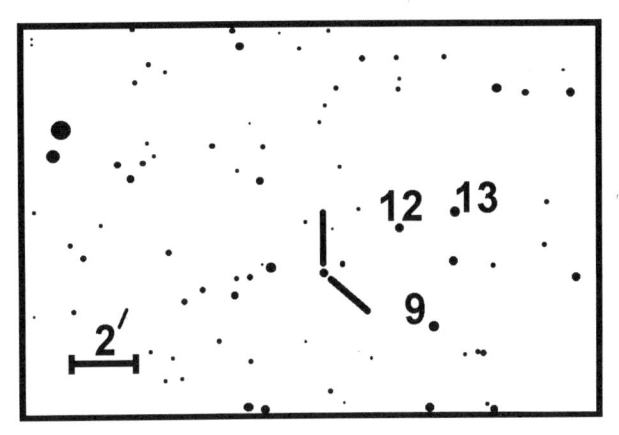

α (2000.0)=$07^h51^m17^s.3$
δ (2000.0)=$14°44'23''$
Type: IP?, NL, XS
V=13.65–14.5
P=$0^d.229167$

RE/RX J0751+14 is an extreme ultraviolet source. Its circular polarization is 4.2(6)% in I band and 2.2(3)% in R band, the linear polarization is 2.2% and 0.8% respectively. There is a clear polarization modulation with the spin period. A longer period, about $0^d.47$, exists. The comparison stars *9, 12, 13* were suggested, differential magnitudes between this stars are:

Stars	ΔV
12-9	−0.453(5)
13-9	−0.17(1)
13-12	0.281(7)

References: [153, 411, 422, 425]

S 193

α (2000.0)=$21^h51^m58^s.0$
δ (2000.0)=$14°06'52''.9$
Type: IP?, NL
V~13
P=$0^d.15$:

References: [102, 132, 229, 436]

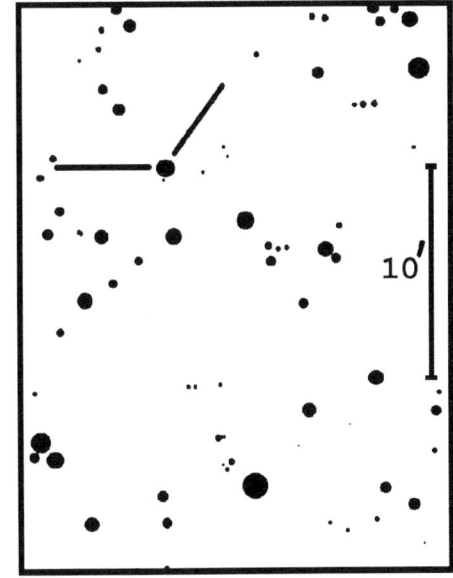

X 0022–7221 in 47 Tuc

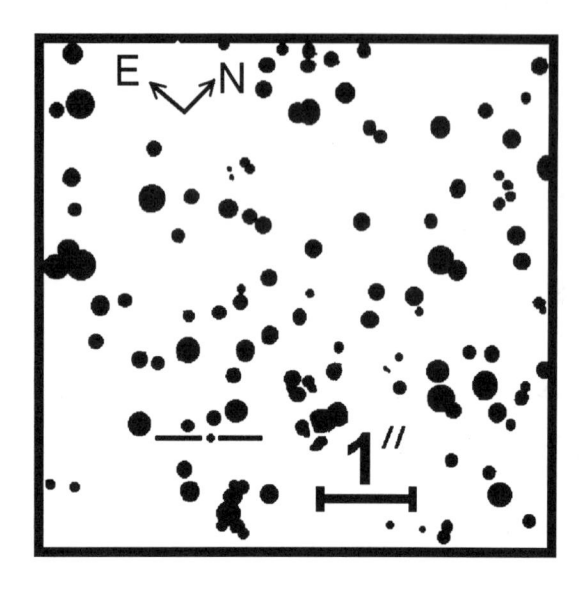

α (2000.0)=$00^h24^m01^s$
δ (2000.0)=$-72°04'$
  Type: IP?, NL, XS
  $V\sim 21$
  $P=0^d.25$:

The orientation for the frame is marked by arrows.

References: [252]

BI Lyn/PG 0900+401

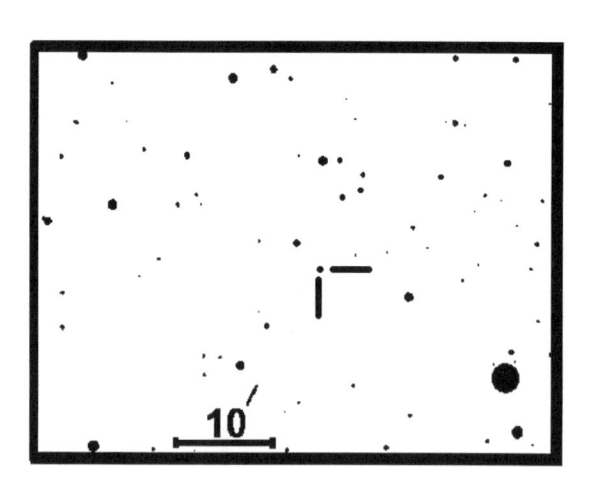

α (2000.0)=$09^h03^m19^s.5$
δ (2000.0)=$39°51'00''.3$
  Type: IP?, NL
  B=12.93–13.13
Spectrum: K3
  $P=0^d.33818$:
   or $0^d.514$:

The comparison stars are [194]:

Star	V	B	U	Star	V	B	U
a	6.34	6.67	6.70	d	10.30	11.01	11.14
b	9.29	10.30	11.0:	e	10.84	11.39	11.33
c	10.10	10.73	10.67	f	10.85	11.61	11.81

References: [115, 194]

AM HERCULIS TYPE STARS (POLARS)

BY Cam

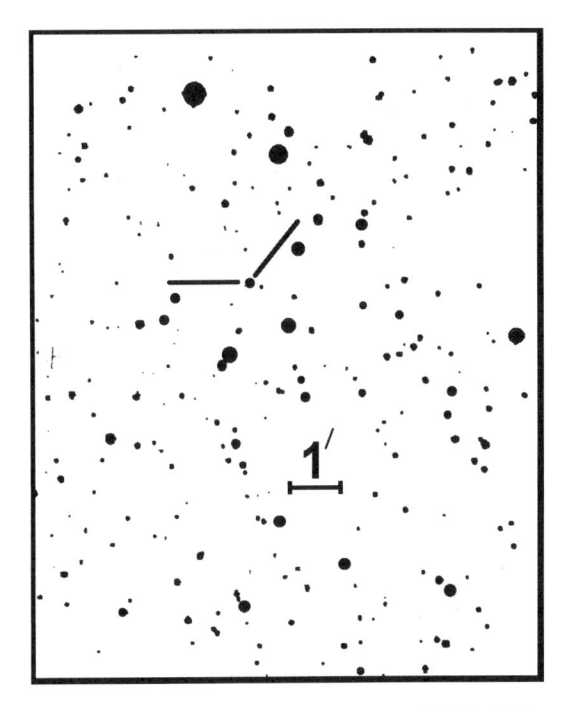

α (2000.0)=$05^h42^m49^s.0$
δ (2000.0) =$60°51'31''$
  Type: AM Her, NL, XS
  V=14.5–17
  P=$0^d.13842380(6)$
Min.phot.=JD 2446586.2559417(1)

BY Cam/H 0538+608 is a polar with asynchronous rotation. The light and polarimetric curves exhibit extreme flickering on time scale 5–10 min and are highly variable from night to night. The profiles show the broad and narrow components, an asymmetry in broad-line profiles may be interpreted as the contribution of the second component.

References: [276, 277, 424]

V 834 Cen

α (2000.0) =$14^h09^m07^s.6$
δ (2000.0) =$-45°17'18''.8$
  Type: AM Her, NL, XS
Spectrum: M6.5 V
  V=14.2–17.2
  P=$0^d.070497235(56)$
Min.phot.=JD 2445048.94238(5)

V 834 Cen /1E 1405-451 is a non-eclipsing binary system, variable soft X-ray source. There are 2 s QPOs of amplitude $3.3 \pm 1.1\%$ in V and $2.5 \pm 0.8\%$ in R. The high-state curve exhibits a broad, deep minimum at phase 0.1, at the low state the deep minimum is absent.

References: [209, 431]

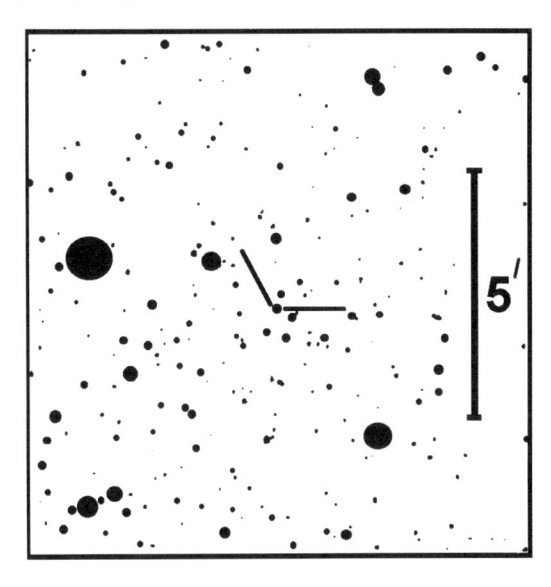

V 1500 Cyg/Nova Cygni 1975

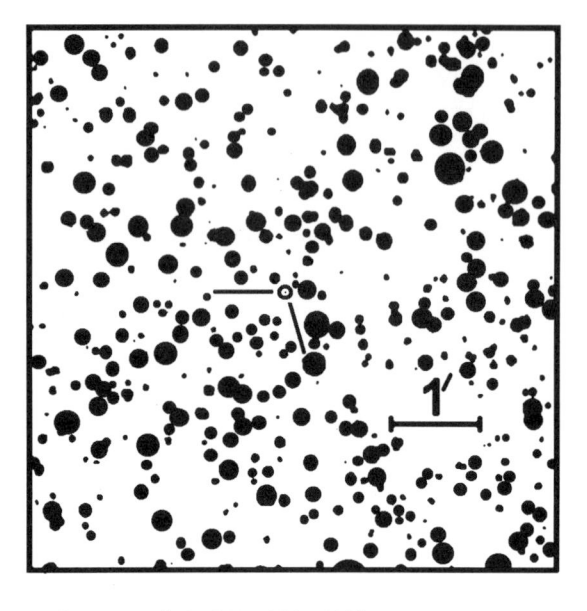

α (2000.0)=$21^h11^m36^s.6$
δ (2000.0)=$48°09'02''$
V=1.85–18.6
Type: AM Her, NA, NL
Spectrum: M4 V
$P=0^d.137164(1)$
Min.phot.=JD 2446875.971(4)

Nova Cygni 1975 was one of the brightest Novae. The system exhibited in 1978–1986 the long-time variability with the period $7^d.69(1)$ [265]. The ephemeris of polarized flux is Max = HJD $244685.992(5)+0^d.137154(4) \cdot E$ [329]. The observed asynchronism is the result of nova outburst.

References: [40, 265, 329, 432]

EF Eri/2A 0311–227

α (2000.0)=$03^h14^m13^s.1$
δ (2000.0)=$-22°35'42''.1$
Type: AM Her, NL, XS
Spectrum: later than M5
B=13.5–17.7
$P=0^d.056266$
Min.phot.=JD 2443894.68241

EF Eri is one of four AM Her systems with the 2–3 s QPOs, the amplitude of QPOs is $A \sim 1.3(1)\%$. This object has very hard X-ray emission. Star a has $V = 12^m.68$, $B - V = 0^m.70$.

References: [41, 402, 403]

UZ For/EXO 0333-255

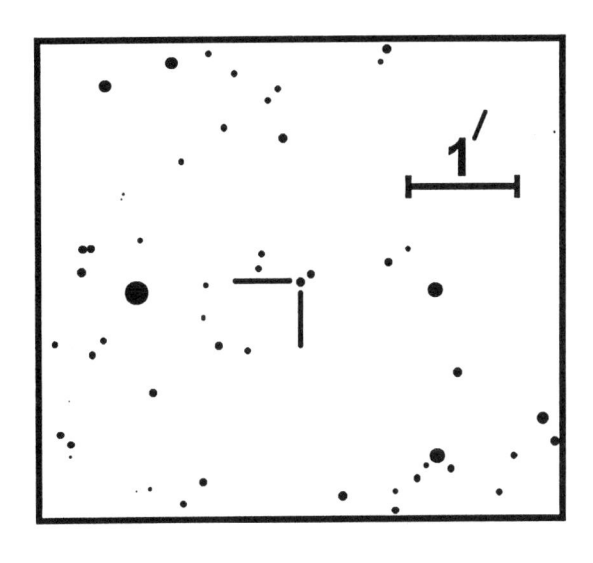

α (2000.0)=$03^h 35^m 28^s.7$
δ (2000.0)=$-25°44'23''.2$
Type: AM Her, NL, XS
Spectrum: M4.5 V
V=17–20.9
P=$0^d.087864(3)$
Min.phot.=JD 2447088.65410

UZ For is a highly inclined *eclipsing* system with a narrow eclipse in the X-rays. The eclipse duration is $0^d.087865$. It is one of massive polars.

References: [102, 413]

CE Gru/Grus V1/Hawkins V1

α (2000.0)=$21^h 37^m 56^s.5$
δ (2000.0)=$-43°42'14''.1$
Type: AM Her?, NL
B=17.5–20.7
P=$0^d.07537(35)$
Max(U)=JD 2446698.977(5)

CE Gru is a faint variable system. The period is derived from the U band observations [347].

References: [102, 145, 347]

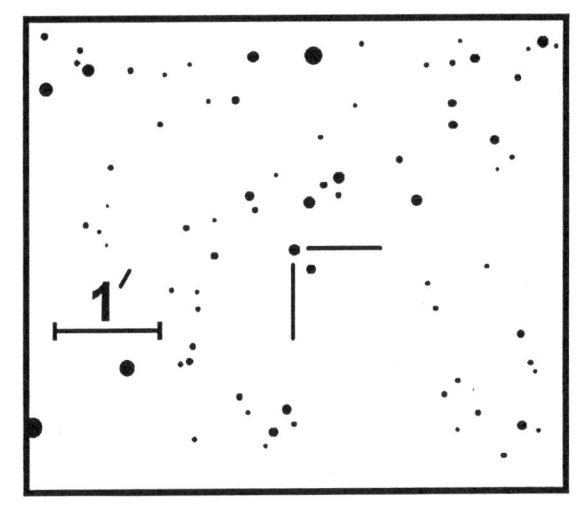

AM Her/2A 1814+4950

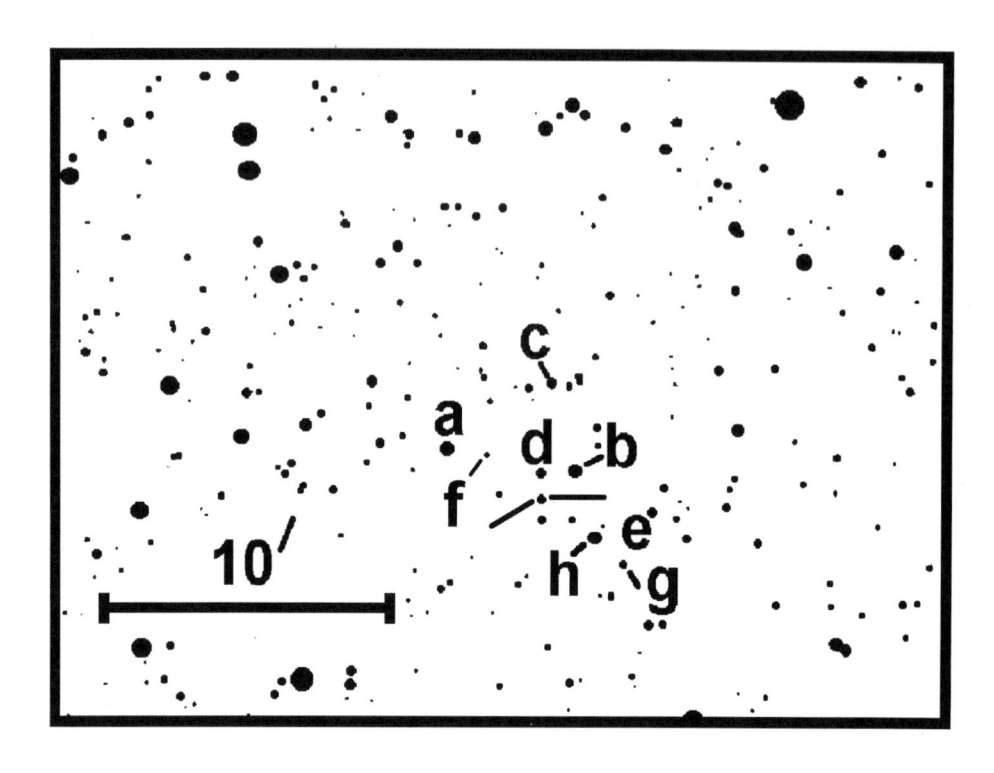

α (2000.0)=$18^h16^m13^s.4$
δ (2000.0) =$49°52'03''.2$
   Type: AM Her, NL, XS
Spectrum: M3 V

V=12.0–15.5
P=0^d.128927041(5)
Min.phot.=JD 2443014.76614

The comparison stars are [10]:

Star	V	B	Star	V	B
a	11.76	...	e	13.13	...
b	12.20	12.79	f	14.64	15.44
c	12.40	12.98	g	14.07	14.84
d	13.10	13.82	h	14.30	14.95

AM Her/3U 1809+50 is an *eclipsing* system. It has two active magnetic poles. Regular oscillations were discovered with the typical timescale of 272 ± 7 s [44].

References: [10, 44, 318, 420]

WW Hor/EXO 0234–5232

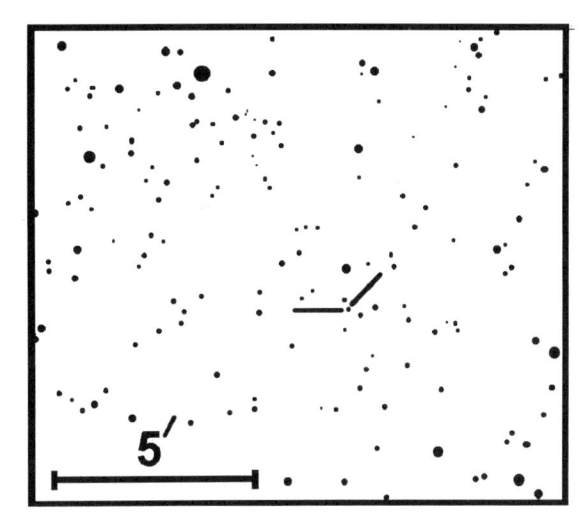

α (2000.0)=$02^h36^m11^s.6$
δ (2000.0)=$-52°19'14''.4$
 Type: AM Her, NL, XS
Spectrum: M6 V
  V=17–21.2
  P=$0^d.080199035(8)$
Min.phot.=JD 2447126.12782

WW Hor is an *eclipsing* system. The system was fainter in 1992 than in 1987. The longitude of the spot is changing.

References: [37, 415]

BL Hyi/H 0139–68

α (2000.0)=$01^h41^m00^s.4$
δ (2000.0)=$-67°53'28''.6$
 Type: AM Her, NL, XS
Spectrum: M3-4 V
  V=14.7–18.5
  P=$0^d.07891518(4)$
Min.phot.=JD 2444884.2176(6)

BL Hyi is a non-eclipsing two-pole system. During the orbital period, there are two linear polarization splashes. The second splash is about 50 min after the first one. Very soft X-rays are observed, there is no hard (≥ 0.4 keV) X-ray emission. "Y" is a constant star.

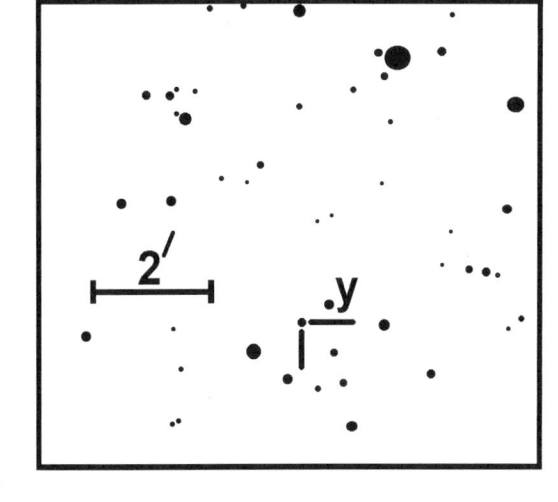

References: [357, 433]

DP Leo/1E 1114+182

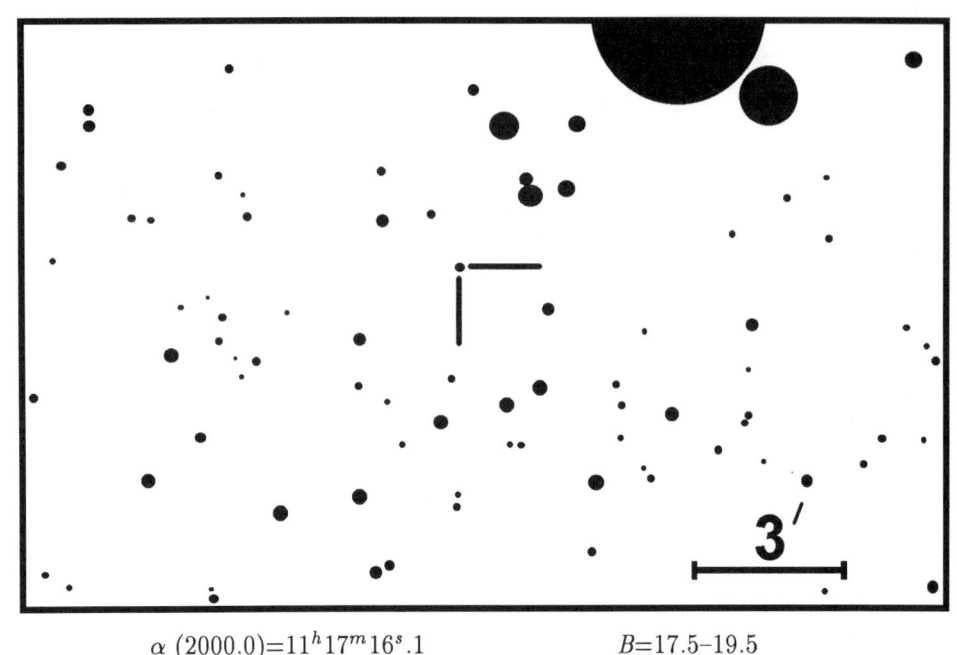

α (2000.0)=$11^h17^m16^s.1$ B=17.5–19.5
δ (2000.0)=$17°57'36''.8$ $P=0^d.06236286(1)$
Type: AM Her, NL, XS *Min.phot.*=JD 2446733.7000
Spectrum: M6 V

DP Leo is a long-period and flickering *eclipsing* system. Two emission poles are seen [90].

References: [38, 90, 435]

ST LMi/CW 1103+254

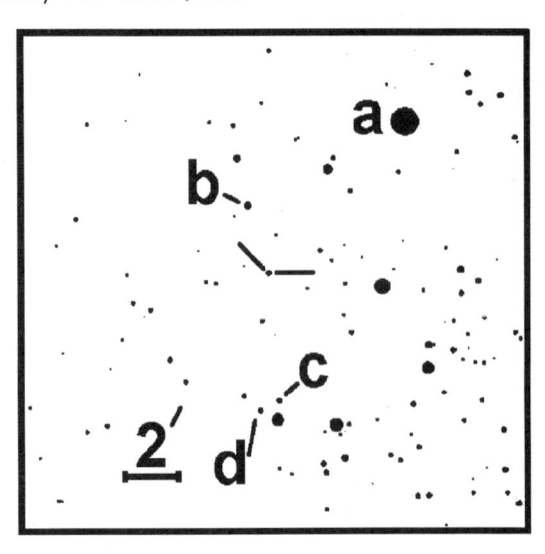

α (2000.0)=$11^h05^m39^s.8$
δ (2000.0)=$25°06'27''.8$
Type: AM Her, NL, XS
Spectrum: M5–6 V
V=15.0–17.2
$P=0^d.07908908(8)$
Max.pol.=JD 2445059.7024

The comparison stars are [130]:

Star	B	Star	B
a	7.5	c	15.49
b	15.07	d	16.22

References: [130, 328, 416]

GQ Mus/N Mus 1983

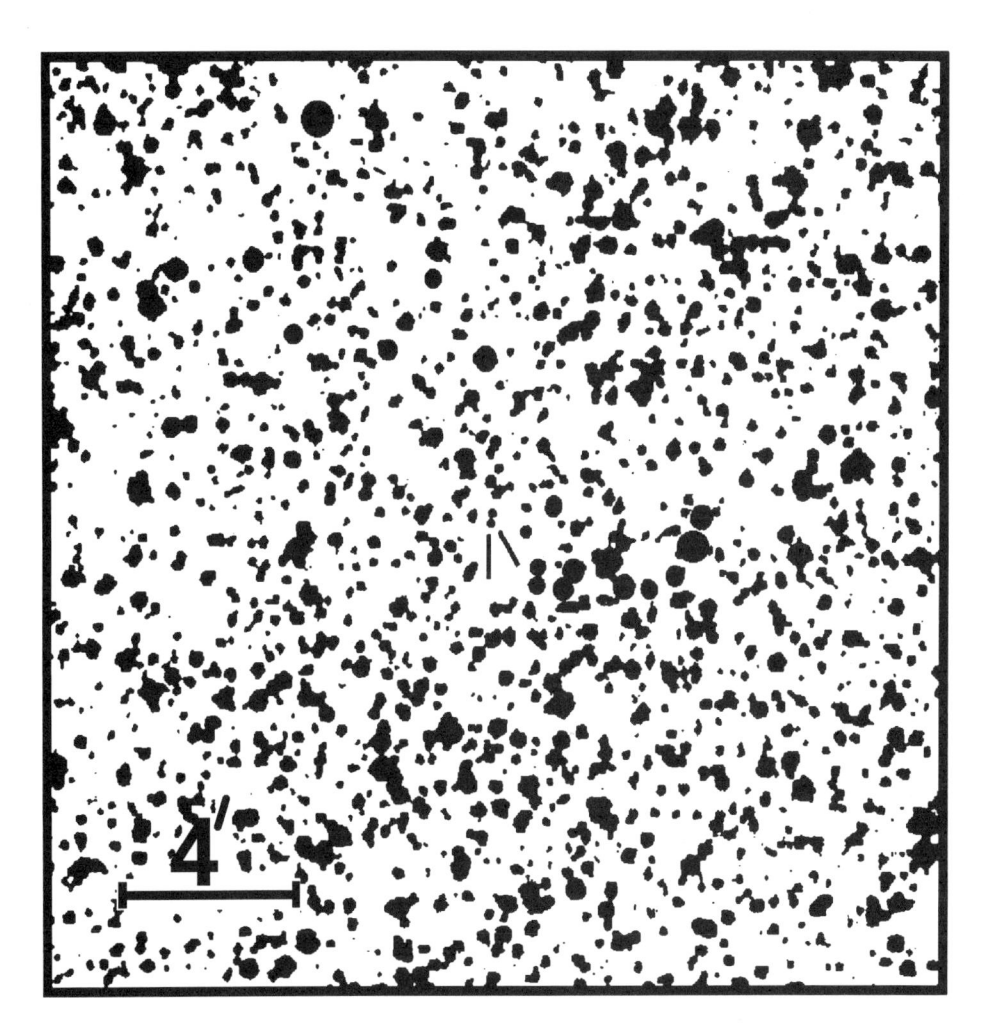

α (2000.0)=$11^h52^m02^s.4$
δ (2000.0)=$-67°12'20''.2$
Type: AM Her?, NA
V=7.2–21
P=$0^d.0593650(2)$
Half-max=JD 2447843.4721(6)

GQ Mus is a moderately fast ($t_2 \simeq 18$ days) classical nova. It erupted in January 1983, amplitude $\Delta m_V \leq 14$. The amplitude of the orbital modulations is $\Delta V = 3.3$. In 1989 the period P was $0^d.0593612(8)$ [96], in 1990 $P = 0^d.05936(8)$, the period derivative $\dot{P} \leq 10^{-6} yr^{-1}$ [97].

References: [40, 96, 97]

V 2051 Oph

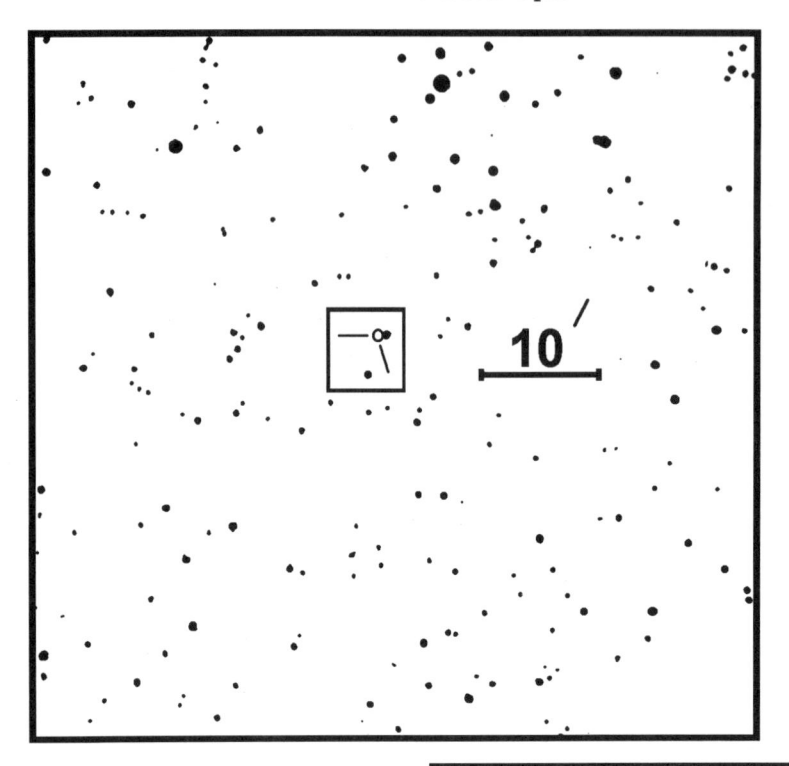

α (2000.0)=$17^h08^m19^s.2$
δ (2000.0)=$-25°48'35''$
  Type: AM Her?, IP?, UG?
    V=13.0–17.5
Spectrum: later than M4 V
    P=$0^d.062427860(2)$
Min.phot.=JD 2444787.32124(5)

The comparison stars are:

Star	V	B	U
a	10.33	10.85	10.96
b	11.87	12.65	12.93
c	13.52	15.45	16.85
d	10.95	11.56	11.72
e	14.13	15.99	17.1:

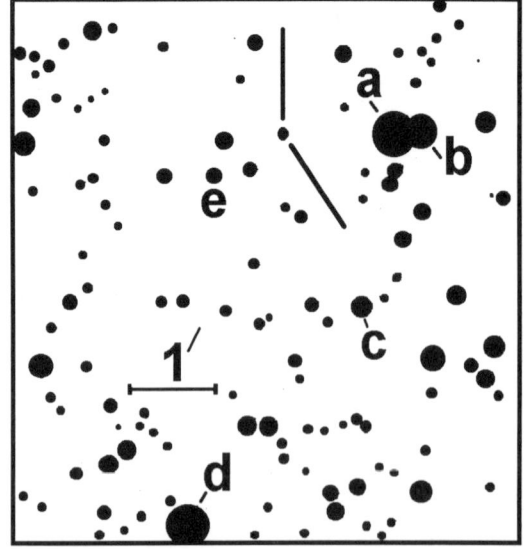

V 2051 Oph is an *eclipsing* system. Usually $\sim 15^m.0$ with strong flickering up to $0^m.4$ in 5 minutes and rare flares of low amplitude. The optical polarization is small. It is not clear yet if V 2051 Oph is a polar or an intermediate polar.
References: [79, 111, 176, 216, 360, 362, 402]

EP Dra/H 1907+690

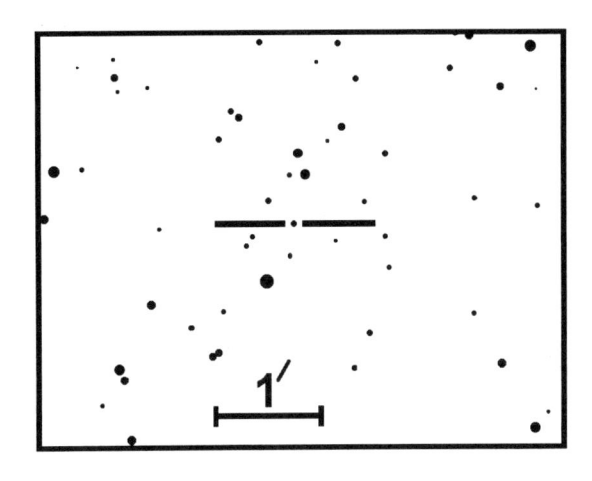

α (2000.0)=$19^h07^m06^s.9$
δ (2000.0)=$69°08'40''$
Type: AM Her, NL, XS
V=17.6–20.8
$P=0^d.0726562$
Min.phot.=JD 2447681.7292(1)

EP Dra is an *eclipsing* binary system with circular polarization of $2.66 \pm 0.09\%$.

References: [277]

VV Pup/RE 0812-1254

α (2000.0)=$08^h15^m06^s.8$
δ (2000.0)=$-19°03'18''$
Type: AM Her
Spectrum: M4–5 V
V=13.9–18.0
$P=0^d.0697468256$

The comparison stars are:

Star	V	B
b	14.76	15.03
c	14.83	15.32
d	12.05	12.48
e	14.19	14.80
f	13.34	13.77
g	13.76	14.36

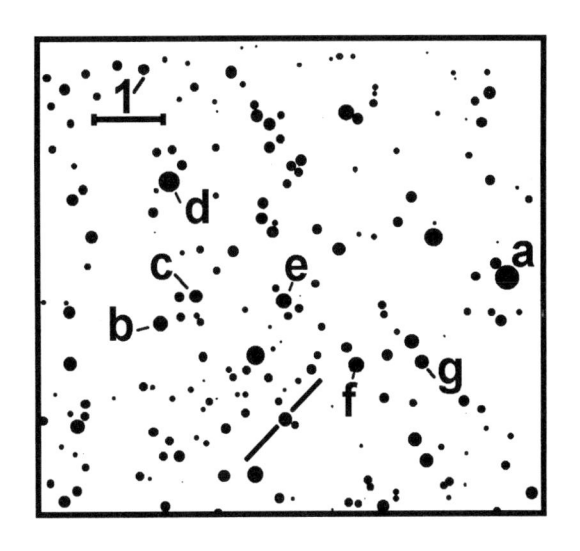

Star "a" is SAO 154020. VV Pup is a two-pole system.

References: [216, 259]

MR Ser/PG 1550+191

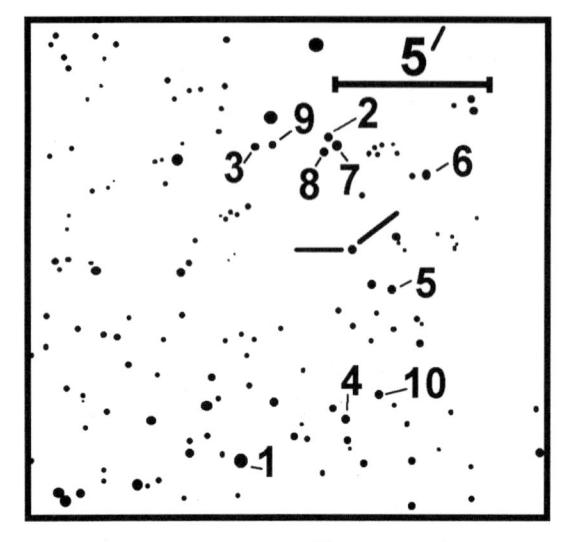

α (2000.0)=$15^h52^m47^s.3$
δ (2000.0)=$18°56'27''.1$
   Type: AM Her, NL, XS
   V=14.7–17
Spectrum: M5–6 V
   P=$0^d.078880(1)$
Min.phot.=JD 2448446.728(2)

MR Ser is a non-eclipsing system, it is fainter than other polars in soft X-rays. It is possible that two accretion poles are visible.

The comparison stars are [318]:

Star	1	2	3	4	5	6	7	8	9	10
B	13.40	16.08	14.75	15.01	15.26	15.52	15.71	15.95	15.99	16.00
V	12.55	15.05	14.20	13.78	14.39	14.57	15.09	14.89	15.12	14.36

References: [4, 318]

DW UMa/PG 1030+590

α (2000.0)=$10^h33^m53^s.1$
δ (2000.0)=$58°46'54''$
   Type: AM?, NL, UX, VY
   V=14.5–17.1
   P=$0^d.13660649(1)$
Mid.ecl.=JD 2446229.0070(1)

DW UMa is a nova-like *eclipsing* system with a deep minimum ($\Delta m \sim 1.5$–2). The duration of the eclipse is about 20 min [151].

References: [151, 178, 417]

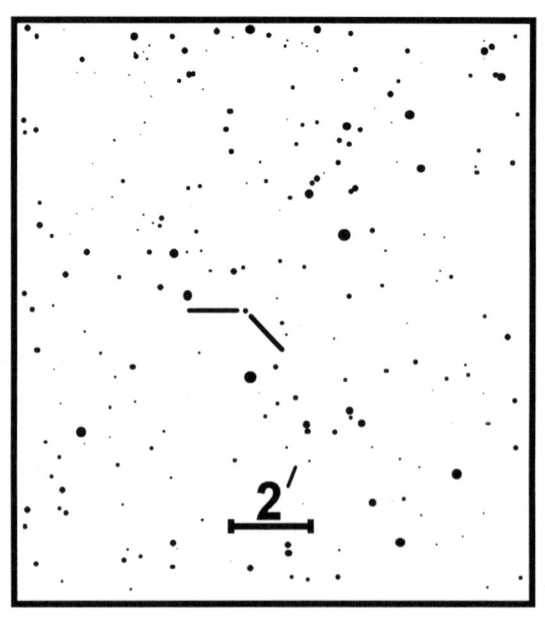

AN UMa/PG 1101+453

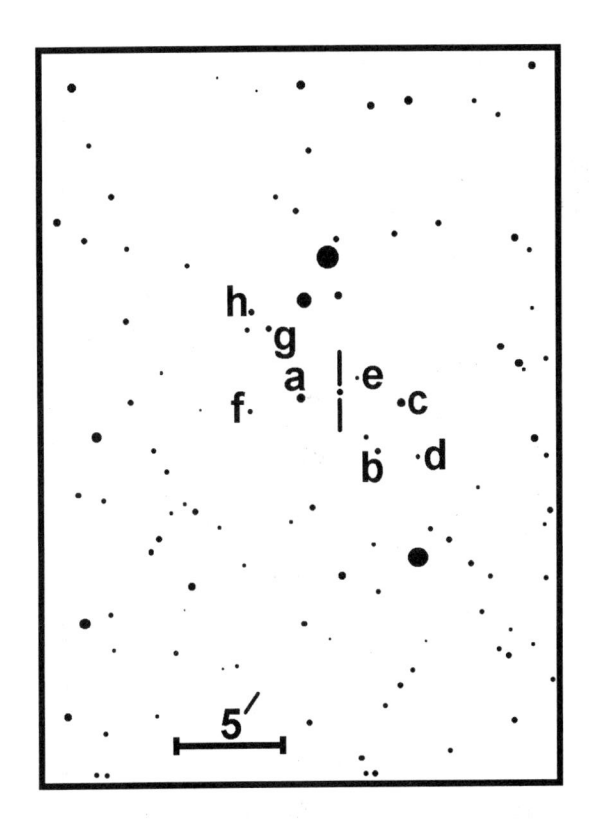

α (2000.0)=$11^h04^m25^s.8$
δ (2000.0)=$45°03'14''$
  Type: AM Her, NL, XS
Spectrum: later than M6 V
  B=14–20.2
  $P=0^d.07975307$
 Min.V=JD 2447126.1278

The comparison stars are [316]:

Star	V	B
a	12.02	12.71
b	14.52	15.50
c	15.55	16.65
d	16.13	16.90
e	16.54	17.44
f	16.29	16.96
g	16.81	17.55
h	17.10	17.83

References: [316]

EK UMa/1E 1048.5+5421

α (2000.0)=$10^h51^m35^s.4$
δ (2000.0)=$54°04'31''.2$
  Type: AM Her, NL, XS
  V=18–20
Spectrum: M5 V
  $P=0^d.0795(1)$
 Min.phot.=JD 2443191.7879

EK UMa is an X-ray source, noneclipsing binary with high circular polarization up to 20%.

References: [228]

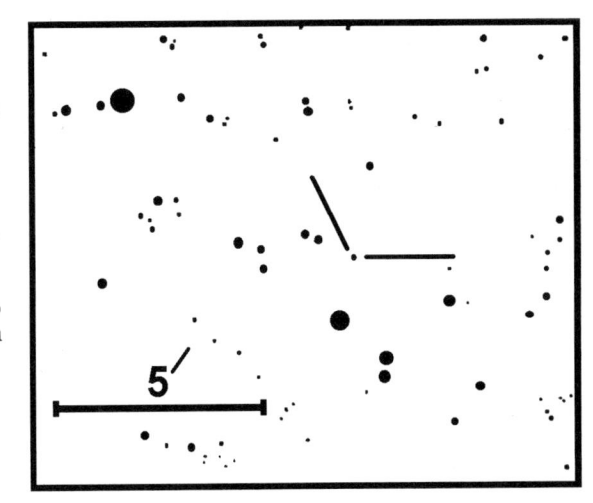

QQ Vul/1E 2003+225

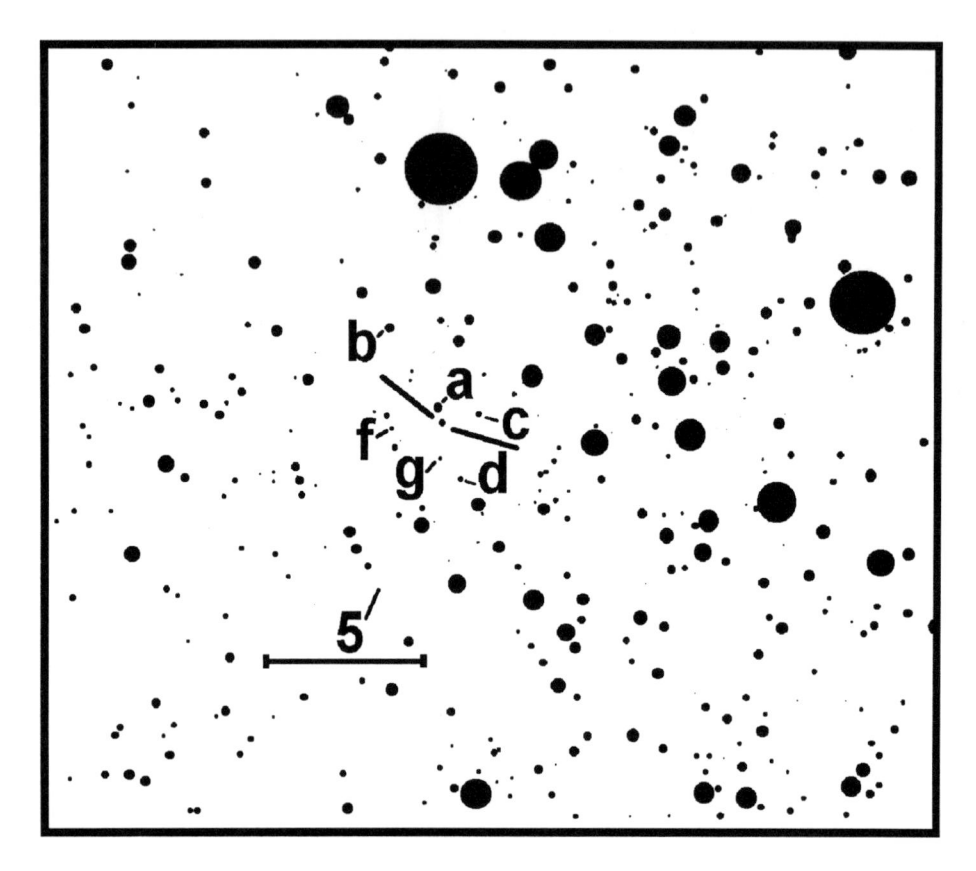

α (2000.0)=$20^h 05^m 42^s.0$ V=14.5–17.0
δ (2000.0) =$22°39'58''$ Spectrum: M2–4 V
     Type: AM Her, NL, XS $P=0^d.15452036(2)$

QQ Vul is a non-eclipsing binary system. The magnitudes of the visual companion with G-K spectrum are: $V = 18.06(5)$, $R = 16.85(2)$, $I = 16(2)$. The long-period irregular changes of the light curve are observed with the characteristic time 4100–4400^d [9].

The comparison stars are [11]:

Star	a	b	c	d	f	g
B	14.8	15.0	15.4	16.0	16.9	17.5

References: [9, 11]

SS UMi/PG 1551+719

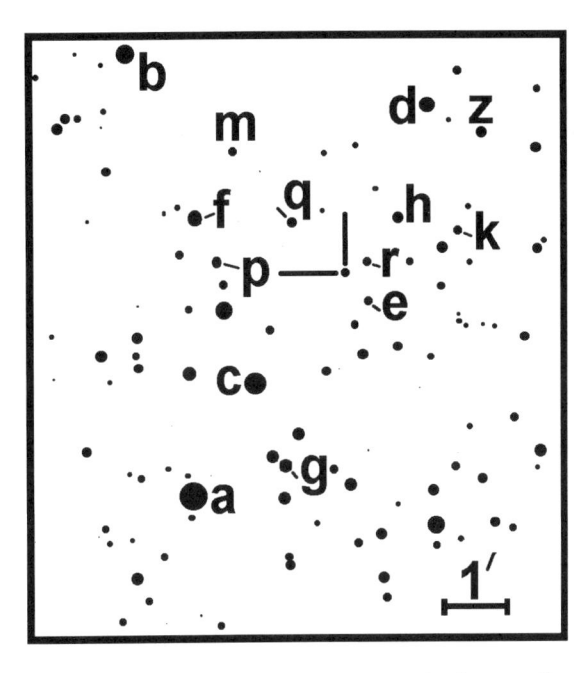

α (2000.0)=$15^h51^m22^s.4$
δ (2000.0)=$71°45'11''.2$
  Type: AM Her?, DN, UG
  V=12.6–17.6
  $P=0^d.0684?$
     $0^d.0701?$
The comparison stars are:

Star	V	B–V
b	12.08	0.03
c	12.70	0.09
d	13.14	0.05
e	17.34	0.13
f	13.96	0.08
g	15.08	0.15
h	15.96	0.11
k	16.06	0.15
m	16.19	0.12
p	17.28	0.24
q	16.38	0.14
r	17.30	0.16
z	14.92	0.10

SS UMi has 10 min superhumps with the amplitude $\Delta m \sim 0.3$. It has two types of maxima: faint and bright, and flickering. The star "a" is variable.

References: [7, 66]

EXO 0329–260/EXO 032957–2606.9

α (2000.0)=$03^h32^m04^s.5$
δ (2000.0)=$-25°56'55''$
  Type: AM Her, NL, XS
  Spectrum: M3.5–4.5 V
  $V \sim 17.5$
  $P=0^d.1586(8)$
  Min.phot.=JD 2447092.6501

References: [102, 414]

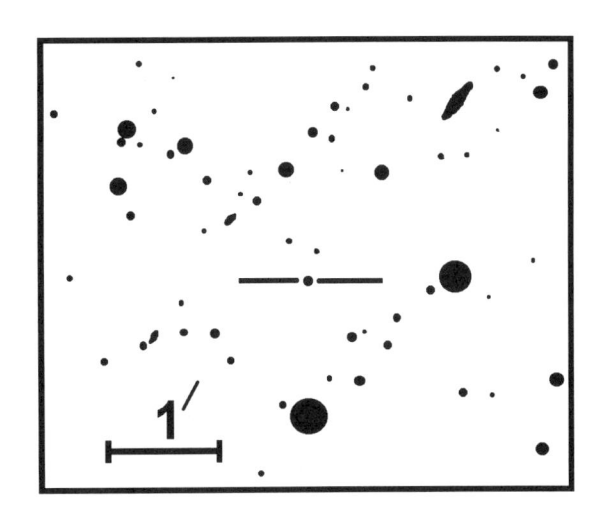

RX J 0531−46

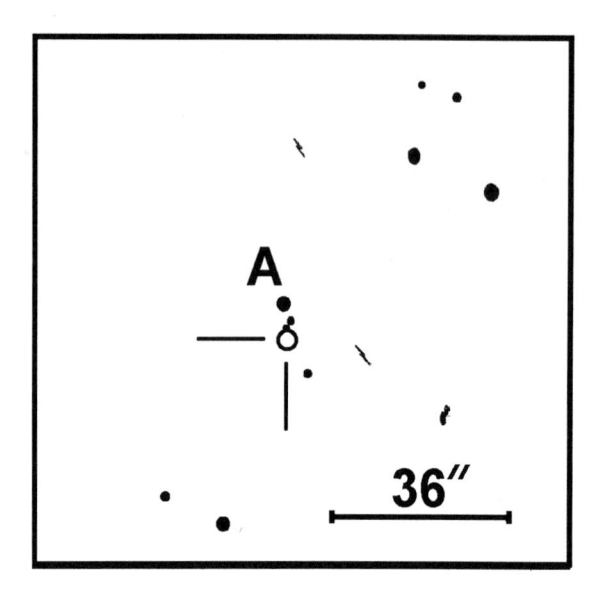

α (2000.0)=$05^h31^m35^s.8$
δ (2000.0)=$-46°24'07''$
  Type: AM Her?, NL, XS
  $V\sim17$
  $P=0^d.0924$:

The V filter CCD image. The uncertainty of the X-ray position is about of 12''. The optical counterpart is the $V \simeq 17$ mag object laying 8'' south of the X-ray position labeled "A".

References: [430]

1E 0830.9−2238

α (2000.0)=$08^h33^m05^s.75$
δ (2000.0)=$-22°48'32''.6$
  Type: AM Her ?
  $V\sim17.7$

References: [149]

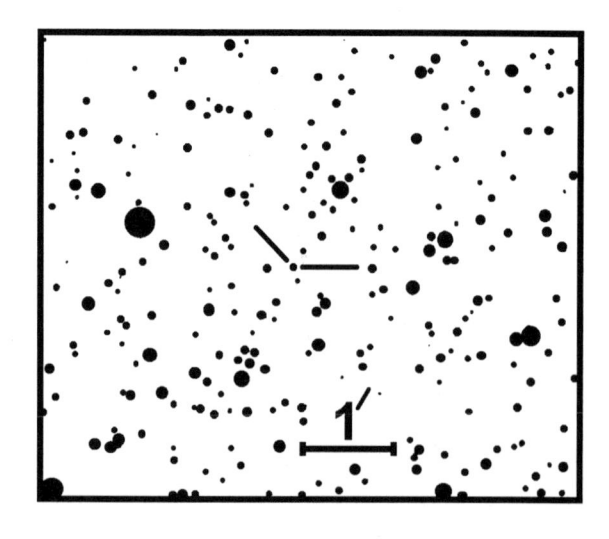

RX J0929.1−2404

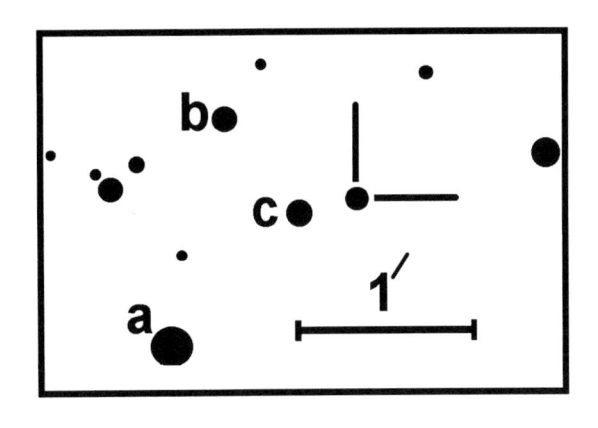

α (2000.0)=$09^h29^m07^s$
δ (2000.0)=$-24°04'56''$
Type: AM Her?
V=17.03
P=0^d.1412(3)
Min.phot.=JD 2449007.588(1)

RX J0929.1−2404 is an *eclipsing* X-ray source. Its orbital period is above the "period gap". The stars marked a, b, c may be used as comparison for differential photometry.

References: [303]

EU UMa/RE 1149+28

α (2000.0)=$11^h49^m55.7^s$
δ (2000.0)=$28°45'07''.5$
Type: AM Her?, NL, XS
B~16.5
P=0^d.062597

RE 1149+28 is a highly variable extreme UV source. It is a non-eclipsing system. The spectrum of the star exhibits strong Balmer lines and He II λ 4686 Å. It has the largest *EUV/Opt* ratio among polars (~ 17.5). Stars "*1*" and "*2*" are the comparison stars used in CCD photometry.

References: [226, 422]

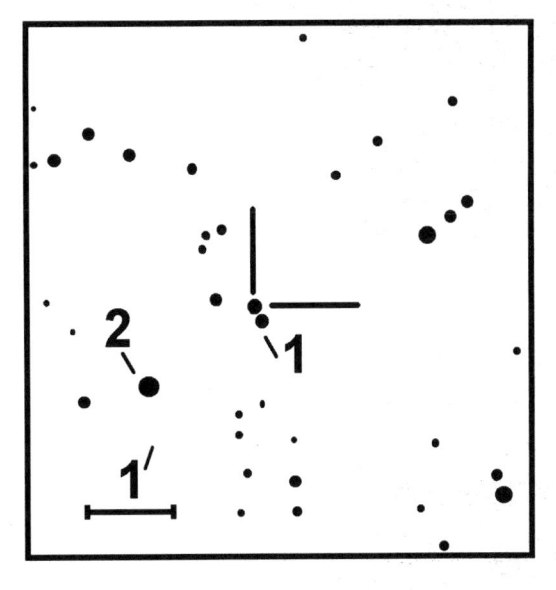

EV UMa/RE 1307+537

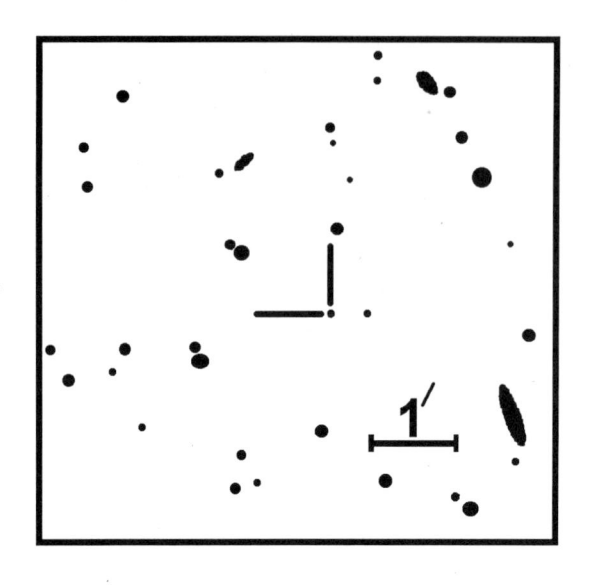

α (2000.0)=$13^h07^m56^s.4$
δ (2000.0)=$53°51'37''$
   Type: AM Her?, NL, XS
   $V\sim17$
   $P=0^d.05534$

RE 1307+537 is one of the faint sources suggested as polars. *ROSAT* EUV observations [248] show that this object is one of the most distant polars at a distance of $z \geq 630$ pc above the Galactic plane. It has the shortest orbital period known for polars.

References: [248]

V 2009-65.5

α (2000.0)=$20^h08^m55^s.8$
δ (2000.0)=$-65°27'43''$
   Type: AM Her
   $V\sim17$
   $P=0^d.1109$

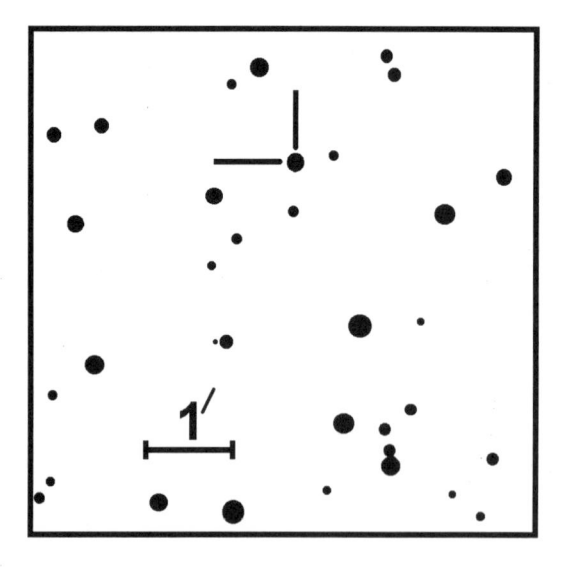

V 2009-65.5 is a new AM Her variable [401]. It has a typical AM Her type spectrum with the broad H I, He I and He II emission lines, and the C-N blend at λ 4650 Å. The spectra show the strong negative circular polarization (to −8%). It is a "two-pole" system, the light curve shows two maxima per orbital period. The first maximum at phase 0.15 is seen at all wavebands, the second one (at phase 0.55) is clearly seen at redder wavebands.

References: [401]

References

1. Abbott T.M.C., Robinson E.L., Hill G.J., Haswell C.A. 1992 *Ap.J.*, **399**, 680
2. Abramenko A.N., Gershberg R.E., Pavlenko E.P. et al. 1978 *MNRAS*, **184**, 27P
3. Ahnert P. 1965 *MVS*, **3**, No 1, 7
4. Allen D.A. 1984 *Proc.ASA*, **5**, No 3, 369
5. Allen D.A., Wright A.E. 1988 *MNRAS*, **232**, 683
6. American Association of Variable Star Observers. Catalog of Variable Star Charts, Cambridge, 1974
7. Andronov I.L. 1986 *Astron. Circ.*, No 1432
8. Andronov I.L. 1991 *IBVS*, No 3645
9. Andronov I.L., Fuhrmann B. 1987 *IBVS*, No 2976
10. Andronov I.L., Korotin S.A. 1982 *Astron.Circ.*, No 1223
11. Andronov I.L., Yavorsky Yu.B. 1983 *Pis'ma Astron.Zh.*, **9**, 556
12. Andronov I.L. 1980 *Astron.Circ.*, No 1417
13. Antokhina E.A., Cherepashchuk A.M., Pavlenko E.P., Shugarov S.Yu. 1993 *Astron.Zh.*, **70**, 804
14. Armstrong J.T., Johnston M.D., Bradt H.V. et al. 1980 *Ap.J.*, **236**, L131
15. Arnold S. et al. 1976 *Ap.J.*, **206**, 790
16. Aslanov A.A., Khruzina T.S. 1988 *Sov.Astr.*, **32**, 100
17. Bailey J. 1979 *MNRAS*, **188**, 681
18. Bailey J., Ward M. 1981 *MNRAS*, **194**, 17P
19. Bailyn C.D. 1992 *Ap.J.*, **391**, 298
20. Bailyn C.D., Grindlay J.E. 1987 *Ap.J.*, **312**, 748
21. Barbieri C. et al. 1985 *Astr.Ap.Suppl.Ser.*, **61**, 163
22. Barwig H., Ritter H., Barnbanter O. 1994 *Astr.Ap.*, **288**, 204
23. Barwig H., Schoembs R. 1981 *IBVS*, No 2031
24. Barwig H., Schoembs R. 1987 *Messenger*, **47**, 19
25. Basko M.M., Goranskij V.P., Lyuty V.M. et al. 1976 *PZ*, **20**, 219
26. Bateson F.M. 1976 *NZAS Publ.*, 4(C76), 39
27. Bateson F.M. 1977 *NZAS Bull.*, Nov. M77/11
28. Bateson F.M. 1977 *NZAS Publ.*, **5**, 27
29. Bateson F.M. 1979 *NZAS Publ.*, **7**, 47
30. Bateson F.M. 1982 *NZAS Publ.*, 10(C82), 12
31. Beavers W.I., Herczeg T.J., Lui A. 1986 *Ap.J.*, **300**, 1
32. Becker R.H. et al. 1982 *MNRAS*, **201**, 265
33. Bell S.A., Hilditch R.W., Pollacco D.L. 1993 *MNRAS*, **265**, 1042
34. Belyakina T.S. 1992 *IK*, **84**, 49
35. Bernacca P.L. et al. 1984 *Astr.Ap.*, **132**, L8
36. Bertola F. 1964 *Ann.Rev.Astr.Ap.*, **27**, 298
37. Beuermann K., Thomas H.-C., Giommi P., Tagliaferri G. 1987 *Astr.Ap.*, **175**, L9
38. Biermann P., Schmidt G.D., Liebert J. et al. 1985 *Ap.J.*, **293**, 303
39. Blair W.P., Raymond J.C., Dupree A.K. et al. 1984 *Ap.J.*, **278**, 270

40. Bode M.F., Duerbeck H.W., Evans A. 1989 In: *Classical Novae*/eds. Bode M., Evans A. Chicester (UK), 249
41. Bond H.E., Chanmugam G., Grauer A.D. 1979 *Ap.J.*, **234**, L113
42. Bond H.E., Liller W., Mannery E.J. 1978 *Ap.J.*, **223**, 252
43. Bond N.E. et al. 1982 *Ap.J.*, **260**, L79
44. Bonnet-Bidaud J.M., Somova T.A., Somov N.N. 1991 *Astr.Ap.*, **251**, L27
45. Borngen F. et al. 1970 *AN*, **292**, H.3, 103
46. Bradt H.V. et al. 1977 *Nature*, **269**, 21
47. Bradt H.V., Apparao K.M.V., Dower R. et al. 1977 *Nature*, **269**, 496
48. Breo A., Pti M. 1959 *PZ*, **12**, 18
49. Brucato R.J., Kristian J. 1972 *Ap.J.*, **173**, L105
50. Bruch A. 1983 *IBVS*, No 2287
51. Bruch A. 1984 *Astr.Ap.Suppl.Ser.*, **56**, 441
52. Bruch A., Fischer F.-J., Wilmsen U. 1987 *Astr.Ap.Suppl.Ser.*, **70**, 481
53. Bruch A. et al. 1987 *Astr.Ap.*, **185**, 203
54. Buckley D.H.A., O'Donoghue D., Kilkenny D. et al. 1992 *MNRAS*, **258**, 285
55. Buckley D.A., Sullivan D.J., Remillard R.A. et al. 1990 *Ap.J.*, **355**, 617
56. Buckley D.A., Tuohy I.R. 1990 *Ap.J.*, **349**, 296
57. Buckley D.A.H., Remillard R.A., Tuohy I.R. et al. 1993 *MNRAS*, **265**, 926
58. Budzinovskaya I.A., Pavlenko E.P., Shugarov S.Yu. 1992 *Sov.Astr.Lett.*, **18**, 201.
59. Burenkov A.N., Voikhanskaya N.F. 1979 *Pis'ma Astron.Zh.*, **5**, 452
60. Caraveo P.A., Bignami G.F., Goldwurm A. 1989 *Ap.J.*, **338**, 338
61. Casares J., Charles P.A., Naylor T. 1992 *Nature*, **355**, 614
62. Casares J., Charles P.A., Naylor T., Pavlenko E.P. 1993 *MNRAS*, **265**, 834
63. Castro-Tirado A.J., Pavlenko E.P., Shlyapnikov A.A. et al. 1993 *Astr.Ap.*, **276**, L37
64. Charles P.A., Naylor T. 1992 *MNRAS*, **255**, 6P
65. Chavira E. 1959 *TTB*, **18**, 3
66. Chen J.-S., Lui X.-W., Wei M.Z. 1990 *Astr.Ap.*, **242**, 397
67. Chevalier C., Ilovaisky S.A. 1977 *Messenger*, **9**, 4
68. Chevalier C., Ilovaisky S.A. 1981 *Astr.Ap.*, **94**, L3
69. Chevalier C., Ilovaisky S.A., Charles P.A. 1985 *Astr.Ap.*, **147**, L3
70. Chevalier C., Ilovaisky S.A. 1987 *Astr.Ap.*, **172**, 167
71. Chevalier C., Ilovaisky S.A., Motch C. et al. 1989 *Astr.Ap.*, **217**, 108
72. Chevalier C., Ilovaisky S.A., van Paradijs J. et al. 1989 *Astr.Ap.*, **210**, 114
73. Chevalier C., Ilovaisky S.A. 1994 Preprint of *Observatoire de Haute-Provence*, No 83
74. Chlebowski T., Halpern J.P., Steiner J.E. 1981 *Ap.J.*, **247**, L35
75. Chochol D. 1988 In: *The Symbiotic Phenomenon, IAU Colloq. No 103*/eds. Mikołajewska J. et al., Dordrecht: Kluwer Acad.Publ., 281
76. Chochol D., Hirc L., Urban Z. et al. 1993 *Astr.Ap.*, **277**, 103
77. Coe M.J., Jones L.R. 1992 *MNRAS*, **259**, 191
78. Cook M.C., Warner B. 1981 *MNRAS*, **196**, 55P
79. Cook M.C., Brunt C.C. 1983 *MNRAS*, **205**, 465

80. Cordova F.A., Garmire G.P. 1979 *Nature*, **279**, 782
81. Cordova F.A. et al. 1981 *MNRAS*, **196**, 1
82. Cordova F.A., Mason K.O. 1984 *MNRAS*, **206**, 879
83. Cowley A.P., Crampton D., Hutchings J.B. 1977 *Ap.J.*, **218**, L3
84. Cowley A.P., Crampton D., Hutchings J.B. 1978 *AJ*, **83**, 1619
85. Cowley A.P., Hutchings J.B., Schmidtke P.C. et al. 1988 *AJ*, **95**, 1231
86. Cowley A.P., Schmidtke P.C. 1990 *AJ*, **99**, 678
87. Cowley A.P., Schmidtke P.C., Crampton D., Hutchings J.B. 1990 *Ap.J.*, **350**, 288
88. Crampton D., Cowley A.P., Hutchings J.B. et al. 1990 *Ap.J.*, **355**, 496
89. Crocker D.A. 1978 *JAAVSO*, **7**, 73
90. Cropper M., Wickramasinghe D.T. 1993 *MNRAS*, **260**, 696
91. Davidsen A., Malina R., Bowyer S. 1977 *Ap.J.*, **211**, 866
92. Davidsen A., Malina R., Smith H. et al. 1974 *Ap.J.*, **193**, L25
93. Davies S.R., Coe M.J. 1991 *MNRAS*, **249**, 313
94. Densham R.H., Charles P.A. et al. 1983 *MNRAS*, **205**, 1117
95. De Freitas Pacheco J.A. 1974 *Astr.Ap.*, **35**, 301
96. Diaz M.P., Steiner J.E. 1989 *Ap.J.*, **339**, L41
97. Diaz M.P., Steiner J.E. 1994 *Ap.J.*, **425**, 252
98. Dokuchaeva O.D. 1976 *IBVS*, No 1189
99. Donoghue D.O. 1987 *Ap.Space Sci.*, **136**, 247
100. Doroshenko V.T., Lyuty V.M., Terebizh V.Yu. 1977 *Pis'ma Astron.Zh.*, **3**, 510
101. Dower R.G., Apparao K.M.V., Bradt H.V. et al. 1978 *Nature*, **273**, 364
102. Downes R.A., Shara M.M. 1993 *PASP*, **105**, 127
103. Downes R.A. 1982 *PASP*, **94**, 950
104. Doxsey R. et al. 1977 *Nature*, **269**, 112
105. Doxsey R.E., Apparao K.M.V., Bradt H.V. et al. 1977 *Nature*, **270**, 586
106. Doxsey R., Grindlay J., Griffiths R. et al. 1979 *Ap.J.*, **228**, L67
107. Duerbeck H.W. 1987 *Space Sci.Rev.*, **45**, 1
108. Dufour R.J., Duval J.E. 1975 *PASP*, **87**, 769
109. Dyck G. 1987 *Sky Tel.*, **73**, 658
110. Echevarria J., Jones D.H.P., Wallis R.E. et al. 1981 *MNRAS*, **197**, 565
111. Echevarria J., Alvares M. 1993 *Astr.Ap.* , **275**, 187
112. Elvius A. 1975 *Astr.Ap.*, **44**, 117
113. Ertan A.Y. 1981 *Ap.Space Sci.*, **77**, 391
114. Feige J. 1959 *Ap.J.*, **129**, 603
115. Ferguson D.H., Green R.F., Liebert J. et al. 1984 *Ap.J.*, **287**, 320
116. Ferguson D.H., Liebert J., Roc Cutri et al. 1987 *Ap.J.*, **316**, 399
117. Forman W., Liller W. 1973 *Ap.J.*, **183**, L117
118. Fuhrmann B. 1985 *MVS*, **10**, No 6, 127
119. Gaposchkin S. 1939 *Harv.Ann.*, **108**, 1
120. Garnavich P.M., Szkody P., Mateo M. et al. 1990 *Ap.J.*, **365** , 696
121. Garcia M.R., Kenyon S.J. 1988 In: *The Symbiotic Phenomenon, IAU Colloq. No 103*/eds. Mikołajewska J. et al., Dordrecht: Kluwer Acad.Publ., 27
122. Giclas H.L., Burnham R., Thomas N.G. 1962 *Lowell Bull.*, No 118

123. Gilliland R.L. 1982 *Ap.J.*, **254**, 653
124. Gilmore A.C., Kilmartin P.M. 1939 *McCormick.Publ.*, **9**, Pt.5
125. Gladyshev S.A., Goranskij V.P., Cherepashchuk A.M. 1987 *Sov.Astr.*, **31**, 541
126. Goranskii V.P. 1972 *PZ(P)*, **1**, 304
127. Goranskii V.P. 1977 *Astron.Circ.*, No 955
128. Goranskii V.P., Samus N.N. 1974 *PZ*, **2**, No 8, 123
129. Götz W. 1983 *IBVS*, No 2363
130. Götz W. 1985 *IBVS*, No 2735
131. Greenstein J.L., Arp H.C., Shectman S. 1977 *PASP*, **89**, 741
132. Green R.F., Ferguson D., Liebert J. 1982 *PASP*, **94**, 560
133. Green R.F., Greenstein J.L. 1976 *PASP*, **88**, 598
134. Green R.F., Richstone D.O., Schmidt M. 1978 *Ap.J.*, **224**, 892
135. Griffiths R.E., Bradt H., Doxsey R. et al. 1978 *Ap.J.*, **221**, L63
136. Griffiths R.E., Gursky H., Schwartz D.A. et al. 1978 *Nature*, **276**, 247
137. Grindlay J. 1977 *IAUC*, No 3104
138. Grindlay J.E. 1979 *Ap.J.*, **232**, L35
139. Grindlay J.E., Liller W. 1978 *Ap.J.*, **220**, L127
140. Gutiérrez-Moreno A., Moreno H., Feibelman W.A. 1992 *Ap.J.*, **395**, 295
141. Hack M., Selvelli P.L. In: 1993 *Cataclysmic Variables and related objects* /eds. M. Hack, C. la Dous, NASA, 511
142. Haefner R. 1988 *IBVS*, No 3261
143. Hanson C.G., Dennerl K., Coe M.J., Davies S.R. 1988 *Astr.Ap.*, **195**, 114
144. Hawkins M.R.S., Veron P. 1987 *Astr.Ap.*, **182**, 271
145. Hawkins M.R.S. 1981 *Nature*, **293**, 116
146. Herbig G.H. 1944 *PASP*, **56**, 230
147. Herbig G.H. 1962 *Harv.Ann.Card*, No 1576
148. Herbig G.H., Neubauer F.J. 1946 *PASP*, **58**, 196
149. Hertz P., Bailyn C.D., Grindlay J.E. et al. 1990 *Ap.J.*, **364**, 251
150. Hertz P., Wood K.S. 1986 *IAUC*, No 4235
151. Hessman F.V. 1990 *IAUC*, No 4971
152. Hildebrand R.H. et al. 1981 *Ap.J.*, **243**, 223
153. Hilditch R.W., Bell S.A. 1994 *MNRAS*, **266**, 703
154. Hinkle K.H., Fekel F.C., Johnson D.S., Scharlach W.W.G. 1993 *AJ*, **105**, 1074
155. Hoard D.W. 1993 *AJ*, **105**, 1232
156. Howarth I. 1977 *J.Brit.Astr.Ass.*, **87**, 176
157. Howarth I., Bailey J. 1980 *J.Brit.Astr.Ass.*, **90**, 265
158. Howell S.B., Hurst G.M., De Young J.A. 1964 *The Astronomer*, **30**, 277
159. Howell S.B., Mason K.O., Reichert G.A. et al. 1988 *MNRAS*, **233**, 79
160. Howell S.B., Szkody P., Kreidl T.J., Dobrzycka D. 1991 *PASP*, **103**, 300
161. Howell S.B., Hurst G.M., Liebert J. et al. 1994 *IBVS*, No 4043
162. Hutchings J.B., Crampton D., Cowley A.P. 1981 *AJ*, **86**, 871
163. Ilovaisky S.A., Chevalier C., Motch C. 1979 *Astr.Ap.*,, **71**, L17
164. Ilovaisky S.A., Chevalier C., Motch C. 1982 *Astr.Ap.*, **114**, L7
165. Iriarte B., Chavira E. 1957 *TTB*, No 16

166. Jablonski F.J., Steiner J.E. 1987 *Ap.J.*, **313**, 376
167. Jankovics I. 1973 *IBVS*, No 840
168. Janot-Pacheco E., Ilovaisky S.A., Chevalier C. 1981 *Astr.Ap.*, **99**, 274
169. Jernigan J.G., Apparao K.M.V., Bradt H.V. *et al.* 1976 *Nature*, **272**, 702
170. Jernigan J.G., Apparao K.M.V., Bradt H.V. *et al.* 1977 *Nature*, **270**, 321
171. Jernigan J.G., Apparao K.M.V., Bradt H.V. *et al.* 1978 *Nature*, **272**, 702
172. Johnston M., Bradt H., Doxsey R. *et al.* 1978 *Ap.J.*, **223**, L71
173. Kato T. *et al.* 1988 *Var.Stars Bull.*, **5**, 18
174. Kazennova E.A., Shugarov S.Yu. 1989 *private communication*
175. Kenyon S.J., Mikołajewska J., Mikołajewski M. *et al.* 1993 *AJ*, **106**, 1573
176. Khruzina T.C., Shugarov S.Yu. 1991 *Atlas of Cataclysmic Variables. U Gem Stars*, Part I,II, Moscow: Moscow University Press
177. Kitamoto S., Tsunemi H., Pedersen H. *et al.* 1990 *Ap.J.*, **361**, 590
178. Kopylov I.M., Somov N.N. 1987 *SCAO*, **56**, 51
179. Kraft R.P., Luyten W.J. 1965 *Ap.J.*, **142**, 1041
180. Krzeminski W. 1974 *Ap.J.*, **192**, 135
181. Krzeminski W., Kraft R.P. 1964 *Ap.J.*, **140**, 921
182. Kubiak M., Krzeminski W. 1989 *PASP*, **101**, 667
183. Kukarkin B.V. 1968 *Astron.Circ.*, No 1479
184. Kurochkin N.E. 1953 *PZ*, **9**, 402
185. Kurochkin N.E. 1959 *PZ*, **12**, 418
186. Kurochkin N.E. 1972 *PZ*, **18**, 425
187. Kurochkin N.E. 1977 *Astron.Circ.*, No 974; 1980, No 1143
188. Landolt A.U. 1970 *PASP*, **82**, 1364
189. Langmeier A., Sztajno M., Vassa W.D. *et al.* 1986 In: *Evolution of Galactic X-ray Binaries. NATO ASI Series*/eds. Trümper J., Lewin W.H.G., Brinkmann W.D., Dordrecht: D.Reidel, 253
190. Lenouvel F., Daguillon J. 1956 *JO*, **39**, No 1, 1
191. Levy D.H., Howell S.B., Kreidl T.J. *et al.* 1990 *PASP*, **102**, 1321
192. Lindsay E.M. 1961 *AJ*, **66**, 169
193. Lipunova N.A. 1988 *Sov.Astr.*, **65**, 99
194. Lipunova N.A., Shugarov S.Yu. 1991 *IBVS*, No 3580
195. Li F.K., van Paradijs J., Clark G.W. *et al.* 1978 *Nature*, **276**, 799
196. Longmore A.J., Tritton S.B. 1980 *MNRAS*, **193**, 521
197. Luyten W.J., Haro G. 1959 *PASP*, **71**, 469
198. Lyuty V.M., Shugarov S.Yu. 1979 *Pis'ma Aston.Zh.*, **5**, 383
199. Lyuty V.M. 1972 *PZ*, **18**, 417
200. Lyuty V.M., Sunyaev R.A., Cherepashchuk A.M. 1973 *Astron.Zh.*, **50**, 3
201. Mardirossian F. *et al.* 1980 *Astr.Ap.*, **85**, 29
202. Margon B. *et al.* 1977 *Ap.J.*, **216**, 811
203. Margon B., Thorstensen J.R., Bowyer S. 1978 *Ap.J.*, **221**, 907
204. Marino B.F., Walker W.S.G. 1983 *South.Stars*, **30**, No 2/3, 241
205. Markaryan B.E. 1967 *AF*, **3**, 511
206. Markaryan B.E. 1968 *AF*, **4**, 144
207. Markaryan B.E., Lipovetzkii B.A. 1971 *AF*, **7**, 511
208. Martynov D.Ya., Kholopov P.N. 1957 *PZ*, **11**, No 3, 222

209. Mason K.O., Middleditch J., Cordova F.A. et al. 1983 Ap.J, **264**, 575
210. Mason K.O., Parmar A.N., White N.E. 1985 MNRAS, **216**, 1033
211. Mattei J. 1983 IAUC, No 3784
212. Mayall M. 1965 KVB, **40**, 241
213. McClintock J.E. 1992 In: Texas/ESO-CERN Symp. on Relativistic Astrophysics, Cosmology, and Fundamental Physics, 1990, Brighton, England/eds. Barrow J.D., Mestel L., Thomas P.A. New York: NY Acad. Sci.
214. McHardy I.M., Pye J.P., Fairall A.D. et al. 1984 MNRAS, **210**, 663
215. McNaught R.H. 1986 IBVS, No 2926
216. Meech K.J. 1983 PASP, **95**, 662
217. Meinunger L. 1972 MVS, **10**, H.3, 56
218. Mendelson H., Mazeh T. 1989 MNRAS, **239**, 733
219. Mendelson H., Leibowitz E.M., Brosch N., Almozina E. 1992 The Astronomer, **29**, No 338
220. Merrill P.W., Burwell C.G. 1950 Ap.J., **112**, 72
221. Middleditch J., Nelson J. 1979 BAAS, **11**, 664
222. Middleditch J., Cordova F.A. 1982 Ap.J., **255**, 585
223. Middleditch J., Pennypacker C., Burns S. 1983 Ap.J., **274**, 313
224. Mikołajewska J., Kenyon S.J. 1992 MNRAS, **256**, 177
225. Miskin N.A. 1970 PZ, **17**, 449
226. Mittaz J.P.D., Rosen S.R., Mason K.O., Howell S.W. 1992 MNRAS, **258**, 277
227. Moneti A. 1989 Messenger, **58**, 7
228. Morris S.L., Schmidt G.D., Liebert J. et al. 1987 Ap.J., **314**, 641
229. Moskalenko E.I. 1995 private communication
230. Motch C., Janot-Pacheco E. 1987 Astr.Ap., **182**, L55
231. Motch C., Pakull M.W., Mouchet M., Beuermann K. 1989 Astr.Ap., **219**, 158
232. MVS, 1957, No 275
233. MVS, 1957, No 311
234. Mufson S.L. et al. 1980 IAUC, No 3471
235. Mukai K. et al. 1988 MNRAS, **234**, 291
236. Mumford G.S. 1966 Ap.J., **146**, 411
237. Mumford G.S. 1971 Ap.J., **165**, 369
238. Munari U. 1992 Astr.Ap. **257**, 163
239. Munari U., Whiteoak P.A. 1989 MNRAS, **239**, 273
240. Murdin P., Branduardi-Raymont G., Parmar A.N. 1981 MNRAS, **196**, 95P
241. Murdin P., Griffiths R.E., Pounds K.A. 1977 MNRAS, **178**, 27P
242. Nassau J.J., Cameron D.M. 1954 Ap.J., **119**, 175
243. Naylor T., Charles P.A., Longmore A.J. 1991 MNRAS, **252**, 203
244. Nevo I., Sadeh D. 1978 MNRAS, **182**, 595
245. Noskova R.I. 1989 Pis'ma Astron.Zh., **15**, 346
246. Okazaki A. 1993 Ap.Space Sci., **210**, 227
247. Oke J.B., Wade R.A. 1982 AJ, **87**, 670
248. Osborne J.P., Beardmore A.P., Wheatley P.J. et al. 1994 MNRAS, **270**, 650
249. Osminkin E.Yu. 1985 PZ, **22**, 261

250. Pakull M. 1979 *Messenger*, **16**, 38
251. Pakull M.W., Beuermann K., van der Klis M., van Paradijs J. 1988 *Astr. Ap.*, **203**, L27
252. Paresce F., de Marchi G., Ferraro F.R. 1992 *Nature*, **360**, 46
253. Parmar A.N., Stella L., White N. 1986 *Ap.J.*, **304**, 664
254. Parmar A.N., Gottwald M., van der Klis M., van Paradijs J. 1989 *Ap.J.*, **338**, 1024
255. Patterson J. 1979 *AJ*, **84**, 804
256. Patterson J. 1980 *Ap.J.*, **241**, 235
257. Patterson J. 1981 *Ap.J.Suppl.Ser.*, **45**, 517
258. Patterson J. 1984 *Ap.J.Suppl.Ser.*, **54**, 443
259. Patterson J. 1994 *PASP*, **106**, 209
260. Patterson J., Nather R.E., Robinson E.L., Handler F. 1979 *Ap.J.*, **232**, 819
261. Patterson J., Eisenman N. 1987 *IBVS*, No 3079
262. Patterson J., Sterner E., Halpern J.P., Raymond J.C. 1992 *Ap.J.*, **384**, 234
263. Patterson J., Halpern J.P., Shambrook A. 1993 *Ap.J.*, **419**, 803
264. Pavlenko E.P., Prokof'eva V.V., Dolgushin A.J. 1989 *Pis'ma Astr.Zh.*, **15**, 611
265. Pavlenko E.P., Pelt J. 1991 *AF*, **34**, 169
266. Payne-Gaposchkin C. 1950 *Harv.Ann.*, **115**, No 14
267. Pedersen H., Ilovaisky S., van der Klis M. 1987 *IAUC*, No 4357
268. Penston M.V., Penston M.J., Murdin P., Martin W.L. 1975 *MNRAS*, **172**, 313
269. Perry M.E., Peterson B.A. 1974 *AJ*, **79**, 1
270. Petit M. 1960 *JO*, **43**, 17
271. Ponman T. 1982 *MNRAS*, **200**, 351
272. Popowa M. 1960 *MVS*, No 463
273. Popowa M. 1960 *MVS*, No 464
274. Popowa M.D., Vitrichenko E.A. 1978 *Astron.Zh.*, **55**, 765
275. Priedhorsky W.C., Terrel J. 1984 *Ap.J.*, **280**, 661
276. Remillard R.A., Bradt H.V., McClintock J.E. et al. 1986 *Ap.J.*, **302**, L11
277. Remillard R.A., Stroozas B.A., Tapin S., Silber A. 1991 *Ap.J.*, **379**, 715
278. Remillard R.A., Bradt H.V., Brissenden R.J.V. et al. 1994 *Ap.J.*, **428**, 785
279. Ritter G.A. 1986 *AN*, **307**, No 4, 221
280. Ritter G.A. 1990 *IBVS*, No 3546
281. Ritter H., Kolb U. 1993 In: *X-ray Binaries*/eds. Lewin W.H.G., van Paradijs J., van der Heuvel E.P.J. Cambridge: Cambridge University Press
282. Robinson E.L. 1973 *Ap.J.*, **183**, 193
283. Robinson E.L. et al. 1978 *Ap.J.*, **219**, 168
284. Robinson E.L., Nather R.E. 1979 *Ap.J.Suppl.Ser.*, **39**, 461
285. Robinson E.L., Shafter A.W., Hill J.A., Wood M.A. 1987 *Ap.J.*, **313**, 772
286. Romano G. 1958 *Mem.S.A.It.*, **29**, 2
287. Romano G., Perissinotto M., 1966, *Padova Publ.*, No 50, 1
288. Romano G., Perissinotto M. 1968 *Padova Publ.*, No 151
289. Rosino L., Romano A G., Marziani P. 1993 *PASP*, **105**, 51

290. Sanduleak N. 1968 *AJ*, **73**, 246
291. Sanduleak N. 1975 *IBVS*, No 1011
292. Sanford R.F. 1947 *PASP*, **59**, 87
293. Schaefer B.E. 1990 *Ap.J.*, **354**, 720
294. Schmidtke P.C. 1988 *AJ*, **95**, 1528
295. Schmidtke P.C., Cowley A.P., 1987, *AJ*, **93**, 374
296. Schmidtke P.C., McGrath T., Cowley A.P., Frattare L. 1993 *PASP*, **105**, 863
297. Schmid H.M., Schild H. 1990 *MNRAS*, **246**, 84
298. Schmutz W., Schild H., Mürset U., Schmid H.M. 1994 *Astr.Ap.*, **288**, 819
299. Schoembs R. 1982 *Astr.Ap.*, **115**, 190
300. Schoembs R., Vogt N. 1981 *Astr.Ap.*, **97**, 185
301. Schoembs R., Zoeschinger G. 1990 Astr.Ap., **227**, 105
302. Schwartz D.A., Griffiths R.E., Thorstensen J.R. et al. 1980 *AJ*, **85**, 549
303. Sekiguchi K., Nakala Y., Bassett B. 1994 *MNRAS*, **266**, L51
304. Semeniuk I. 1980 *Astr.Ap.Suppl.Ser.*, **39**, 29
305. Seregina T.M., Shugarov S.Y. 1988 *private communication*
306. Seward F.D., Charles P.A., Smale A.P. 1986 *Ap.J.*, **305**, 814
307. Shafter A.W. 1983 *IBVS*, No 2377
308. Shafter A.W. 1985 *AJ*, **90**, 643
309. Shafter A.W., Ulrich R.K. 1982 *BAAS*, **14**, 880
310. Shafter A.W., Szkody P. 1984 *Ap.J.*, **276**, 305
311. Shafter A.W., Harkness R.P. 1986 *AJ*, **92**, 658
312. Shafter A.W., Szkody P., Thorstensen J. 1986 *Ap.J.*, **308**, 765
313. Shahbaz T., Naylor T., Charles P.A. 1993 *MNRAS*, **265**, 655
314. Shakun L.I. 1988 *Astron.Circ.*, No 1525
315. Shara M.M. et al. 1984 *Ap.J.*, **282**, 763
316. Shugarov S.Yu. 1978 *PZ*, **20**, 507
317. Shugarov S.Yu. 1995 *Astron.Circ.*, in press
318. Shugarov S.Yu. 1995 *private communication*
319. Shustov B.M., Tutukov A.V. 1985 In: *Recent Results on Cataclysmic Variables ESA SP-236*, 113
320. Silber A., Remillard R.A., Horne K., Bradt H.V. 1994 *Ap.J.*, **424**, 955
321. Smak J. 1967 *Acta Astr.*, **17**, 255
322. Smak J., Stepien K. 1975 *Acta Astr.*, **25**, 379
323. Smale A.P., Corbet R.H.D., Charles P.A. 1988 *MNRAS*, **233**, 51
324. Stauffer J. et al. 1979 *PASP*, **91**, 59
325. Steiner J.E. et al. 1984 *Ap.J.*, **280**, 688
326. Stepanyan D.A. 1980 *AF*, **16**, 187
327. Stepanyan D.A. 1982 *PZ*, **21**, 691
328. Stockman H.S., Foltz C.B., Schmidt G.D., Tapia S. 1983 *Ap.J.*, **271**, 725
329. Stockman H.S., Schmidt G.D., Lamb D.Q. 1988 *Ap.J.*, **332**, 282
330. Stolz B., Schoembs R. 1981 *IBVS*, No 1955
331. Stover R.J. 1981 *Ap.J.*, **249**, 673
332. Szkody P. 1976 *Ap.J.*, **207**, 190
333. Szkody P. 1982 *BAAS*, **14**, 881

334. Szkody P. 1987 *Ap.J.Suppl.Ser.*, **63**, 685
335. Szkody P., Shafter A.W., Coeley A.P., 1984 *Ap.J.*, **282**, 236
336. Szkody P., Mateo M. 1986 *AJ*, **92**, 483
337. Szkody P., Feinswog L. 1988 *Ap.J.*, **334**, 422
338. Sztajno M. 1979 *IBVS*, No 1710
339. Thorstensen J.R., Ringwald F.A., Wide R.A et al. 1991 *AJ*, **102**, 272
340. Thorstensen J., Charles P., Bowyer S. et al. 1979 *Ap.J.*, **233**, L57
341. Thorstensen J.R. 1986 *AJ*, **91**, 940
342. Thorstensen J.R. 1987 *Ap.J.*, **312**, 739
343. Thorstensen J.R., Freed I.W. 1985 *AJ*, **90**, 2082
344. Thorstensen J.R., Vennes S., Shambrook A. 1994 *AJ*, **108**, 1924
345. Tsunemi H., Kitamoto S., Okamura S., Roussel-Dupre D. 1989 *Ap.J.*, **337**, L81
346. Tuohy I.R., Buckley D.A.H., Remillard R.A. et al. 1986 *Ap.J.*, **311**, 275
347. Tuohy I.R., Ferrario L., Wickramasinghe D.T., Hawkins M.R.S. 1988 *Ap.J.*, **328**, L59
348. Tsesevich V.P. 1952 *PZ*, **8**, 412
349. Tsesevich V.P., Dragomirezkaya B.A. *RW Aurigae type stars*, Kiev: Naukova Dumka, 1973
350. Tsesevich V.P., Goranskii V.P., Samus N.N., Shugarov S.Yu. 1979 *Astron. Circ.*, No 1043
351. Tsesevich V.P., Kazanasmas M.S. *The Atlas of finding charts of variable stars*, Moscow: Nauka,1971
352. Udalski A., Szymanski M. 1988 *Acta Astr.*, **38**, 215
353. Udalsky A., Kaluzny J. 1991 *PASP*, **103**, 198
354. Van Buren D., Charles A., Mason K.O. 1980 *Ap.J.*, **242**, L105
355. Van Woerd H. et al. 1988 *Ap.J.*, **330**, 911
356. Van der Klis M., Clausen J.V., Jensen K. et al. 1985 *Astr.Ap.*, **151**, 322
357. Visvanatan N., Pickles A.J. 1982 *Nature*, **298**, 41
358. Vogt N. 1976 In: *Structure and Evolution of Close Binary Systems. IAU Symp. No 73*/eds. Eggleton P.P. et al. Dordrecht: D.Reidel, 147
359. Vogt N. 1977 *VSS*, **5**, 1
360. Vogt N. 1980 *NZAS Publ.*, **8**(C80), 8
361. Vogt N. 1982 *Astr.Ges.Mitt.*, **57**, 79
362. Vogt N. 1983 *Astr.Ap.Suppl.Ser.*, **53**, 21
363. Vogt N. et al. 1981 *Astr.Ap.*, **94**, L29
364. Vogt N., Bateson F.M. 1982 *Astr.Ap.Suppl.Ser.*, **48**, 383
365. Voloshina I.B. 1986 *Pis'ma Astron.Zh.*, **12**, 219
366. Voloshina I.B., Lyuty V.M. 1987 *PZ*, **22**, 551
367. Voloshina I.B., Shugarov S.Yu. 1989 *Pis'ma Astron.Zh.*, **15**, 723
368. Voloshina I.B., Lyuty V.M. 1993 *Astron.Zh.*, **70**, 61
369. Von Hoffmeister C. 1929 *Sonn.Mitt.*, No 16
370. Von Hoffmeister C. 1963 *VSS*, **6**, 1
371. Von Hoffmeister C. 1967 *AN*, **289**, Pt.5, 205
372. Von Löchel K. 1965 *MVS*, **3**, 107
373. Vorobjeva V.A. 1960 *PZ*, **13**, 72

374. Voikhanskaya N.F., Nazarenko I.I. 1984 *Pis'ma Astron.Zh.*, **10**, 439
375. Wachmann A.A. 1961 *Bergd.Abh.*, **6**, 1
376. Wachmann A.A. 1963 *Bergd.Abh.*, **6**, 99
377. Wade R.A., Quintana H., Horne K., March T.R. 1985 *PASP*, **97**, 1092
378. Wade R.A. et al. 1987 *BAAS*, **19**, 548
379. Wagner R.M., Kreidl T.J., Bus S.J., Williams W. 1989 *Ap.J.*, **346**, 971
380. Walker A.D., Olmsted M. 1958 *PASP*, **70**, 495
381. Walker M.F. 1981 *Ap.J.*, **248**, 256
382. Walter J. et al. 1959 *Ric.Astr.*, **6**, No 28
383. Wargan W., Vogt N. 1982 *Astr.Ges.Mitt.*, **55**, 77
384. Warner B. 1976 *Observ.*, **96**, 49
385. Warner B. 1982 *MNASSA*, **41**, 15
386. Warner B. 1988 *MNASSA*, **47**, No 3/4, 30
387. Warner B., Brickhill A.J. 1974 *MNRAS*, **166**, 673
388. Warner B., Brickhill A.J. 1978 *MNRAS*, **182**, 777
389. Warren P.R., Penfold J.E. 1975 *MNRAS*, **172**, 41P
390. Watson M.G., Sherrington M.R. 1978 *MNRAS*, **184**, 79P
391. Weber R. 1966 *IBVS*, No 123
392. Weber R. 1967 *BSAM*, No 13, 51
393. Weekes T.C., Geary J.C. 1982 *PASP*, **94**, 708
394. Wegner G. 1975 *MNRAS*, **171**, 637
395. Wenzel W. 1979 *IBVS*, No 1720
396. Wenzel W. 1980 *MVS*, **8**, 133
397. Wenzel W. 1983 *IBVS*, No 2430
398. Wenzel W. 1984 *MVS*, **10**, 51
399. Westerlund B. 1968 *Ann.Uppsala Ast.Obs.*, **4**, 1
400. White N.E., Mason K.O., Huckle H.E. et al. 1976 *Ap.J.*, **209**, L119
401. Wickramasinghe D.T., Ferrario L., Bailey J. et al. 1993 *MNRAS* **265**, L29
402. Williams G. 1983 *Ap.J.Suppl.Ser.*, **53**, 523
403. Williams G., Johns M., Price C. et al. 1979 *Nature*, **281**, 48
404. Wood J.A., Drew J.E., Verbunt F. 1990 *MNRAS*, **245**, 323
405. Wu C.C., Panek R.J. 1982 *Ap.J.*, **262**, 244
406. Wyckoff S., Wehinger P.A. 1978 *PASP*, **90**, 557
407. Young P., Schneider D.P., Shectman S.A. 1981 *Ap.J.*, **245**, 1035
408. Yudin B.F. 1982 *Astron.Zh.*, **59**, 302
409. Zhang E.-H. et al. 1986 *Ap.J.*, **305**, 740
410. Zverev M.S., Makarenko E.N. 1979 *PZ(P)*, **3**, No 17, 431

Additional references

411. Andronov I.L. 1991 *IBVS*, No 3828
412. Andronov I.L., Pavlenko E.P., Seregina T.M. et al. 1992 In: *Stellar magnetism*. St.Petersburg: Nauka, 160
413. Beuermann K., Thomas H.-C., Schwope A. 1988 *Astr.Ap.*, **195**, L15
414. Beuermann K., Thomas H.-C., Giommi P. et al. 1989 *Astr.Ap.*, **219**, L7
415. Beuermann K., Thomas H.-C., Schwope A.D. et al. 1990 *Ap.J.*, **238**, 187

416. Cropper M. 1994 *MNRAS*, **222**, 853
417. Dhillon V.S., Jones D.H.P., Marsh T.R. 1994 *MNRAS*, **224**, 859
418. Diaz M.P., Steiner J.E. 1991 *AJ*, **103**, 964
419. Garlick M.A., Rosen S.R., Mittaz J.P.D. et al. 1994 *MNRAS*, **267**, 1095
420. Heise J., Verbunt F. 1988 *Astr.Ap.*, **189**, 112
421. Horne K., Welsh W.F., Wade R.A. 1993 *Ap.J.*, **410**, 357
422. Kazarovets E.V., Samus N.N. 1995 *IBVS*, No 4140
423. Kenyon S.J., Mikołajewska J.1995 *AJ*, **110**, 391
424. Mason K.O., Liebert J., Schmidt G.D. 1989 *Ap.J*, **346**, 941
425. Mason K.O., Watson M.G., Ponman T.J. 1992 *MNRAS*, **258**, 749
426. Mukai K. 1990 *MNRAS*, **245**, 385
427. Nussbaumer N., Schmutz W., Vogel M. 1986 *Astr.Ap.*, **169**, 154
428. Rosen S.R., Clayton K.L., Osborne J.P., McGale P.A. 1994 *MNRAS*, **269**, 913
429. Patterson J., Steiner J.E. 1983 *Ap.J.*, **264**, L61
430. Reinsch K., Burwitz V., Beuermann K. et al. 1994 *Astr.Ap.*, **291**, L27
431. Sambruna R.M., Chiappetti L., Treves A. et al. 1991 *Ap.J.*, **374**, 744
432. Schmidt G.D., Stockman M.S. 1991 *Ap.J.*, **371**, 749
433. Schwope A.D., Beuermann K. 1989 *Astr.Ap.*, **222**, 132
434. Skopal A. 1994 *IBVS*, No 4096
435. Stockman H.S., Schmidt G.D., Liebert J., Holberg J.B. 1994 *Preprint STSI*, No 819
436. Szkody P., Garnavich P., Castelaz M., Manino F. 1994 *PASP*, **106**, 615
437. Tuohy I.R., Remillard R.A., Brissenden R.J.V. 1990 *Ap.J.*, **359**, 204
438. White II J.C., Honeycutt R.K., Horne K. 1993 *Ap.J.*, **412**, 278
439. Zhang E.-H., Robinson E.L., Ramseyer T.T. 1991 *Ap.J.*, **381**, 534